Springer Series in Physical Environment

5

Liu Tungsheng

Loess in China

Second Edition

With 80 Figures

China Ocean Press Beijing

Springer-Verlag Berlin Heidelberg New York
London Paris Tokyo

Professor Dr. LIU TUNGSHENG
Institute of Geology
Academia Sinica Qijiahuozi
Beijing, China

First Edition Published by China Ocean Press 1985
and Distributed in the People's Republic of China

ISBN-13: 978-3-642-82834-8 e-ISBN-13: 978-3-642-82832-4
DOI: 10.1007/978-3-642-82832-4

Library of Congress Cataloging-in-Publication Data. Liu, Tung-sheng. [Huang t' u yü huan ching. English] Loess and the environment / Liu Tungsheng et al. p. cm. — (Springer series in physical environment ; 5) Translation of: Huang t' u yü huan ching. ISBN 0-387-16717-X (U.S.) 1. Loess. 2. Loess—Environmental aspects—China. I. Title. II. Series. QE579.L5813 1988 551.3'54—dc19 88-15943

Softcover reprint of the hardcover 2nd edition 1988

Typesetting: K+V Fotosatz GmbH, Beerfelden
Offsetprinting and Bookbinding: Konrad Triltsch, Graphischer Betrieb, Würzburg
2132/3130-543210 – Printed on acid-free paper

Foreword

Around 1825, a young professor at the University of Heidelberg, Karl Caesar von Leonhard (1779–1862), included a sample of yellow silty loam, from the locality Haarlass near Heidelberg in the valley of the Neckar, in his collection of minerals and rocks (cf. Kirchheimer 1969). He used the local word "loess" and introduced it (1824, p. 722–724) to the scientific world. At that time he could not have imagined that the term "loess" would become a ubiquitous term to describe a wide field of research. Karl Caesar von Leonhard was not aware of the aeolian nature of loess. This was recognized much later by Ferdinand von Richthofen (1833–1905) during his famous journeys in China, published, for instance, in the first comprehensive book on geomorphology in German (cf. Richthofen 1886).

Today, loess is considered a fascinating aeolian sediment mainly deposited in or around periglacial landscapes. Most of the great loess deposits are therefore of Pleistocene age but, under special conditions, Holocene loesses are also known. Where loess occurs, it dominates the geoecology of an area: it is fertile and stimulates agriculture on the one hand; on the other, it is subject to serious erosional hazards and – as a consequence – is one of the sources of pollution and siltation of adjacent rivers. In a scientific view, loess contains an exciting archive of regional data on climatic and geomorphic history based on fossil soils, pollen records, tephra etc.

Regional information on loess are, therefore, of great value to all interested in the Earth's physical environment. Thus, we are glad to present this book on *Loess in China* to the Scientific Community. Volume 5 of the Springer Series in Physical Environment displays an enormous amount of very interesting regional data and information hitherto unavailable to most geoscientists in the Western World.

The present edition is based on a revised English version of the Chinese book. The first English version had only a limited distribution. The publisher, the editors, and the members of the advisory board of this Series feel that this edition of *Loess in China* demonstrates the present state of loess research in the classic area of loess deposits. Besides information not published in the West before, it presents data which will be very welcome in the context of international scientific exchanges. It is hoped that this publication will further strengthen international cooperation by overcoming language barriers in this field of our physical environment.

Heidelberg, June 1988 DIETRICH BARSCH

References

Kirchheimer F (1969) Heidelberg und der Löß. Ruperto Carola vol. 46, p. 201–205 (Heidelberg)

Leonhard KC v (1824) Charakteristik der Felsarten. Dritte Abtheilung: Trümmer-Gesteine. Lose Gesteine. Kohlen. Heidelberg (Engelmann), p. 599–772

Richthofen F v (1886) Führer für Forschungsreisende. Hannover 734 pp.

Preface

"Loess and the Environment" is an abridged translation of the monograph under the same title recently published in Chinese. It is a sequel to the three monographs: "The Loess Along the Middle Reaches of the Huanghe (Yellow) River", "The Loess Deposits of China" and "The Composition and Texture of Loess". Despite the fact that a great number of books and papers have appeared in China in the last 30 years little was known to scientists abroad due mainly to linguistic obstacles. This English version has been prepared for the forthcoming International Symposium on Loess to be held in Xian, 1985, facilitating the scope of exchange between the scientists of China and other countries.

It is no problem for the Chinese readers to understand the type localities of the loess profiles and the connotation of the terms – such as "Yuan", "Liang" and "Mao". These will, however, be somewhat unfamiliar to overseas scientists. Therefore, some technical devices have been introduced into the translation to meet their needs. The map in the inside front cover shows the important locality names mentioned in this book. Some necessary annotations were also added in the text. Even so, it is still difficult to resolve all the problems, and readers are referred to the above-mentioned Chinese literature for further explanation information.

The English translation was done by Drs. Sun Qixian, Guo Qiti, Xu Zhonglun, Feng Liangyuan, Shen Defu, Zhang Shouxin, Li Huhou, Zheng Honghan, Shi Jiyang, Zheng Shaohua, Ai Yang, Jiang Hui, and Ms. Zhou Weijian. The author is grateful to them for their efforts, without which it would have been impossible to have this book ready in such a short time. Owing to pressure of time, the authors fear that there are some inconsistencies in context in the articles and in phraseology in English. Fortunately, the Canada Energy Delegation headed by Dr. H. Stafford visited China in May, and the authors are indebted to Dr. H. Stafford and Drs. A. N. Hutton and N. R. Fischbuch, the members of the delegation, who found time in the midst of pressing affairs in visiting China to critically review the manuscript of the Foreword and Chapters I to III. Later Prof. G. J. Kukla, Drs. Warc Serreze, Joyce Gavin, Dorothy Peteete and Tom Janecek of the Lamont

Doherty Geological Observatory of the Columbia University, New York, reviewed whole the manuscript. The authors wish to express the deepest thanks to them for their linguistic revision of the manuscript and for their critical comments, many of which were incorporated into the final version. The author has benefited greatly from discussions with Prof. Kulka. His knowledge of the loess of the world and his first knowledge expertise in Chinese loess has rendered the discussion very profitable. The authors wish to thank Prof. Zhou Mingzhen and Dr. Sun Qixian for reviewing the manuscript. However, the authors alone are responsible for any errors or mistakes that still remain.

In the technical preparation of this book, the authors were assisted by Zhao Xitao, Han Jiamao and Zheng Honghan in Editing. Cui Jiuxu typed the manuscript, Xia Juanjuan rearranged the references. The authors wish to extend their appreciation to them all, as well as to the publishers, for their unfailing helpfulness during the realization of this technically difficult book.

Beijing, China, 1988 LIU TUNGSHENG

Contents

Chapter 6. Loess and the Environment

Introduction

The book on "Loess and the Environment" is now in your hands. Since 1980 personnel of the Quaternary Research Laboratory of the Institute of Geology, the Quaternary Geochemical Research Laboratory of the Institute of Geochemistry, and the Xian Laboratory of Loess and the Quaternary Geology of Academia Sinica have been investigating loess at Luochuan, Shaanxi Province, in continuation of the past work. This book is the result of a great deal of cooperation.

By now, 20 years have elapsed since the book on "The Composition and Texture of Loess" was written jointly by the personel mentioned above in 1964, with Luochuan loess as the major object of the study. It has been over 30 years since the study of loess was begun and the first Laboratory of Quaternary Research in China was established at the Institute of Geology. At that time, none of us thought that we could continue studying loess and jointly publish monographs after 20, even 30 years. We are very glad to be able to study loess as our life work, and to continue in our collective long-term cooperation, which has enabled our recent progress. We now understand what 20 and 30 years of continuing research mean. The hardship and significance of scientific research lies in continuity of this sort.

Time has deepened our understanding of loess, has enabled us to learn from past work, and time also forces us to look further ahead. The content of this book is a review of our loess research and of our future plans.

When the Sanmenxia Reservoir Project was begun in 1953, we carried out loess research and made a comprehensive investigation of water and soil conservation along the middle reaches of the Huanghe (Yellow) River under the guidance of the late Prof. Zhu Kezhen (Kuochin Chu), the former vice president of Academia Sinica. After this investigation in 1958, we made a reconnaissance along ten major geological profiles on the Loess Plateau. From 1958 to 1961, we had surveyed loess outside the Plateau in Shandong, Qinghai and Hebei Provinces, and compiled "The Map of Loess Distribution in China". At that time, many young men worked on the Plateau night and day. During 1962–1964, two monographs, "The Loess Along the Middle Reaches of the Huanghe (Yellow) River" and "The Loess Deposits of China" were completed with the support offered by the late Prof. Hou Defeng, the Director of the Institute of Geology, and Yu Xiangchun and Peng Hui. Afterwards, we made an investigation of Luochuan loess for the first time and then published the book "The Composition and Texture of Loess".

Over 20 years have elapsed. With the progress in scientific study, some of our coworkers have become involved in starting new research fields. On the occasion of the publication of this book, we first think of them. This book contains their work and thoughts. Their wisdom and their style are manifested in this book by the lasting imprint they made on the thinking of the authors.

Besides this group, many people in many institutions have been engaged in loess research in China over the past two decades, especially the personnel working in agricultural and engineering departments. Their understanding and comprehension of loess has inspired our ideas and aroused our attention. Their work has helped us to study more rigorously and act with more caution, thus avoiding possible mistakes. This mutual encouragement is difficult to express merely by thanks.

The results obtained from 1954 to 1966 are as follows:

1. Before participating in the VIth International Quaternary Congress held in Poland in 1961, under the guidance of the late Prof. Li Siguang (J. S. Lee), the former vicepresident of Academia Sinica, we had divided the loess series of China into Malan Loess (still using the old name), Lishi Loess and Wucheng Loess. Combined loess stratigraphic and biostratigraphic subdivision and nomination have been gradually adopted by geological circles in this country. The definition of loess deposits in China, consequently, has been expanded and includes sedimentary regions less than 20 m thick to over 200 m thick. The geological age of loess in China is from Late Pleistocene to the entire Quaternary. From the point of view of engineering geology, Malan Loess belongs to collapsible loess, while Lishi Loess and Wucheng Loess are basically non-collapsible. This is of significance for construction in the Loess Plateau. Meanwhile, the recognition of older loess in China is of worldwide paleoclimatic significance. In recent years, scholars in Europe and America have begun to pay increased attention to old loess strata.

2. As far as particle composition and microtexture of loess is concerned, Wang Tingmei, Su Lianyi and Zhu Haizhi classify loess into three belts: sandy loess, loess and clayey loess, progressing from northwest to southeast. The belts reflect the dynamic action of wind transportation of loess, and demonstrate that sandy loess is the main source of silt along the lower reaches of the Huanghe River, which is of great importance in its control.

3. Through continuous work on loess and paleosol sequence, Wang Kelu, Wu Zirong, Wen Qizhong, Yang Lihua, Guo Xudong (Guo Jinluan), Lu Yanchou and An Zhisheng recognized at Luochuan the first, second and third sandy loess layer (also called the upper, lower and bottom sandy loess layer). Coarse-grained sandy loess is a good aquifer in the loess "Yuan" area. It represents dry and cold climate in terms of depositional environment.

All the above results have already been described in the three books mentioned earlier. In this book the authors discuss several problems raised by the work done from 1966 to 1984, especially between 1980 and 1984.

1. Considering chronostratigraphy, the timescale of deposition of loess and paleosol sequence has been obtained by using paleomagnetic, thermoluminescence and radiocarbon dating techniques. From the top downward, the dated layers are: Modern loess (contemporary windborne dust deposits), and loess of the past 10000 years (Holocene loess, dated by ^{14}C); loess of the past 100000 years (Malan Loess, dated by thermoluminescence); paleosol about 500000 years old (5th paleosol, corresponding to the time of the 10th layer with *Homo erectus pekinensis* at Zhoukoudian); the first sandy loess about 800000 years old (deposited before the B/M boundary); the second sandy loess about 1150000 years old (deposited before the Jaramillo subchron); paleosol assemblage about 1800000 years old (in the middle part of Wucheng Loess, in the Olduvai subchron); basal loess deposited about 2400000 years ago (after M/G boundary).

The chronologic framework makes it possible to study accurately the sequence of geologic events from the time when loess began to accumulate, to the time it reached a 200-m thickness. It shows us how the climate of the Loess Plateau changed, how ancient humans, other fauna (especially rodents in the loess regions, and vegetation evolved and how gullies and ravines formed over the past 2.4 million years. Thus one can trace back the character and range of environmental changes, and find out their duration. In this sense, the Loess Plateau in China may present unique opportunities for the study of Quaternary environmental change.

2. Environmental assessment of the Loess Plateau. The control of Plateau environment is a task of top priority. What is the process of soil erosion? How did gullies and ravines form? Did the arid environment of the present steppe evolve from the former forest vegetation? What was the history of the Loess Plateau, and what will be its future? Appraising the Loess Plateau requires considering it as a system which includes subsystems from inorganic (soil composition) to organic (organism and its activities), scales from macroscopic (geomorphy, structure of strata, etc.) to microscopic (common and trace elements, heavy minerals, clay minerals and distribution of amino acids), times from past to present, zones

from source through transportation area to accumulation and redeposition belts (from western China to eastern China), and so on.

Loess and paleosol sequences in China contain rich geologic information which can be studied as a natural system. Information on the geological aspects of the loess deposits is treated in the first part of this book, with further comprehensive analysis to be made in future work.

3. The process of loess formation can be deduced by analyzing the processes of the transportation and deposition of modern silt in dust storms. Such studies may provide the basis for models of lifting, transportation and deposition of loessial deposits. After the study of drifting sand by Bagnold (1941), attention was focused on sand and sand dunes, less on eolian silt. One of the reasons is that it is difficult to find dust deposition areas in nature, whereas in the oceans, lakes, rivers, or in sand dunes, study of depositional processes is relatively simple. However, the vast area from Xinjiang in the west to the North China Plain in the east is unique and the complete process of transportation and deposition of silt, i.e., the formation material of loess, can be studied here. In addition, there are historical records spanning several thousand years, and more weather and climate observations of the dust storms (recent equivalent of loess formation) are added each year.

The study of all the processes leading to loess deposition will probably answer many unsolved problems related to loess; for example, the source of kaolinite in loess, the reason of the high $CaCO_3$ content in loess, and the formation mechanism of silt size, sharp angular quartz particles. All these problems are of major importance in the protection of the current environment, in particular the knowledge of mechanism of wind transport of dust and a correct comprehension of the propagatioin of solid particles in the atmosphere.

4. Loess-paleosol sequences record the evolutionary history of geological events leading to the formation of loess deposits and the development of paleosols during the past 2.4 million years. In the loess-paleosol sequence geologic and climatic events are inseparable. The well-developed polygenetic paleosol of the 5th and the 2nd paleosol assemblage (W_{s-2}) records the episodes of optimal climate; Malan Loess and 1st and 2nd sandy loess layers record significant dry-cold climatic events. Such extremes happening in the long process of geological history should be regarded as abrupt occurrences of global significance. Loess and paleosols are treated in this book as geologic units sensitive to climate. According to the probable climatic signature of loess and paleosols, the history of long-term climatic fluctuations in the Loess Plateau has been semi-quantitatively reconstructed, and correlated with the oxygen isotope record of the deep-sea sediments. The information

on global climatic fluctuations provided by loess-paleosol sequences over the past 2.4 million years is more detailed, in a sense, than that in deep-sea sediment. By the multispectral analysis of loess-paleosol sequences over the past one million years, the periods of different oscillations have been identified, enabling the prediction of future geological and climatic events in the Loess Plateau and of the trend of future environmental changes. The analysis of short-term periodicities is the major remaining task in the research of environmental trends at the Loess Plateau.

To sum up, various environmental processes and the depositional process of loess, closely related to climatic fluctuations, are reflected by the loess-paleosol sequence of 2.4 million years, and constitute a perfect three-dimensional system. This is a new orientation for further study of loess.

Naturally, different views of individual problems are expressed in parts of this book. It is considered better to present all the different views to the reader and let him draw his own conclusion rather than to nip off the young shoots of a possible breakthrough only for the sake of coherence of the material presented in this book. Some results described in this book are tentative, leaving much work to be done. In this respect readers are requested to make their comments and present their criticism to the authors.

The book presents collective results which were jointly discussed but separately written. Chapter 1 was written by Lu Yanchou, section I of Chapter 2 by Zheng Honghan, section II by Wu Zirong and Yuan Baoyin, section I of Chapter 3 by Liu Tungsheng, Wen Qizhong, An Zhisheng and Han Jiamao, section II by An Zhisheng and Liu Tungsheng, section III by Qiao Yulou, Huang Baolin, Shen Chengde and Zhou Mingfu, section IV by Zheng Honghan, section V by An Zhisheng and Liu Tungsheng, section I of Chapter 4 by Zheng Shaohua, Yuan Baoyin, Gao Fuqing and Sun Fuqing, section III by Chen Deniu, Gao Jiaxiang, Gao Fuqing and Liu Tungsheng, section III by Zhou Kunshu, Lin Shaomeng, Liu Ruiling and Zhu Yizhi, section I of Chapter 5 by Hu Biru and Lu Yanchou, section II by Wei Lanying, section II by Han Jiamao, Zheng Honghan and Gu Xiongfei, section IV by Wen Qizhong, Diao Guiyi and Sun Fuqing, section V by Wen Qizhong and Yu Suhua, section VI by Xiang Mingju, Sun Fuqing and Wen Qizhong, section VII by An Zhisheng and Lu Yanchou, section I of Chapter 6 by Chen Mingyang and Liu Tungsheng, section II by An Zhisheng, Geng Ansong and Zheng Honghan, section III by Liu Tungsheng, An Zhisheng, Wen Qizhong, Zheng Honghan and Lu Yanchou, section IV by Lu Longhua and An Zhisheng, section V by Wen Qizhong, Diao Guiyi and Yu Suhua, section VI by Yu Zhicheng and Gao Fuqing, section VII by Wei Lanying and Lu Yanchou, section VIII by Chen Mingyang, section I of Chapter 7 by Zhang Shouxin, section II by Li Huhou.

Our cordial thanks go to the several institutions that have made contributions to the study of loess and given us support and help. They are the Northwest University, Institute of Hydrogeology and Engineering Geology of the Chinese Academy of Geological Sciences, Hydrogeological Teams No. 1 and 2 of the Bureau of Geology and Mineral Resources of Shaanxi Province, the Huanghe River Water Conservancy Committee of the Ministry of Water Conservancy and Electric Power, the Research Academy of Meterology of the National Bureau of Meteorology, the Ministry of Foundation Base of the Academy of Architecture, the Institute of Water and Soil Conservation, and the Nanjing Institute of Pedology, Academia Sinica. We are indebted, for their concern and assistance in the preparation of this book, to leading members of the Division of Earth Sciences, Xian Branch of Academia Sinica. Participating in the study of clay minerals of loess, the late Gu Xiongfei had also prepared the manuscripts and collected the material, and on his decease we lost a very responsible editor. Wu Zelin did much editing. An Zhisheng and Han Jiamao did a lot in organizing the collective efforts which led to the writing of this book. Shao Xingya, Yu Zhicheng, Shen Xiaodong, Chen Aihua, Wang Zhiyu, Gong Shuhua and Zhang Ying prepared the illustrations. Gui Wenli, Zhang Yaguang, Xu Yun and Cheng Yingliang prepared the photographs. Gao Fuqing, Zhu Yizhi and Zhou Yonghang did much work in arranging the materials and checking the data. Han Jiamao, Cui Jiuxu and Bater were responsible for the arrangement of references. Li Huamei, Wang Junda, Lu Liangcai, Huang Renliang, Wang Mingliang and Li Haibin helped us in loess research work. Zhu Xianmo, Wang Yongyan, Zhang Zonghu, Xi Chengfan, Dai Yingsheng, Ding Guiyu, Zhu Haizhi, Wang Kelu, Zhang De'er, Qiu Shihua, Ma Xinghua, Liu Chun, Liu Chengzuo, Zhu Bingquan, Zhu Fukang, Wang Deqian and Chen Yun gave us support and reviewed sections of this book. We hereby express our thanks to all of them.

CHAPTER 1

A Brief History of Loess Studies in China

1.1 The Significance of the Study of Loess

Loess[1] covers about 10% of the land surface of the Earth and is concentrated in the temperate zones and in semi-arid desert margins: in the so-called "world wheat belt", which are the highly developed industrial and agricultural areas as well as the densely populated areas of the world. The economy as well as the history of mankind is closely linked to the distribution of loess.

In China, loess occupies 440000 km^2 of which thick loess amounts to 273000 km^2 in the middle reaches of the Huanghe (Yellow) River (Liu Tunsheng et al. 1965). Together with the secondary loess (including the North China Plain), the distribution area exceeds 1000000 km^2, accounting for over 1/10 of the total area and more than 1/5 of the tillable land in China. Now, more than two hundred million people are living in the loess and secondary loess belts, amounting to 1/5 of the national population. Since ancient times, the cultural development of China has been closely related to loess. The Lantian Man, the Dali Man, the Dingcun Man and the Hetao Man in the Paleolithic created magnificent prehistoric cultures in the loess region. In the Neolithic, the splended Painted Pottery culture flourished 6000 or 7000 years ago in the loess region. People at that time, created the earliest dry farming of the world by exploiting the unique natural environment of the Loess Plateau, which consequently became one of the first agricultural zones of the world (He Bingti 1969), during the 3000 years from the Shang and Zhou to the Tang and Song dynasties. It was not a fortuitous phenomenon that the political, economic and cultural center of China was located in the loess and secondary loess region. The loess, being loose, porous and homogeneous, is easy to cultivate, and turns into fertile farmland favorable for agricultural production, consequently affecting the cultural development of ancient societies. Loess, however, is subject to erosion. Severe soil erosion in the Loess Plateau not only destroys the ecological equilibrium, but also endangers the lower reaches of the Huanghe River. The annual mean silt discharge of the Huanghe River amounts to 16 hundred million tons or more, 80% of which comes from the Loess Plateau. The key to harnessing the Huanghe River lies in the water and soil conservation on the Plateau. This involves an understanding of the distribution of loess, its geological age, topographic condition, composi-

[1] Loess here refers to loess, secondary loess and loess-like deposits.

tion and the physico-mechanical properties of its components, as well as the formation and evolution of the Loess Plateau itself. In addition, to solve engineering problems, economically and reasonably such as saturation slumping and collapse encountered in the loess region, it is necessary to study the basic properties of loess.

R. F. Flint (1974) pointed out that a major theoretical subject in the earth sciences will be the study of the evolution of the atmosphere, including long-term climatic changes. Loess is the product of atmospheric activity. Abundant biological remains and well-developed loess-paleosol sequences are preserved in the Loess Plateau, recording major geological events that occurred in Eurasia over the past two million years. Bioclimatic changes, the evolution of ancient man, etc. are all recorded. Recent studies (Liu Tungsheng and Yuan Baoyin 1982) have shown that the record of biostratigraphic change preserved in the Luochuan loess profile is comparable to that preserved in the profile of Quaternary deep-sea sediments. The correlation between them will provide clues for resolving the mechanism and dynamic forcing of regional and global long-term climatic changes.

In recent years, many scholars considered that the major subject of earth sciences in the 1990's will be the study of interaction between the geosphere and biosphere, of the processes of global evolution, and understanding of the trends of natural environment in which mankind lives. Since loess lies in the belt in which much of the interaction between geosphere and biosphere is taking place, further study of loess will provide valuable material for solving the above-mentioned tasks.

1.2 The Terminology and Concept of Loess

"Huangtu" (yellow earth) is a general term used by the Chinese people for loose earthy deposits, its range is different from that of "loess" as used in geology. The deposit referred to as "Huangtu" by the Chinese people is an earth, yellow in color, neither sand, nor mud. Hence "Huangtu" should be an accumulation of silt or dust, yellow in color.

The term "Huangtu" was found in Chinese literature as early as some 2000 years ago. For example, "in the morning of one day in the fourth month of the third year of Yuanfeng, Jianshi, in the sovereignty of the Emperor Cheng Di, Han Dynasty (78 B.C.), strong winds blew from the northwest. Red and yellow thin, floating clouds were hanging over the sky and yellow dust (Huangtu) fell on the ground from morning till night". (Ban Gu: *The Books of the Western Han Dynasty*, Wuxingzhi, vol. VII). This account by Ban Gu (A.D. 32–92), a great historian, reflects that the ancient Chinese people had an understanding of the relationship between "Huangtu" and wind-borne dust and of the principal properties of "Huangtu". Ban Gu related the term "dust" to loess, vividly describing the particle size of loess.

The German, English, Russian and French equivalents of the Chinese "Huangtu" are Löss, Loess, Лёсс and Leuss, respectively, which are all derived from German Loß or Loeß. K. Leonard used the word "Loeß" based on the word

"Loesch" for loose soil as named by the local people of the Rhine Valley. The words "Loesch" or "Losch" and "Los", "Losen" are similar in meaning, and all imply loose and/or loose texture. According to the original meanings of Löss, Loess, Лёсс and Leuss, their relation to loess is not as precise as that of the Chinese word "Huangtu".

F. v. Richthofen (1877), carrying out geological investigations in China from 1866 to 1872, made a comparison between "Huangtu" in China and Loss of the Rhine Valley in Germany, and held that loesses in China and Germany are basically the same, all being light, greyish yellow calcareous loam, loose, porous, thick-layered, nonstratified, with vertical joints, precipice-forming, and often containing terrigenous mollusks. F. v. Richthofen gave the definition of loess with these characteristics and considered that loess is generally believed to be eolian in origin, thus distinguishing it from lacustrine loess-like deposits. Later, V. A. Obruchev (1933, 1958) discussed the problem as "What is loess?", etc. Other scholars such as L. S. Berg (1928, 1946), R. J. Russell (1944), R. F. Flint (1947), Ma Rongzhi (Ma Yungchi 1945), S.S. Morozov (1951), J. Dylik (1954, 1961), V. V. Popov (1957), Zhang Zonghu et al. (1959), N. I. Kriger (1965) and I. J. Smalley (1971) have discussed the subject from various points of view. Liu Tungsheng (1958, 1959), Liu Tungsheng and Chang Tsunghu (1962), Liu Tungsheng et al. (1964, 1965, 1966) have made a systematic review of the problem.

According to the results of the study of loess and loess-like deposits in various parts of China and referring to the idea of Obruchev (1933), Liu Tungsheng et al. pointed out in 1965 that loess is believed to be wind-derived, not secondarily disturbed, unstratified, calcareous and porous yellow silty earthy deposit. The Loess Plateau in Shanxi, Shaanxi and Gansu provinces is taken as representative. They further pointed out that besides eolian deposits, yellow, stratified earth silts of other origin with sand and gravel interlayers are called loess-like deposits, and are now called secondary loess. Most of the latter are loess reworked by water and redeposited in arid and semi-arid zones. Their lithologic characters and physico-mechanical properties are similar to those of loess.

1.3 The Process of Loess Formation

The correct concept of loess formation should consider all the processes interacting in the production, transportation and deposition of loess material, as well as in its subsequent changes.

1.3.1 The Source of Loess Material

The composition of loess is commonly characterized by a predominance of silt-sized particles and enrichment of carbonates. Quartz grains compose over half of the silt-sized particles. How are the quartz grains formed? Where does the carbonate come from?

Quartz in loess is in irregular, subangular and angular grains (Wang Kelu and Pei Jingxian 1964), with characteristic surfaces resulting from mechanical fragmentation, such as typical conchoidal fracture, sharp edges, ruptures, etc. (Smalley and Vita-Finzi 1968; Smalley 1971; Lu Yanchou et al. 1976; Liu Tungsheng et al. 1978). This shows that the grains cannot result from in-situ weathering and soil-forming in loess accumulation area as proposed by L. S. Berg (1932).

It is commonly considered that glacial grinding is a favorable mechanism for producing large amounts of silt-sized quartz grains (Smalley 1971). Desert processes, i.e., great temperature differences, particularly under the influence of soluble salts (Na_2SO_4), can also produce fine particles from coarse quartz grains in desert and gobi (rocky desert) regions. The distribution of loess is closely related either to desert and gobi, e.g., the loess in China (Liu Tungsheng and Chang Tsunghu 1962), or to continental glaciation, e.g., loess in Central Europe and North America (Flint 1947; Kriger 1965). The Cenozoic and Mesozoic detrital sediments form the deserts and gobi (about 1.5 million km^2 in area) northwest of the loess area in China. They are probably the source material of the loess accumulation in China.

There are two sources of carbonate in loess: (1) Carbonate which is carried from the source area. Rounded calcite in silt size, called primary calcite, commonly exists in loess in China, and its amount is directly proportional to the $CaCO_3$ content of the loess (Liu Tungsheng et al. 1966). In desert and gobi of arid regions the surface of silt-sized quartz and feldspar grains is frequently coated with $CaCO_3$. This coating is partly carried with the grains, when blown to the loess accumulation area. $CaCO_3$ occupies 3% of the dust sample which was collected at Tempe, in Arizona State on April 22, 1972 and July 21, 1973 by Pewe et al. (1981). (2) Carbonate is believed to be brought into loess by atmospheric precipitation, ground water and biological activities, i.e., the secondary carbonate. High evaporation relative to precipitation in arid and semi-arid regions is favorable for the enrichment of carbonates in loess deposits. For the middle reaches of the Huanghe River, it has been estimated that about 3% – 10% of $CaCO_3$ content is derived from atmospheric precipitation and ground water in the loess-forming process.

1.3.2 Transportation and Sedimentation of Loess

F. v. Richthofen (1877) considered loess a subaerial deposit. V.A. Obruchev (1933) held that silt from Central Asia could be blown by anticyclonic winds, even to the remotest distant regions where it settles and becomes loess. Judging from the records of "dust falls" in Chinese history (Wang Jiayin 1965; Zhang De'er 1982) and the study of a dust storm event which occurred on April 17–21, 1980 in Northern China (Liu Tungsheng et al. 1981, 1982), one can see that the transport and accumulation of dust in China are related to atmospheric activities on various scales closely linked with the development of cold fronts. With the emergence of a strong cold front in Siberia and southern Mongolia and the accompanying strong winds, dust in gobi and desert can be readily lifted and carried in suspension by turbulent eddies and, consequently drift southeastward with a cold front in the westerly current, and eventually settle in the middle reaches of the Huanghe River and in North China. This process is also related to strong anticyclonic systems (see Chap. 6, Sect. I). Most of the "dust fall" areas described in Chinese history lie in the region near the East Asian substationary trough, where the prevailing subsidence favors the fallout of dust and the deposition of loess (Zhang De'er 1982). This is an important condition for the development of thick Chinese loess deposits, that rank first in the world. Since the geographic position, topographic conditions and atmospheric setting of loess areas throughout the world are quite different, the transportation and depositional processes of loess materials are certainly highly diverse, being controlled by atmospheric circulation on several scales and related to site-specific weather processes.

1.3.3 The Biochemical Stage of Loess-forming and Loess Evolution

Dust which settles on the surface of arid and semi-arid regions is affected by rain water, frost, snow and biological activity which proceed in an oxidizing environment in weakly alkaline medium. Secondary carbonate turns the dust into a buff, soft, porous and homogeneous earth with a "mottled" texture. This is the "loess-forming process", biochemical in nature, but including the contribution of the physical acts of congelifraction and evaporation. The above-mentioned process is also called "loessification". Obviously, our concept of loessification is different from that proposed by L.S. Berg (1932).

In arid and semi-arid regions, while experiencing the secondary impregnation by calcium carbonate, the dust accumulation is also affected by other weathering processes, whose intensity depends mainly on the prevailing bioclimatic environment. In the environmental condition of high aridity and low biological production rate, secondary carbonate forms rapidly, while other weathering is weak. However, in the environmental conditions of low aridity and high biological production rate, secondary carbonate develops slowly, while other weathering processes at a significant rate. In such conditions the dust is directly affected by a strong soil-forming process, becoming incorporated into a soil. This is one of the important factors leading to regional differences in the lithologic character of loess series from the same period.

Once deposited, the loess formations develop in different ways under the influence of diverse geological agents in various environments. Liu Tungsheng et al. (1964, 1984) discussed this topic, and further details are given in Section VII of Chapter 5.

The formation and evolutionary processes of loess are diagramatically shown in Fig. 1.1.

1.4 A Brief History of Loess Studies

It is certain that in the progress of history people want to understand loess, the character of the natural environment of loess regions and the history of its devel-

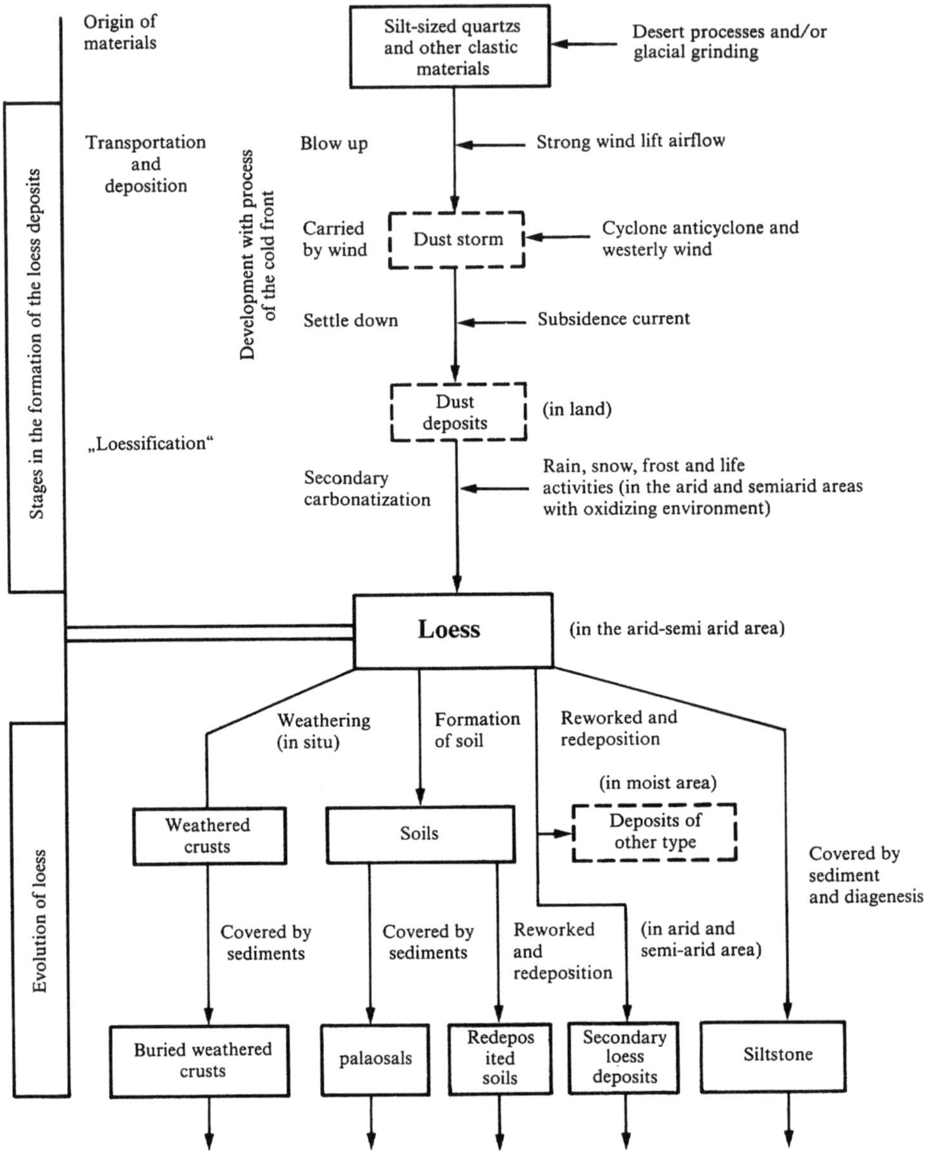

Fig. 1.1. Diagram of the formation and evolutionary processes of loess

opment, and want to utilize this knowledge to transform the environment and exploit it to the greater advantage of mankind.

Loess is closely related to the cultural development of China. There are accounts of topography, river systems, earthquakes, dynamic-geological phenomena and "dust fall" events in ancient Chinese books and records, as well as in local chronicles (Wang Jiayin 1965). These records have not yet been systematically studied.

Since the middle of the last century, with the development of geology, loess has become an object of scientific research. Viewed in the worldwide perspective study of loess has passed through the following stages.

1. Period of Dominant Catastrophism (Before ca. 1840). At the end of the 18th century and at the beginning of the 19th century, the theory of catastrophism, advanced by G. Cuvier, spread widely in geological circles. The study of loess, at that time, was focused on determining whether the loess was formed in the pluvial age.

2. Period of Dominant Neptunism (ca. 1840–1870). The realistic methods of C. Lyell and uniformitarianism exerted a new influence on the study of loess. It was commonly believed, in this period, that loess was water laid and of alluvial, proluvial or lacustrine origin. It is worth pointing out that the studies of mollusks in loess in the Rhine and the Danube Valley made by A. Braun (1847) had a great impact on the study of the Quaternary. Braun's study has shown that xerophilous mollusk shells dominate in loess, while aquatic mollusks are rare. Later, it was found that the majority of mollusk species are now living in coniferous forest belt above the 2100 m elevation in the Alps and the Jura Mountains. These findings provided an important basis for the eolian theory of loess and the understanding of its climatic environment.

3. Development of the Principle of Atmospheric Dust Origin (ca. 1870–1930). Various schools prevailed in this period, most of them approving the principle of atmospheric dust origin, advanced by F. v. Richthofen and further developed by V.A. Obruchev.

After investigating the loesses in North China and in the Rhine Valley, as well as personally experiencing a dust storm in North China, F. v. Richthofen clearly pointed out in his geological book *China* (vol. 1, 1877) that loess is a subaerial deposit. In his opinion loess can only form when sand and dust particles are lifted by the wind and deposited in an arid grassland or in an inland salt lake basin by action of wind and rain. Accordingly, Richthofen classified loess into land loess and lake loess.

After making investigations in Central Asia and North China, V.A. Obruchev (1933) came to a deeper understanding of atmospheric dust (eolian) deposits, pointing out many times that loessial deposits can be classified into typical loess (or primary loess or typical atmospheric dust loess) and secondary loess, which is the loess-like sediment produced by retransportation of typical loess by running water. Being aware of the fact that gobi, desert and loess are arranged in successive belts, he related loess formation with desert and gobi, considering the loess in North China to be a desert loess (or the so-called hot loess), whereas that in western Europe and Central Europe was glacial loess (or the so-called cold loess), genetically related to glaciations and formed in the periglacial zone. He stressed the role of anticyclonic activity in the transportation and sedimentation of loess material (Obruchev 1933, 1951, 1958).

Since the 1960s, Liu Tungsheng and Chang Tsunghu (1962), Liu Tungsheng et al. (1964, 1965, 1966, 1978, 1981, 1982) have carried out further research on the dynamics of loess formation and on its relationship to the Quaternary bioclimatic evolution. They also studied the significance of loess accumulation on the present ecological environment of the loess region.

Simultaneously with the prevailing eolian theory, the principle of in-situ weathering and the soil-formation hypotheses of loess formation were taking shape, and played an important role in the understanding of loess formation. After surveying the loess in China, J. Thorp (1935) proposed that the loess is a kind of arid zone soil. L. S. Berg (1932, 1946, 1947) advanced the concept of loessification. He noted that loess is concentrated in the arid and semi-arid steppe and desert belt, and that it is rich in $CaCO_3$, microaggregates. He explained the high $CaCO_3$ content in the loess by the concept of loessification. However, he neglected the basic character of loess, its great thickness and sorting of mineral particles, disproving his idea of in-situ weathering and eluvial origin of loess. In the present understanding, loessification is the evolutionary process which takes place in the atmospheric dust accumulated in steppe environment and is the final stage of the formation of loess (namely, the biochemical stage of the "loess-forming" process).

In addition, J. Dylik (1954, 1961), a Polish researcher, suggested that loess in Poland is formed by an eolian accumulation of ice-melting and weathering products in a periglacial environment. This hypothesis deepened and broadened the understanding of the loess formation process.

4. Period of the Study of Material Composition of Loess and of Its Engineering Properties (ca. 1940–1970). After the 1930s, the methodology of the composition and of the physical and mechanical properties of loose soils gradually developed, thus providing conditions for better understanding of material composition and basic properties of loess. In the 1940s, especially after World War II, problems of engineering geology such as slope stability in reservoirs and irrigation ditches, subsidence of foundations, saturation collapse, etc. have become part of various engineering studies carried out in the areas of loessial deposits in many countries, especially in the Soviet Union. Since the founding of the People's Republic of China, especially in the 1950s, along with the large-scale economic construction in the loess region of Northwest China, the harnessing of the Huanghe River and the water and soil conservation in the Loess Plateau have been carried out. All of these have greatly promoted the studies of the material composition and physico-mechanical properties of loess and of its engineering characteristics. In this respect, Soviet scholars such as N. Ya. Denisov (1946, 1953a, b) and A.K. Larionov et al. (1959), and others have made an extensive study of loessial deposits in the U.S.S.R. The conditions of the saturation-collapse of loess were analyzed and classifications proposed according to the engineering requirements. Zhang Zonghu et al. (1959), Zhang Zonghu (1962), Huang Qiang (1963) and Lin Zaiguan and Liang Weiming (1982) have carried out research on the physico-mechanical properties of loessial deposits and on their engineering character which guided the construction in the loess belt.

The physico-mechanical properties of loessial deposits and their engineering character are closely related to material composition and texture of the deposits, their geological age, geomorphology and genesis. Hence, many scholars are engaged in the studies of material composition, stratigraphy, geomorphological and genetic types, and in this way greatly enhance our understanding of loess formation.

In view of the needs of economic reconstruction in China and considering the loess in the middle reaches of the Huanghe River to be important of the world loess, Liu Tungsheng (1958), Liu Tungsheng and Chang Tsunghu (1962), Liu Tungsheng et al. (1965, 1966), from 1956 onward, have systematically researched the distribution, stratigraphy, material composition (including grain-size, mineral and chemical composition) and texture of loess and investigated relationship of these properties to the engineering characteristics (such as saturation collapse and slope stability). They compiled "The Map of Loess Distribution Along the Middle Reaches of the Huanghe (Yellow) River" and "The Map of Loess Distribution in China". They classified the loess in China into Wucheng Loess of Early Pleistocene age, Lishi Loess of Middle Pleistocene and Malan Loess of Late Pleistocene. They proposed the differentiation of the loess and the loess-like deposits, and found that the loess in China has a characteristic zonal distribution of thickness, stratigraphy sequence, material composition and physico-mechanical properties. From northwest to southeast, the thickness decreases, the particle size, the $CaCO_3$ content, pH value, void ratio and tendency to saturation collapse also decrease, and there is an increase in dry bulk density as well as an improvement in mechanical properties. Through the study of biological remains (mammal fossils and spores and pollens) preserved in loess and paleosol, they have confirmed that loess in China is the airborne dust accumulated in arid and semi-arid steppe environment. All these results have not only deepened the understanding of the process of loess formation, but also laid the foundation for resolving engineering geological problems of the loess belt in China. Recent studies by Zhai Lisheng (1983) and Gao Guorui (1981) also confirmed the above findings.

5. Period of the Study of Loess Ecology and Environment (from the Late 1960s Onward). From the mid-1960s to beginning of the 1970s, a significant advance took place in the earth sciences, with the emergence of the theory of continental drift, of sea floor spreading, and of plate tectonics. Geology became an important part of environmental sciences. At the end of the 1960s and at the beginning of the 1970s extensive studies on the paleoecology and paleoclimates of deep-sea sediments have been carried out.

The studies mentioned above successfully revealed the paleoclimatic fluctuations and paleoecological changes during the last one million years (Emiliani 1955, 1966, 1972; Hays et al. 1976; Shackleton and Opdyke 1973, 1976), thus providing important leads for probing the causes of long-term climatic variations. However, the deep sea is, after all, far away from the region inhabited by mankind, where perfect paleoclimatic and paleoecologic records are preserved in the continental loess deposits. The studies of loess-paleosol sequences in the Loess Plateau by the authors (Liu Tungsheng et al. 1965, 1978, 1981, 1984; Lu Yanchou and An Zhisheng 1979) and of the loess in Central Europe by G. Kukla (1961, 1970, 1977) indicated that the records of paleoclimatic and paleoecological variations are comparable to those of the deep-sea sediments. More significantly, loess is closely connected with human health. In view of this understanding, specialists in earth sciences and in medical sciences have jointly investigated the relationship between endemic diseases in the loess region (such as Keshan disease, Kashin-Beck disease, local fluourine poisoning, etc.) and loess geochemistry and biochemistry, studied the ecological systems of the loess region, and examined the relationship between the fertility of loess region (including soil elements and micromorphology) and increase in crop production, as well as the improvement of the environmental quality of the loess region.

Today, many scientists are concerned with the influence of the shift of deserts and the increase of atmospheric dust on health. The loess in China shows a close reaction to the expansion of deserts. The study of loess sections in the Loess Plateau has revealed many advances and retreats of deserts in geologic and historical times (Liu Tungsheng et al. 1984), thus providing the basis for the study of future expansions of deserts in China and the necessary preventive measures. But the researches mentioned above are still in an initial stage of investigation.

The study of the paleoenvironment and the modern environment of the loess region has also revealed that the understanding of loess formation has been deepening along with the progress of civilization and the development of science and technology. The scientific concept of loess involves the understanding of the complete cycle of loess formation, of the basic properties of loess, the environmental evolution and forecast, and the practical application of this information to production in order to speed up the progress of human society. All the abovementioned findings are the result of studies made during the past 100 years or so. The authors of this book try to describe different aspects of loess formation and of the evolution of bioclimatic environment in ancient and modern times, using the Luochuan profile as an example. They also discuss some geological and environmental projects related to the economic development of the Loess Plateau.

CHAPTER 2

Regional Distribution of the Aeolian Products (Loess)

2.1 Loess Distribution

Study of loess distribution is closely related to the study of loess origin. Results obtained from such study form an important basis for protecting environment and planning industrial and agricultural production. The study of loess distribution in China can be divided into three stages as follows:

First stage (from the latter half of the last century to the founding of the People's Republic of China). Some foreign scientists working in China at that time, such as B. Pumpelly, B. Willis, F. v. Richthofen and V. A. Obruchev, put forward views on loess origin based on geologic investigations. At the same time, maps of loess distribution were compiled by V.A. Obruchev (1958), H. Schmithener (1932), G.B. Cressey (1934), Masukenkichi and Todatatsu (1944), respectively. However, all these maps are mere sketches, outlining loess distribution theoretically but without detailed information. It is worth noting that the results of loess stratigraphic and paleontological studies done by P. Teilhard de Chardin and C.C. Young (1930) at that time, especially their study on fossil rodents from loess, are of great importance in distinguishing different layers of loess (1931). Additionally, the study of the Malan terrace by J.G. Andersson (1939), the Sharawusu Formation by P. Teilhard de Chardin and E. Licent (1930), and the observation of buried soil from loess by J. Thorp (1936) have deepened the understanding of loess development and distribution at different stages as well as the recognition of its environmental history.

Second stage (from 1949 to the 1970s). After the founding of the People's Republic of China, personnel from production units, universities and scientific research institutions did much work in loess regions from geological, geographical, pedological, hydrological, textural and mechanical aspects, oriented toward the water and soil conservation, engineering construction and land rehabilitation. This work not only provided many new data important in production, but also advanced the scientific knowledge of loess, revealing more about loess distribution. Several papers and maps were published at that time by Liu Tungsheng et al. (1964, 1965, 1966) of the Institute of Geology, Academia Sinica. They summarized systematic regularities in slope orientation and loess zonation, as well as the relationship between loess and fluvio-lacustrine strata, loess and paleosols, loess and underlying relief, and so on. The results have provided an important scientific basis for understanding the source of loess, and for distinguishing the engineering qualities and water-bearing properties of loess at different localities.

Engineering construction, land utilization and rehabilitation were greatly aided by the results.

Third stage (since the 1970s). With the extensive application of radiocarbon chronology, magnetostratigraphy and thermoluminescence dating in loess research, a preliminary chronological scale was obtained for exploring the loess depositional cycles, their regional record and variations. The results of biostratigraphy, pedostratigraphy, mineral stratigraphy and geochemistry, the depositional cycles, the intercalations of loess paleosol, or of loess and fluvio-lacustrine strata can now be placed into a reliable chronologic framework. Furthermore, we can make extensive climatostratigraphic correlations. Various authors have discussed climatic records of the past 2.4 million years, the beginnings of loess deposition in China (Liu Tungsheng et al. 1980a), 1.1 million years (Lu Yanchou and An Zhisheng 1979), 120 thousand years (An Zhisheng and Lu Yanchou 1984), the last 10 thousand years (Zheng Honghan 1984) and the last 5 thousand years (Zhu Kezhen 1973). In addition, meteorological and climatological methods were adopted to study historic and modern fallen dust (Wang Jiayin 1965; Zhang De'er 1982, 1984). All these have deepened our understanding of loess distribution (Liu Tungsheng et al. 1982). With the application of meteorological methods and satellite images in the study of dust falls, it has been further confirmed that loess is an aeolian deposit and its material is transported by wind.

The third stage of the study of loess distribution found that:

1. The horizontal loess distribution was controlled not only by the agent which transported the loess material, but also by the mode of rock disintegration, aeolian transportation, sedimentation and weathering. This understanding led to a development of a sedimentary model valid throughout the loess belt – from the source area through the accumulation region to the erosion belt.
2. The vertical loess distribution goes from the Pleistocene through the Holocene to the present, and the study expands into the future.

2.1.1 Area of Loess Cover

Loess in China is widespread between the northern foothills of the Kunlun Mts. and Qinling Mts. and the southern foothills of the Altay, Alashan and the Da Hingan Mts., forming a belt elongated from NWW to the SEE. The eastern part of the loess belt extends from the Songhuajiang – Nenjiang rivers in the north (typical loess starts from mountains of the western Liaoning Province and the Rehe mountains), to the middle and lower reaches of the Changjiang (Yangtze) River in the south, running between 30 and 49 °N. The best-developed loess lies between 34 and 45 °N, along the middle reaches of the Huanghe River, where it attains its greatest thickness. Middle reaches of the Huanghe River form the center of loess sedimentation in China (Fig. 2.1).

As for the area of loess cover, varying figures have been given by different authors due to different concepts and accuracy of the studies. According to our data, the area of loess cover, as estimated by Zheng Honghan from *The Map of Loess Distribution in China* (1/2,000,000) compiled by Prof. Liu Tungsheng et al.

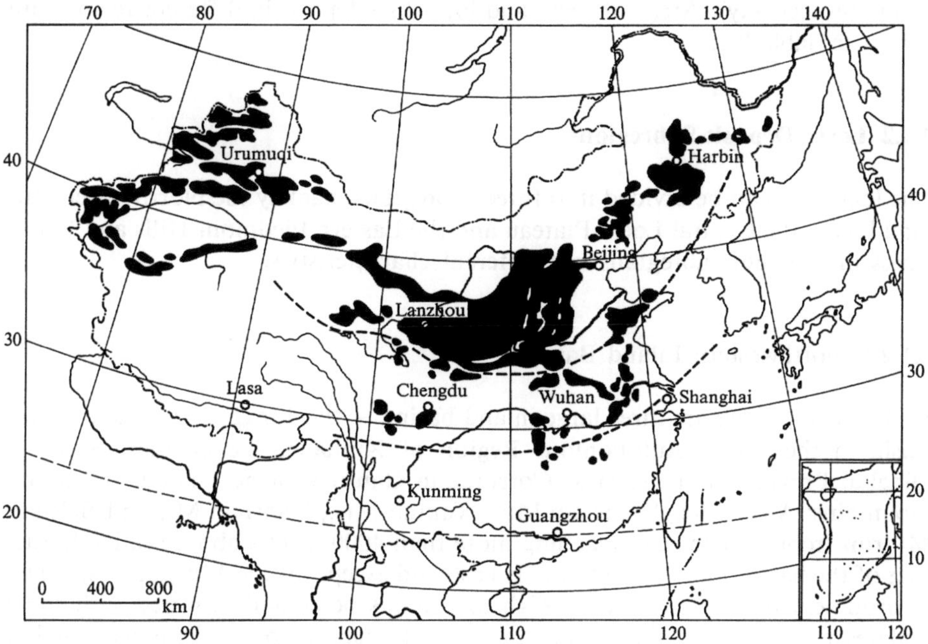

Fig. 2.1. Sketch map showing the loess distribution in China

Table 2.1. Areas of Loess and Loess-like Deposits in China[a]

	Loess		Loess-like deposits	
	Area km²	Percent of total loess area	Area km²	Percent of total loess-like deposit area
Songhuajiang-Liaohe Plain	11,800	3.09	81,000	31.83
Huanghe valley lower R.	26,000	6.82	3,880	1.56
Huanghe valley middle R.	275,600	72.36	2,400	0.94
Huanghe valley Qinghai	16,000	4.20	8,800	3.45
Hexi-Corridor of Gansu	1,200	0.35	15,520	6.09
Uygur of Junggar	15,840	4.15	91,840	36.09
Xinjiang Tarim	34,400	9.03	51,000	20.04
Total area	380,840	100.00	254,440	100.00

[a] Liu Tungsheng et al. (1965).

(1966), is 440,000 km^2. The loess-like deposits cover an area of 191,840 km^2. By comparing the loess area and loess-like deposits area with the total land area of China, loess occupies 4.4%, and loess-like deposit approximately 1.9%. Later, based on *The Map of Loess Distribution in China* (1/4,000,000) Liu Tungsheng and his coworkers estimated that the loess area is 380,840 km^2 and the loess-like deposit area is 254,440 km^2 (occupying 3.9% and 2.65% of the total land area of

China respectively). Areas of loess and loess-like deposit in different regions are shown in Table 2.1.

2.1.2 Loess Deposit Subregions

The loess belt can be divided into three subregions, namely the Northwestern Inland Basin, the Central Loess Plateau and the Eastern Piedmont Hills and Plain. Loess in different subregions has different characteristics.

2.1.2.1 Northwestern Inland Basin

In this region there are three large inland basins – the Junggar Basin and Tarim Basin in the Uygur Autonomous Region of Xinjiang, the Qaidam Basin in Qinghai Province and the Hexi Corridor in Gansu Province. The surrounding mountains; Altay Mts., Tianshan Mts., Kunlun Mts., Nanshan Mt. and Beishan Mt. run from west to east, reaching more than 4000 meters above sea level. Surface of the basin floors dips gradually eastward except for the Junggar Basin. The environment is characterized by high mountains perennially covered with snow and deserts and saline lakes in the center. Quaternary deposits of alluvial, eluvial, deluvial and proluvial origin lie between the deserts (or saline lakes) and the mountains, appearing as a belt, with loess covering the mountain foot hills. In the Junggar Basin, loess cover lies at the foot of the western mountains and south of the mountains, in the Kupu Valley. In the southern part, loess is widespread at the northern foot of the Tianshan Mts., appearing as a belt best developed in the eastern Tianshan Mts. Loess of the Tarim Basin is distributed mainly near the Yengisar and Keiyang regions at the northern foot of the Kunlun Mountains in the western and southwestern parts of the basin. Loess is also developed in the area between Haquibuha and Ala Gou at the southern foot of the Tianshan Mountains. Loess of the Qaidam Basin is distributed in Mingshan region in the eastern part of the basin and along the northern slope of the Kunlun Mountains. Loess covers are mainly in the piedmont belt north of Xiligou, Qagan Us, Xiangride and Burhan Budai Mountain. Loess in the Hexi Corridor is mainly distributed in the eastern part of Jiuquan, ranging from Minle through Shandan, Wuwei, Gulang and Tianzhu counties, to Wushao Mountain.

Loess in the inland basin of northwestern China is located basically in a windswept belt. Loess is liable to be retransported and redeposited by wind or by glaciers, forming different kinds of loess-like deposits under conditions of dry climate, high surface radiation and great daily and annual temperature differences. According to the correlation of lithologic characteristics of the loess of Tuomur Peak in the Tianshan Mts. done in the last several years, the well-preserved original loess formed possibly from Late Pleistocene to Holocene, lying mainly in the eastern part and western part of the basin and on the northern slopes of the mountains. Generally speaking, loess contains some glacial detritus and rich soluble minerals in varying amounts at different localities. Samples from Tuomur Peak show large differences in particle size and mineral composition at different localities (Chap. 5). Loess of Qaidam Basin, Qinghai Province and at other places is also rich in salt and mirabilite.

2.1.2.2 Central Loess Plateau

The valley of the Huanghe River from Longyang Gorge to Sanmen Gorge belongs to its middle reaches, and it is the center of loess deposition in China. The surrounding mountains are Helan Mt. in the west, Yinshan Mt., including Langshan Mt., Ura Mt. and Daqing Mt. in the north, the Taihang Mts. in the east, and the Qinling Mts. in the south. The Loess Plateau is distributed in the area from the southern edge of the desert near the Great Wall of China in the north to the northern slopes of the Qinling Mts. in the south. Additionally, loess possibly extends to the south of the Qinling Mts. through mountain passes at some localities.

Loess of the central plateau has a great thickness. It covers the Tertiary or other old strata, forming different kinds of loess landforms – "Yuan", "Liang" and "Mao"[2]. The Loess Plateau can be divided into three subregions by mountains trending in approximate N-S directions. The western subregion is bounded by the Wushao Mts. and the Liupan Mts. Within this subregion, loess covers the mountain slopes, the intermontane basin and the higher terraces. The central subregion is located between the Liupan Mts. and the Luliang Mts. Here loess appears in the form of continuous cover overlying the Pliocene Red Clay, filling most of the ancient valleys and basins. The loess strata are about 100 m thick; bedrock is exposed in some deep cut valley bottoms, loess and paleosols of different ages in loess "Yuan" being an important characteristics of this subregion. However, loess of different ages shows angular unconformities or is intercalated with fluvio-lacustrine deposits. All these features suggest the existence of sedimentary hiatuses in loess sections. Loess and aeolian sand are superimposed alternately in the northern part of the Loess Plateau near the desert edge, showing the rise and fall of wind action since early Quaternary. In the eastern subregion between the Luliang and Taihang Mts., the basin and the surrounding mountains form a sharp relief contrast. Here loess covers mainly the rim of the basin or the river terraces. A thin loess layer covers the dividing ridges of intermontane basins.

Although the landscapes in the three subregions are quite different, the stratigraphy of loess sections from Early to Late Pleistocene is similar. Regional differences in stratigraphy at different localities is mainly manifested by: paleosol layers of different intensity, occurrence of fluvio-lacustrine strata, disconformities of loess of different age, and intercalations of aeolian sand. In general, paleosols and fluviolacustrine strata are well developed in the eastern part, whereas the intercalations of aeolian sand are more common in the western and northern parts.

Shown in Fig. 2.2 is a tentative comparison between the loess sections in the Datong basin of Shanxi Province, Luochuan "Yuan" of Shanxi Province and the Longxi basin of Gansu Province. It shows varying paleosol development during the same pedogenic period. The general tendency is that the thickness of paleosol

---

[2] "Yuan", "Liang" and "Mao" are local names. "Yuan" means a high table land consisting of thick loess, "Liang", a flat ridge covered by loess, and "Mao", a round hill with loess as its cap.

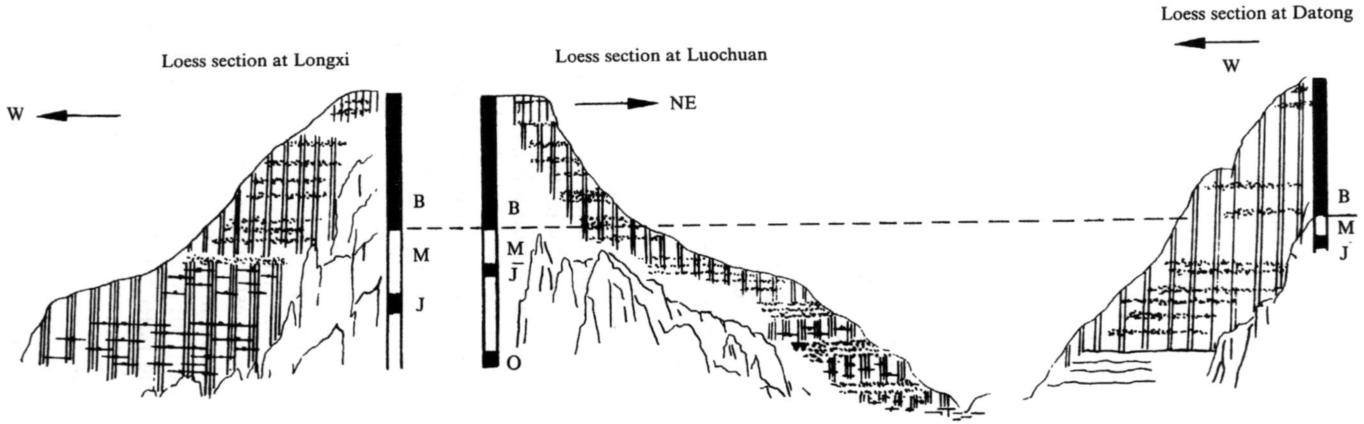

Fig. 2.2. Comparison of paleosol layers in different pedogenic stages

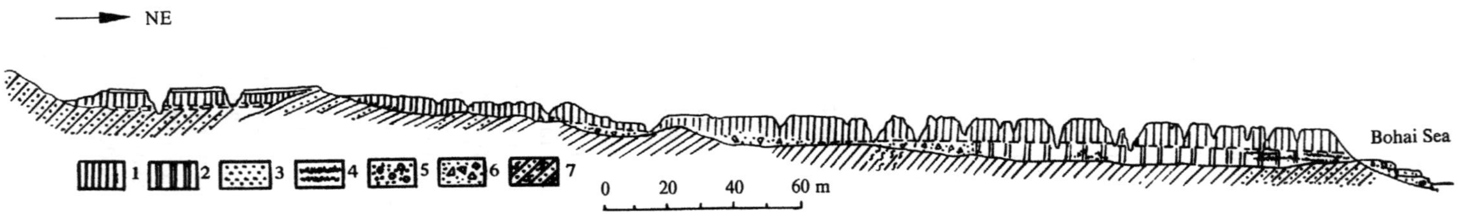

Fig. 2.3. The Quaternary geological section at Toji Island, Shandong Province. *1* Malan Loess; *2* upper part of Lishi Loess; *3* detritus; *4* paleosol; *5* sand and gravel; *6* weathering layer; *7* bedrock

layers increases eastward and southward with regional variation in different stages. Luochuan is a center of loess deposition in China, and the loess sections there contain the best pedogenetic records.

2.1.2.3 Eastern Piedmont Hills and Plain

This region is bound by the Da Hinggan Mts., Yanshan Mt., Taihang Mts., the eastern part of the Qinling Mts. and Dabie Mts. Geomorphically, the loess units cover the surrounding piedmont hills, the Great Plain, Miaodao Islands, lower mountains and hills in the eastern and south-central Shandong Province, and the hills near Nanjing. Loess distribution is discontinuous. Except for the piedmont hills and the surrounding mountains, most of the area is occupied by the alluvial plain, which received thick loess-like deposits with fluvio-lacustrine interlayers of sand, gravel and clays. Loess cover is separated by extensive alluvial plains and forms isolated islands. The northern boundary of typical loess lies near Dabaling, north of Bairin and Tuchengzi. Loess spreads from there eastward and southward, to Ongniud, Chifeng, Jianping, Bianqiang, Chaoyang, Lingyuan, Pingquan, Chengde and Longhua counties to the west of the Rehe mountains and west to the Liaohe River plain. Loess covers the top of gentle hills, mountain slopes and higher terraces. According to the loess section found at Harqin by Ding Menglin, at least seven layers of paleosol underlying Malan Loess can be distinguished, and most of them contain *Prosipheus intermedius* indicative of Early Pleistocene Wucheng Loess.

It has been found in the field by Liu Tungsheng and his coworkers recently that the loess on Miaodao Islands seems to be influenced by the retreat of sea wave action, and is mixed with marine gravel and elluvial-dilluvial detritus. But the loess itself has characteristics to common continental loess with loose texture, porous structure and rich carbonate content. Loess material was obviously transported by wind and deposited there before the separation of the islands from the continent. The loess section contains at least three layers of paleosol, and a typical section began its development at least about 300,000 years ago, continuing up to Late Pleistocene. A few vertebrate fossils have been found – *Coelodonta antiquitatis, Mammuthus primigenius, Cervus canadiensis, Pseudaxis hortulonum, Struthiolithus anderssoni,* etc. at Toji island. These fossils are all closely similar to those from Malan Loess in northern Shanxi Province. Han Jingtai[3] has found that most of redeposited marine foraminifera in Malan Loess of Miaodao Islands, are about 100,000 years old. The lithological character of loess, including its physical and heavy mineral characteristics, excludes marine origin. A reasonable interpretation of such findings is that an extensive marine plain formed after the sea level dropped in the Malan stage or some other colder stages, and the Miaodao Islands received some material transported by wind from the marine plain with the majority of detritus coming from northwestern China as a windblown dust. Such a loess can be recognized as another subtype, different not only from that produced by glaciation (such as the Tuomur Peak loess), but

[3] Han Jingtai, 1984, A discussion on the genesis of the loess in Penglaige, Shandong.

also from the plateau loess surrounding deserts. Aeolian loess with admixture of marine is best developed in Miaodao Islands and nearby Penglaige, so it can be called "Penglaige Loess" (Fig. 2.3).

Within the boundary of Shandong Province, loess is mainly distributed in intermontane basins and piedmont belt on the northern slope of the Mount Taishan and the Lushan Mt. with a thickness from 5 to 7 m to up to 20–30 m. A thin loess cover can be found in the northwestern part of Shandong Province and at some sites in front of the mountains in Shandong Peninsula.

Xiashu loam section near Nanjiang at about 32 °N was regarded as a Late Pleistocene "Malan Loess". However, An Zhisheng has recently found that the middle and lower parts of the section are different from the Malan deposits and that the four layers of reddish buried soil can be compared with the upper part of "Lishi Loess" covering some 500,000 years (Chap. 5, Sect. VII).

2.1.3 Some Aspects of the Loess Distribution in China

New observations on the loess distribution in China have been made in recent investigations since the late 1970s and especially since the 1980s.

1. The aeolian origin of loess was further supported by observed records of "dust rain" distribution in historical times, by the modern meteorological and atmospheric physical analysis of weather processes leading to dust storms and by the detailed geological study of loess distribution and its environment. The conclusion can be drawn that loess is undoubtedly a silt transported by wind from inland basins under westerly winds and deposited on marginal grasslands under dry conditions.

2. Loess distribution from west to east reflects the development of loess from detritus production through aeolian transportation to sedimentation and secondary variations. Inland basins in western China are important source areas for dust material, rims of the basins and the central Loess Plateau are dust deposition regions, and the eastern piedmont hills and plains are areas of resedimentation and erosion. Careful recording of the whole process of silt formation and transportation by westerlies suggests that the environment of China is especially suitable for the development of loess. The loess belt in China is an unparalleled natural laboratory for the study of aeolian dust formation.

3. The loess in China is widespread over an extensive area encompassing different geomorphic units and climatic zone. Besides loess cover at the periphery of deserts in the middle reaches of the Huanghe River, other loess has been found recently in the western area and the eastern area – Tuomur Peak loess and Penglaige loess. The occurrence of loess with regional characteristics indicates the existence of various types of loess in China.

4. It can be seen that loess distribution has a clear regional character. Loess belt shifted longitudinally in different stages, alternating warmer and colder climatic conditions lead to cyclical loess sedimentation, and loesses of different stages show continuous superposition from Early to Late Pleistocene.

By study of loess distribution, not only are the characteristics of natural environmental processes revealed, but also more valuable results achieved.

2.2 Geological Structure of Basins with Loess Sediments

2.2.1 Basins with Loess Sediments and Their Strata

The Loess Plateau is developed on the basis of a few large and relatively shallow basins with the Mesozoic basement. The thick loess deposited in these basins in Quaternary and the landscape of the Loess Plateau emerged. A number of large loess "Yuan" are distributed east of Liupan Mountains to west of the Luliang Mountains at about 36 °N. Several examples are the Dongzi "Yuan", the Luochuan "Yuan" the Jixian "Yuan". Because of their complete stratigraphic sequences, research on them will allow a systematic recognition of the natural environment of loess evolution. Such knowledge may also be used in addressing production problems of the Loess Plateau, such as soil erosion, agricultural regionalization and natural resources exploitation. The Luochuan "Yuan" is an ideal area for research on formation and geological structure of loess "Yuan", not only because of its thick loess sediments, complete strata and clear contact relation, but also for its mark beds for dividing the strata and reflecting climatic character.

Luochuan loess deposition basin ranges from Huanglong Mountain 1748 m east, to Ziwuling Mountain 1800–1900 m west, and Yijunliang Mountain 1500 m south, and to the Laoshan Mountain divide at 1450 m north, with an average height of 1100 m. Luochuan "Yuan" is situated at the center part of the basin. Its total area is about 3500 km^2 (length-70 km, width-50 km). Red Clay of the Late Pliocene about 20 m thick and loess over 100 m thick are deposited there.

1. Red Clay of the Late Pliocene. This widely distributed sediment overlies a Pre-Tertiary base. It ranges in thickness from 5–10 m to the southeast to a maximum of 30–40 m to the northwest. It is mainly composed of brown-red clay containing calcareous nodules with yellow-brown sand and gravel at the bottom.

2. Wucheng Loess of the Early Pleistocene. In the Wucheng Loess, which is widely found in Luochuan with a thickness of 40–50 m, loess is interbedded with fossil soils and calcareous nodule beds. A sandy loess is found at the base of the sequence, and is known as the bottom sandy loess, suggesting that the climate was drier and colder when it was deposited. It thus has an important implication for stratigraphical division and the study of climatic change.

The Wucheng Loess is occasionally underlain by a "mixed bed", 1–2 m thick, which is characterized by mixture of loess and Red Clay of the Late Pliocene. A "pill structure" is also occasionally present. Sometimes fluvio-lacustrine sediments, 5–15 m thick can be found in one or two sections, such as in the Qingchi section, southeast of Luochuan.

3. Lishi Loess of the Middle Pleistocene. Lishi Loess overlapes the Wucheng Loess and is about 70 m thick. 14–15 loess and paleosol sequences can be found in Lishi Loess. Each individual paleosol is well developed and has a thickness of more than 1 m. The fifth paleosol is the best, often ranging between 4 and 6 m, which has the strongest red color and is in fact a complex paleosol consisting of 2 or 3 paleosols. Because of its characteristics, the fifth paleosol can easily be recognized in the field and is regarded as a key for distinguishing the upper and lower parts of Lishi Loess.

The upper part of Lishi Loess, about 25–30 m thick, contains 3–4 thick paleosols, which are dark in color, well developed and contain calcareous nodules. The loess is greyish yellow to light brown in color, and in it vertical joints and large pores are developed.

The lower part of Lishi Loess contains 9–10 paleosols and 2 sandy loesses (the first and the second sandy loesses) with a thickness of 40–50 m. The lower portion of the Lishi Loess is darker in color than its upper portion, but the paleosols are light in color as compared to the upper part of the Lishi Loess. The loess of the lower part is harder, with fewer large pores and occasionally intercalated with a few calcareous nodules. According to paleomagnetic research (Heller and Liu Tungsheng 1984), the boundary of B/M is situated near the first sandy loess (between the seventh and the eighth paleosols). Obviously, the first sandy loess is a marker bed subdividing the Lishi Loess and the second sandy loess is the boundary of Lishi and Wucheng Loess. It was recognized as such for a long time, indicating that the sandy loesses can also mark stratigraphic divisions.

4. Malan Loess of the Late Pleistocene. Malan Loess is rather thin, no more than 10 m. It is distributed from the local erosion basis to the piedmont slope, and thus is wider in distribution than the Wucheng Loess and Lishi Loess. It overlaps unconformably onto the older loess and other strata. The lithological character of Malan Loess is similar to that of the upper part of Lishi Loess, being light grey-yellow silt, loose, with large pores, plant roots and vertical joints. Instead of brown-red paleosols, black loam with a thickness of 0.5–1 m can usually be found in Malan Loess. Loess of less than 1 m thick on the black loam in the Luochuan region may be Holocene in age.

2.2.2 Division of Ages for Luochuan Loess-Paleosol

Based on the study of the various dating techniques, the chronological sequence of the Luochuan loess has been established. The ages of loess strata obtained from the above-mentioned research provide a base for better understanding of loess. According to these age data, together with characters of mammalian fossils, spores and pollen, and lithology, etc., a scheme of important geological events has been proposed (Fig. 2.4). There are five geological boundaries especially suited for the loess of the Luochuan "Yuan".

1. The First Paleosol (S_1) with an Age of More Than 100000 yr B. P. (by Thermoluminescence Dating and Accumulation Rate of Loess). The first paleosol underlying Malan Loess is the obvious boundary easily identified in the place where Malan Loess is deposited, and dates from about 100000 yr B. P. Loess has been accumulated rapidly since then.

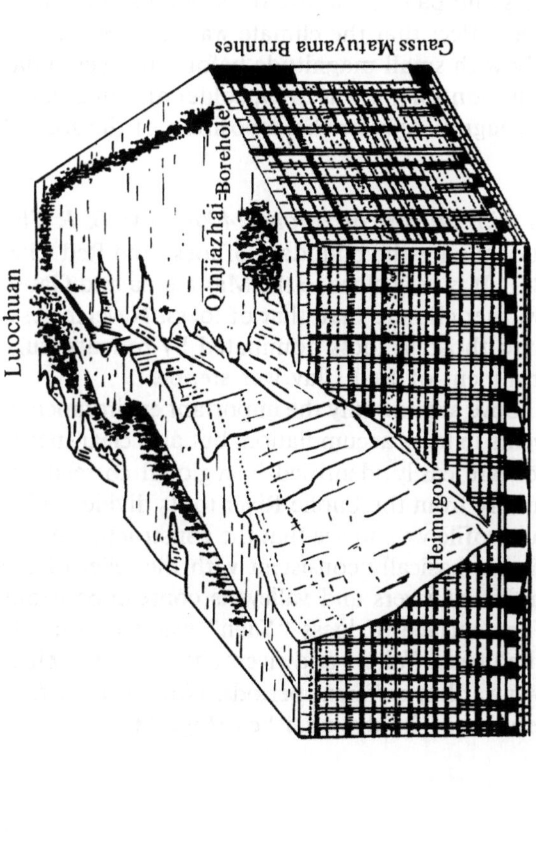

Fig. 2.4. Block diagram of loess strata at Heimugou, Luochuan

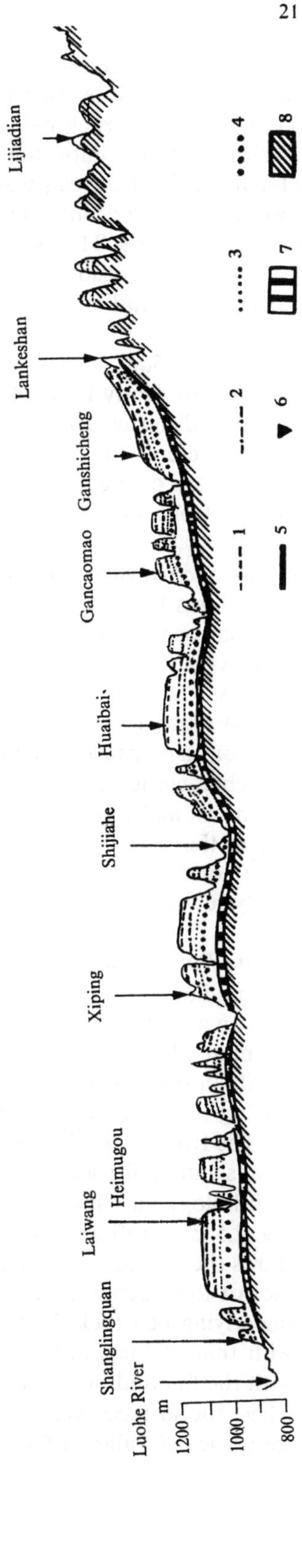

Fig. 2.5. The profile showing loess strata of Luohe River-Lijiadian (formation phases and ages of river course and valley in the Loess Plateau). *1* S_1; *2* S_5; *3* upper sandy loess; *4* lower sandy loess; *5* lower boundary of loess; *6* site of profile; *7* Red Clay; *8* bedrock

2. The Fifth Paleosol (S_5) with an Age of About 500000 yr B. P. From magnetostratigraphy, thermoluminescent dating and the rate of sedimentation of loess, the age of each paleosol has been calculated. The S_5 is the most stable. It is a distinct and important geological boundary dated at 500000 yr B. P. The climate at that time was warm and moist. The sedimentation rate of the loess obviously decreased and the sedimentary gaps and erosion surfaces appear near rivers and valleys. This boundary has an important implication in the climatic and paleontological study. It can be compared with the tenth bed of the Locality 1 in the Zhoukoudian region where fossils of Peking Man were first found.

3. The First Sandy Loess with an Age of About 800000 yr B. P. In the Luochuan region, this sandy loess (4–6 m thick), covered by the eighth paleosol, is easily recognized in the field because of its stable horizon and obvious characteristics. This represents the most intense period of loess. A paleomagnetic survey shows that the age of the bed is about 800000 yr B. P. (the boundary of B/M is located not far from it).

4. The Second Sandy Loess with an Age of About 1150000 yr B. P. The second sandy loess has the same characteristics as the first one. It is situated under the 14th paleosol with an age of about 1150000 yr. B. P. The importance of this sandy loess is that it represents a period of great changes in the loess depositional process. The crowed brown-red paleosols appear in the loess sediments under the second sandy loess. Also the loesses between the paleosols are all thin and dark in color, in contrast to the thick loess and paleosol above the second sandy loess. The characteristics mentioned above reflect that the climate was relatively warm and moist and fluctuated frequently with small magnitude before the accumulation of the second sandy loess. On the contrary, it was drier, colder and fluctuated less frequently with relatively large magnitude after the deposition of the second sandy loess.

5. The Bottom Boundary of Loess with an Age of About 2400000 yr B. P. The bottom boundary of the loess in Luochuan is the boundary of loess and Red Clay with an age of about 2400000 yr B. P. It is very close to the Matuyama and Gauss boundary. Red Clay represents a warmer climate than that of loess. Nevertheless, fossils in the Red Clay are similar to those in loess dominated by rodents, suggesting that the sedimentary environment of Red Clay was steppe.

The five boundaries mentioned above are not only the important environmental breaks and significant geological events in loess accumulation, but also clear marks of lithology and stratigraphy, which are easily identified in regional correlation. Fig. 2.5 is a profile measured essentially from the Luohe River to its divide, which shows the five boundaries. From the profile we can see that the constructional surface of loess deposits of every period is basically consistent with the relief of the underlying bed rock. The development of rivers and valleys is contemporaneous with that of loess accumulation. The surface of loess "Yuan" was conformable with the flat and broad surface of bed rock. Valleys and rivers shaped in the place where the surfaces were cutting downward in different periods. Now, at least four categories of valleys of various periods can be distinguished (Fig. 2.6).

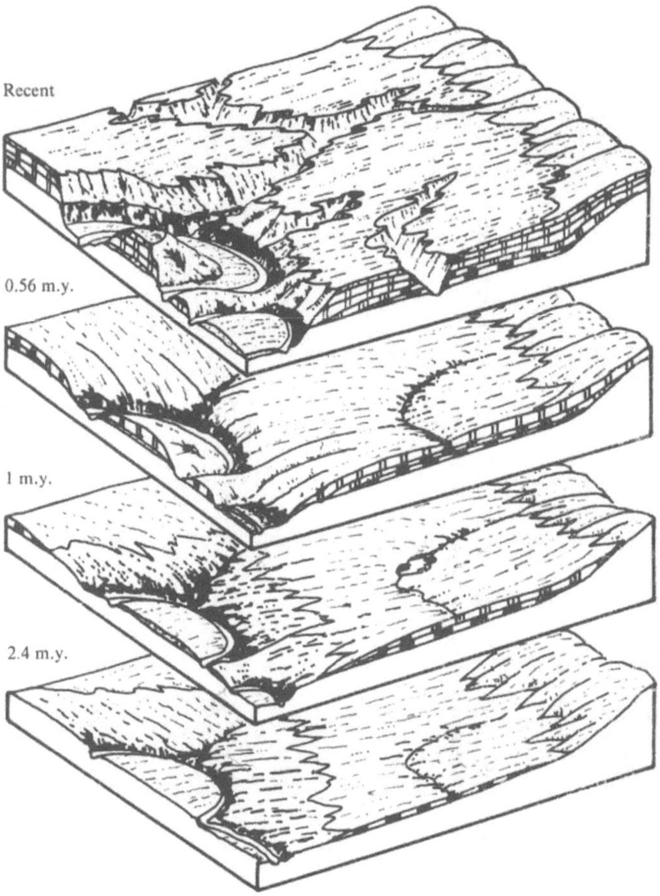

Fig. 2.6. Developing process of valleys in loess "Yuan" region of Luochuan

1. The early stage valleys include the Luohe River and its first tributaries. Figure 2.6 shows the Quaternary geological structure of the Luohe River; an incised meanders in the Loess Plateau with a height difference of about 200 m from the surface of loess "Yuan" to the river bed. The rock is exposed on both sides of the valley 80 – 90 m higher than the river bed. It forms a fairly steep slope. The bed rock surface inclines gently to the valley, forming the original bedrock basin of Luochuan. This is where the Red Clay of the Late Pliocene and loess were accumulated and the geomorphology of loess "Yuan" was finally shaped. Red Clay of Late Pliocene was absent in most areas near Luohe River. The terraces of the valley are covered with loess, but only 4 paleosols were developed in it (Fig. 2.7). The fifth paleosol dips abruptly (21°) to the river at the place where loess "Yuan" and valley meet. Here, the fifth paleosol is thick due to redeposition caused by the slope erosion. This indicates that in the development of the fifth paleosol there was a rather strong erosion. According to the previously mentioned data, we suggest that the Luohe River had been in its embryonic stage after sedimenta-

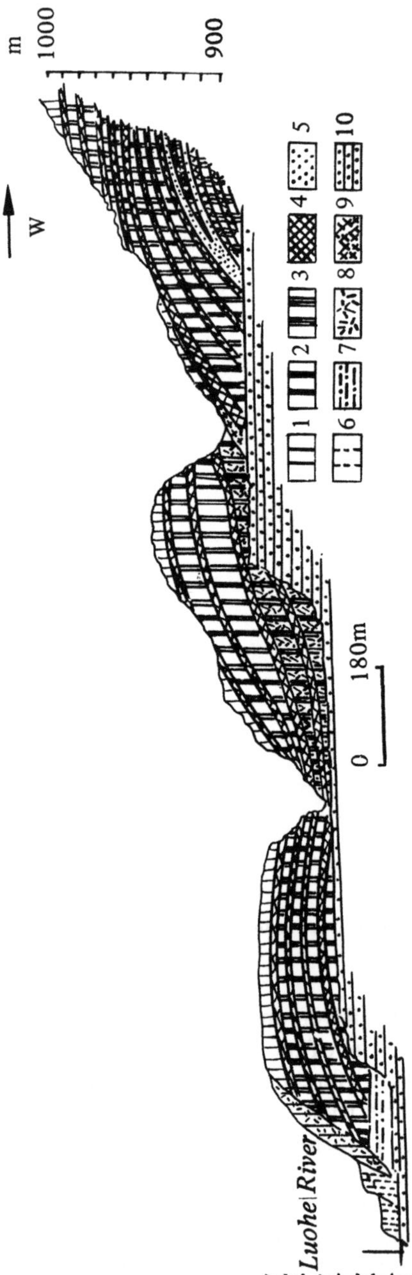

Fig. 2.7. Geological profile of the Quaternary sediments at Xialingquan, Luochuan. *1* Malan Loess; *2* Lishi Loess; *3* Wucheng Loess; *4* palesol; *5* sandy loess; *6* loess-like sediments; *7* silty clay; *8* slope wash; *9* redeposited paleosol; *10* bedrock

tion of the Red Clay of the Late Pliocene and developed continually until the fifth paleosol formed. Because of the long period of erosion, a steep slope of bedrock with a height of 80–90 m can be found in the Luohe River within the form of incised meanders. There are three terraces in the valley: the first and the second are accumulational terraces formed in the Holocene and in the late Malan stage respectively. The third is a terrace with bedrock basement of about 2 m high. On the surface of the rock basement, there are, in ascending order, gravel and sand beds of fluvial facies about 8 m thick and wind-borne loess with four paleosols. Thus the third terrace was first formed roughly at a time simultaneous with that of the development of the fifth paleosol, or even earlier. It means that at the time when the fifth paleosol was formed the river had nearly the same drainage pattern as that of today. Correlated with the geological age of the fifth paleosol the Luohe River might have already existed at 500000 yr B. P. The primary tributaries entering the Luohe River directly, such as the Jiezi Stream and Xiangu Stream, were developed at the same ages as that of the Luohe River.

2. The middle stage valleys mainly include secondary tributaries of the Luohe River such as the Heimugou. On the slopes of the middle and lower reaches of these valleys there are Malan Loess and the first paleosol (S_1) which covered an ancient valley with a wide and shallow profile. From the paleosol (S_1) and the Malan Loess deposited on these ancient valleys it is supposed that the valleys of this period had been formed before the second paleosol was developed, at about 200000 yr B. P.

3. The late stage valleys are characterized by further dissection of the early and middle stage valleys. Along the slope of the middle stage valleys, many small gullies were developed perpendicular to the main valley. They developed at the late Malan stage.

4. The latest stage valleys formed in historic time. An example is west of Luochuan city where a relic of Painted-Pottery Culture dated 5000–6000 yr B. P. was found in a large area along the road near Luochuan Middle School. A gully cut right through the middle of the relic and advanced headward about 50 m. This headward erosion suggests that the gully has extended 50 m during the past 5000–6000 years. The valley of this period is characterized by steep, straight walls and pointed ends. It developed more rapidly than the others. The development of the latest stage valley is directly related to human activities.

CHAPTER 3

Time Sequence of Loess Deposits

3.1 Luochuan Loess-Paleosol Sequence

In the past few years, considerable progress has been made in the study of stratigraphic division of the loess in China (Liu Tungsheng et al. 1980; Wang Yongyan et al. 1982; Zhang Zonghu 1983; An Zhisheng and Lu Yanchou 1984). Investigation of the loess-paleosol sequence and its geological consequences appears to show three developing tendencies. First, the division of loess strata has become increasingly detailed, and the chronology has been employed in stratigraphy. Second, the study of loess has been developing in the approach from regional environments towards global climatic changes. Third, it has entered upon a new phase of dealing with the Quaternary geological events. It has now been recognized that the Luochuan stratigraphical series, developed in the past 2.4 m.y., is one of the most suitable objects for investigations of the Quaternary stratigraphy, chronology and relevant geological events.

3.1.1 Geological Events Recorded by Loess-Paleosol Sequence

It is of interest to note that clear boundaries can be observed between loess and paleosol layers, indicating that abrupt changes occurred during the transition between the two facies. Locally, the deposition of loess and the formation of paleosol should proceed progressively, but if viewed as a scale of climatic changes represented by loess deposition reflecting dry-cold climate and by paleosol formation reflecting warm-humid climate, they should clearly differ from each other. On the other hand, such a transition, taking place in an area as wide as hundreds of thousand km^2, has also recorded an areal and clearly variational geological event that is of common significance to that area (i.e., an event of transition from loess to paleosol genesis, or an event of superimposition of loess upon paleosol). The question is "Why did such nearly repeated jumping events take place in the Loess Plateau, while it could hardly be observed in lake or marine sediments?" The reason is plausibly related to the fact that the loess-paleosol sequence itself is an integral, relatively consecutive, well exposed and stable continental facies. One of the current tasks, however, is to probe into the occurrence and developmental history of geological events by investigating the loess-paleosol sequence. At this point, a question might arise: "What would happen if one compared the geological events recorded by the loess-paleosol sequence to those in areas apart from the Loess Plateau?" This underscores an important problem concerning the correlation among Quaternary geological events. The character and extent of the events, such as the occurrence and development of fossil-man in northern China, the transgression and regression of the sea, the development and contraction of alluvial fans, changes in the river system and the appearance and disappearance of ancient lakes, can all be compared with the loess-paleosol sequence. Therefore, its is extremely important to establish a loess stratigraphical sequence on the basis of geological events. The geological events recorded by the Quaternary strata in China closely reflect the global climate change and at the same time are closely related to regional environmental change. Thus they necessitate a detailed study of the loess-paleosol sequence developed over 2.4 m.y.

3.1.2 Luochuan Loess Section

The Luochuan area is a large basin with Pleistocene loess deposits, which constitute a stable and consecutive loess-paleosol sequence, and may serve as a representative of the whole loess area.

Wang Kelu and Wen Qizhong first mapped a loess stratigraphical section in West Gully, Shijiazhuang, Luochuan in 1957. From 1962–1973, Wen Qizhong, Zhu Haizhi and Guo Jinluan (Liu Tungsheng et al. 1966) divided the loess strata of Heimugou, Luochuan into Malan Loess of the Late Pleistocene, Lishi Loess of the Middle Pleistocene and Wucheng Loess of the Early Pleistocene. During the second half of the 1970s, Wen Qizhong, An Zhisheng and Lu Yanchou continued the survey of the Heimugou section. In 1980, Liu Tungsheng, Wen Qizhong, Wu Zirong, Yuan Baoyin, Sun Fuqing and Zhu Yizhi mapped for the second time the Luochuan loess section with the assistance of the Hydrogeological Team No. 2 of Bureau of Geology and Mineral Resources of Shaanxi Province, which performed the drilling work of the Qun 22 borehole and categorized the cores with some complementary descriptions made by An Zhisheng.

The Heimugou section is located in the Potou, 4 km away from Luochuan city. The gully is 140 m deep with the bottom composed of Triassic sand stone and sandy shale. A Late Pliocene Red Clay layer of 10–15 m thick overlies the bedrock. The loess layer is 130 m thick, and overlies the Red Clay layer. The individual loess-paleosol layers are named from the top to the bottom: S_0, L_1, S_1, L_2, S_2, ... L_{14}, S_{14}, L_{15}, W_{S-1}, W_{L-1}, W_{S-2}, W_{L-2}, W_{S-3}, W_{L-3} (Figs. 3.1 A and 3.2).

The Qun 22 borehole is located on the east side of Yaotun, near a branch of the Heimugou gully, in Qinjiazhai, Luochuan County. The Qun 22 hole is 144 m in depth; the Quaternary loess-paleosol deposits occur above the depth of 136 m, and the Pliocene Red Clay lies from 136 m down to 144 m (Fig. 3.1 B).

As can be seen from the above figures, the structure of the Luochuan section is characterized by the following aspects:

1. Paleosol or weathering layers, horizontally distributed or slightly inclined, occur alternately with loess, indicating that there were multiple hiatuses or slow dust deposition intervals during the entire loess deposition period.

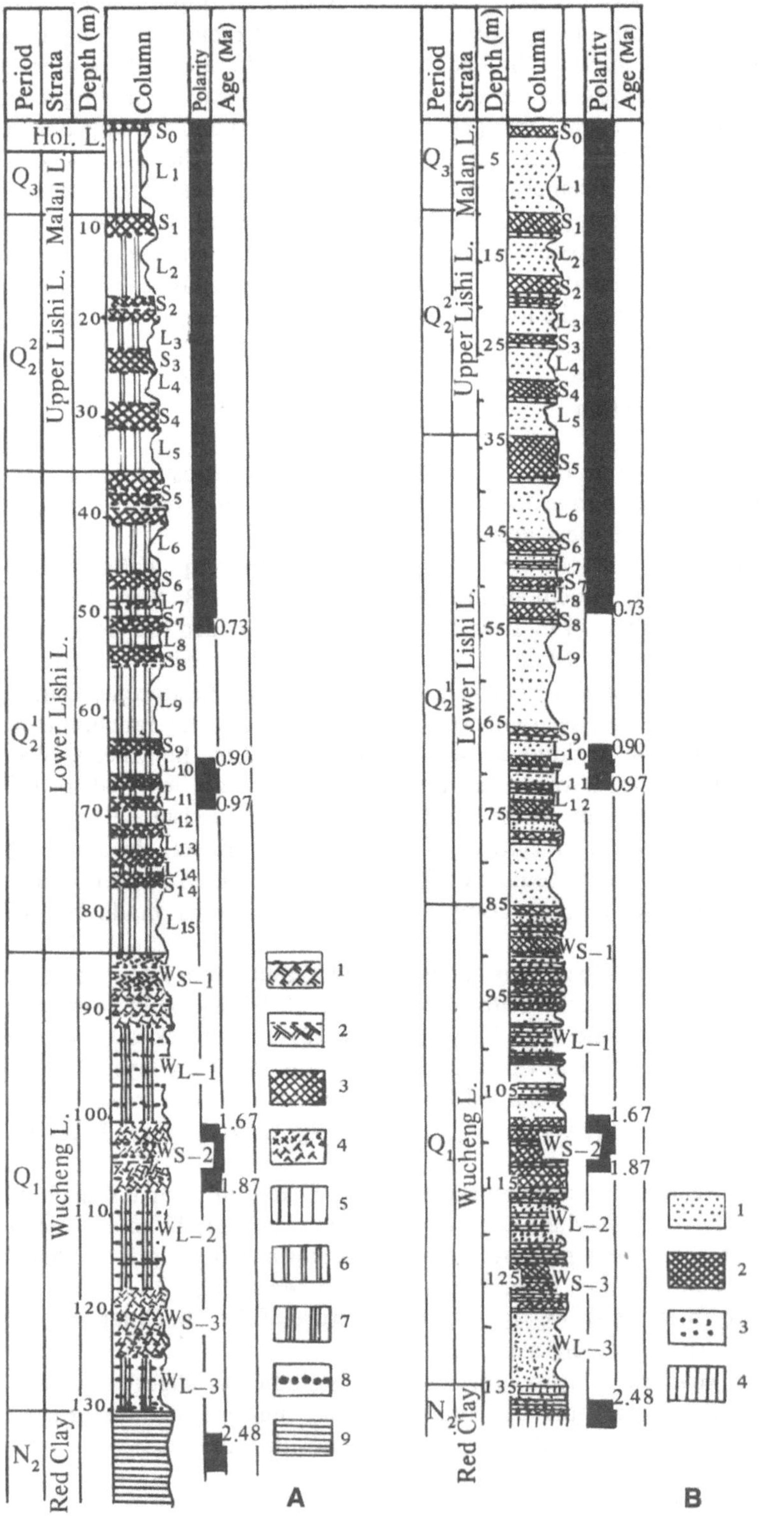
Period
Strata
Depth (m)
Column
Polarity
Age (Ma)
Hol. L.
Malan L.
Upper Lishi L.
Lower Lishi L.
Wucheng L.
Red Clay
0.73
0.90
0.97
1.67
1.87
2.48
A
B

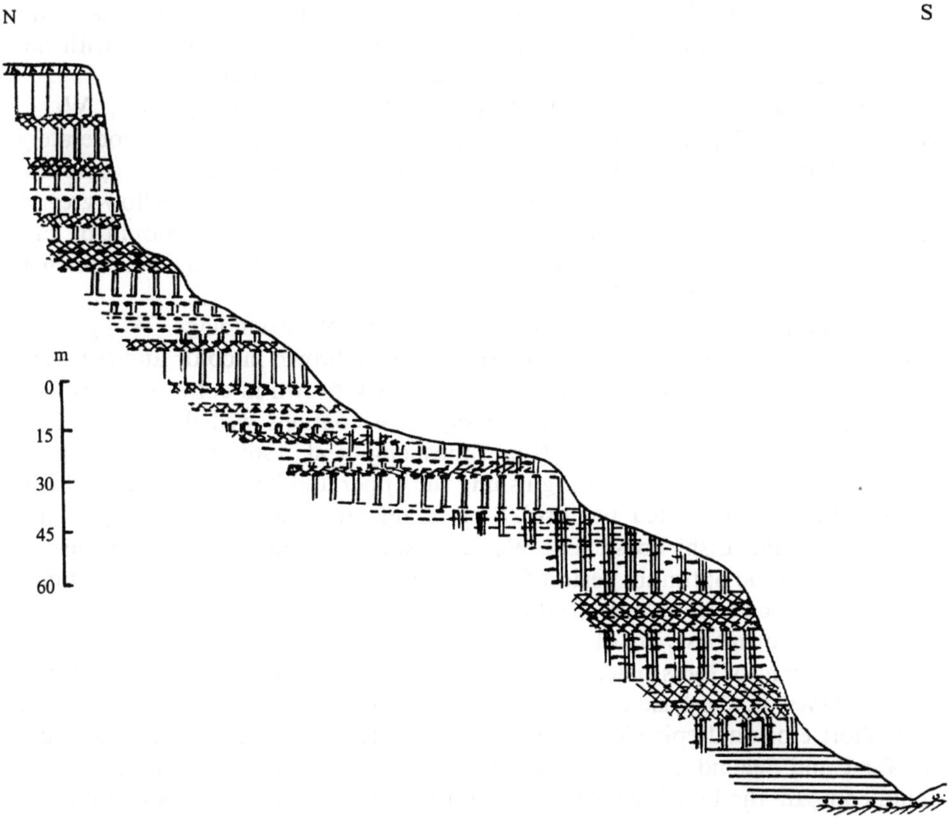

Fig. 3.2. Loess section in Heimugou, Luochuan. (The explanation is the same as that in Fig. 3.1 A)

2. Loess-paleosol assemblages of different periods bear different features. The Wucheng Loess of the Early Pleistocene contains three fairly concentrated paleosol assemblages (W_{S-3}, W_{S-2}, W_{S-1}). The paleosol or weathering layers of each assemblage overlie each other. Among them, some thin loess layers of less than 1 m thick were observed. Each paleosol assemblage alternates with a thick loess layer (W_{L-3}, W_{L-2}, W_{L-1}). Calcareous nodule layers are sandwiched in each thick loess layer.

The lower part of Lishi Loess formed in the early period of Middle Pleistocene. There are three densely distributed paleosol assemblages ($S_{14}-S_9$, S_8-S_6, S_5) occurring within the portion from L_{14} to S_5. Assemblages $S_{14}-S_9$ and S_8-S_6 contain a loess layer of 1–2 m thick. The three assemblages are separated by two thick loess layers (L_9, L_6). In the upper part of Lishi Loess

Fig. 3.1. A Stratigraphical section of loess-paleosol in Heimugou, Luochuan, *1* Black loam; *2* buried black loam; *3* paleosol; *4* paleosol; or weathering layer; *5* Malan Loess; *6* Lishi Loess; *7* Wucheng Loess; *8* calcareous nodule; *9* clay. **B** Luochuan Qun 22 core section (revised and supplemented from the data provided by the Hydrogeological Team No.2, Shanxi Province). *1* Loess; *2* paleosol and weathering layer; *3* calcareous nodule; *4* red clay

(L_5-S_1) and within the portion of Late Pleistocene Malan Loess L_1, there are four paleosol-loess assemblages with the paleosol regularly alternating with the loess, indicating a clear cycle of soil-forming and dust accumulation.

3. There are 11 loess layers of more than 2 m thick, namely W_{L-3}, W_{L-2}, W_{L-1}, L_{15}, L_9, L_6, L_5, L_4, L_3, L_2, L_1, which were weakly or moderately weathered and have recorded 11 prolonged dry-cold dust accumulation episodes. Among them, the upper and lower sandy loess layers (L_9, L_{15}) are yellow-grey in color, thick and less compact or porous, contain relatively large amount of carbonate, and stand in nearly vertical walls. Thus they can serve as an ideal reference for distinguishing stratigraphic positions.

4. Characteristic paleosol layers: The paleosol assemblage W_{S-2}, best developed in the middle of Wucheng Loess, is dark brown in color and coherent in character. The complex paleosol layer S_5 is dark brown in color and also has a clear prismatic structure with ferromanganese coatings well developed. It is known as "Luochuan soil layer" (Liu Tungsheng et al. 1965) and has been accepted as a marker layer for distinguishing the upper and lower parts of the Lishi Loess. The well-developed paleosol S_1 is a reference layer for distinguishing Malan Loess and Lishi Loess. The W_{S-2}, S_5 and S_1 paleosol or assemblages all indicate a climatic optimum period for pedogenesis, and are of prime importance for the identification of loess strata.

No unconformities were observed in the section, indicating the continuity of the development of loess-paleosol sequence. Some redeposited red soil aggregates were observed in the base of Wucheng Loess, indicating that there might have been short erosional episode following Gauss chron. If S_5 and S_1 paleosols and the First and Second sandy loess L_9 and L_{15} are taken as reference layers, one could compare the Luochuan section with the other sections of loess region, also demonstrating lack of hiatuses (Fig. 3.3).

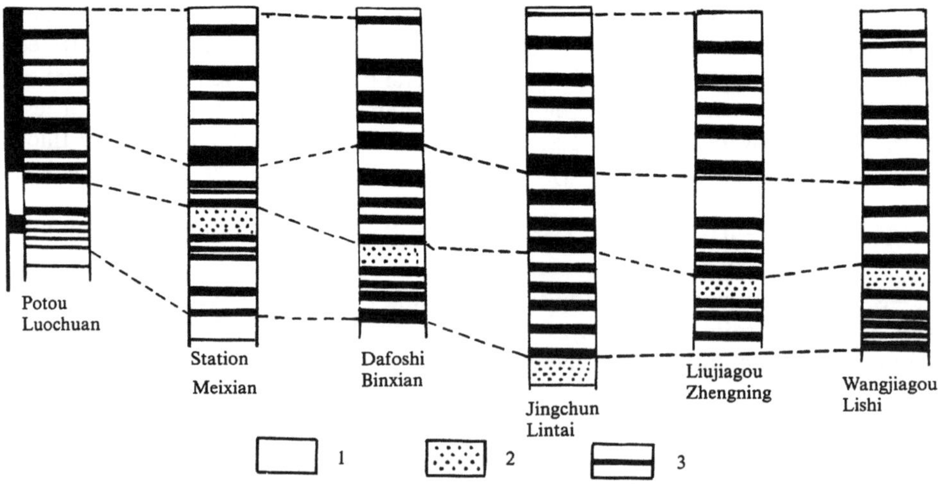

Fig. 3.3. Comparison between several loess sections in the Loess Plateau. *1* Loess; *2* sandy loess; *3* paleosol. (Data from Lui Tungsheng, Wang Yongyan, Wang Kelu, Su Lianyi and others)

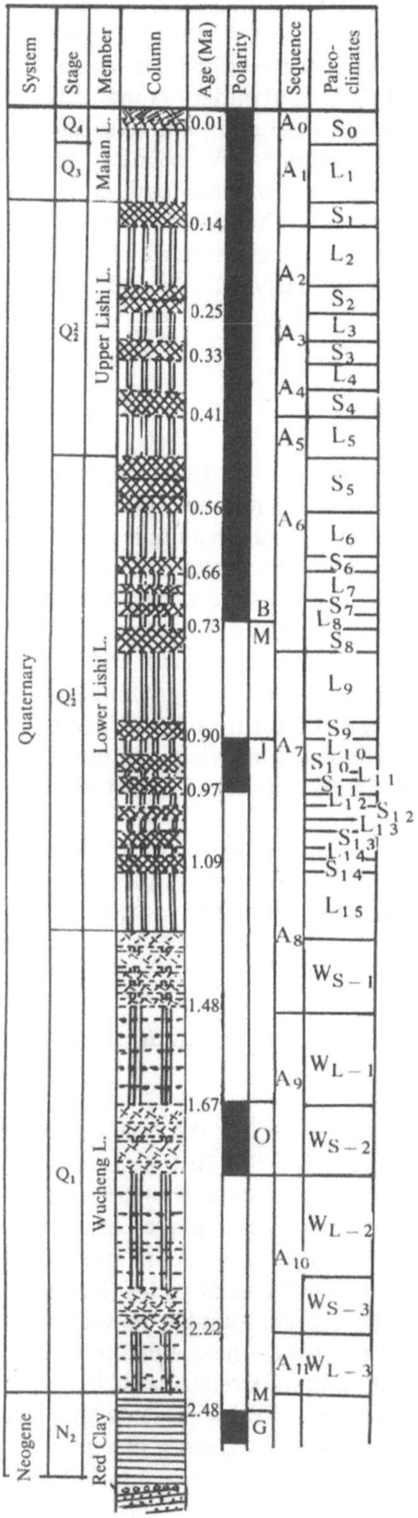

Fig. 3.4. Stratigraphical division of Luochuan loess-paleosol sequence

The Malan, Lishi and Wucheng Loesses in the Luochuan section bear their own petrological characteristics (Liu Tungsheng et al. 1965). A number of fossil mammalian faunas have recently been discovered within the sections. For example, the Lishi Loess contains fossil species typical of the Zhoukoudian fauna, and the upper part of Wucheng Loess contains *Myospalax* fossil comparable to the Nihewan fauna. The age of the *M. omegodon* in the lower part of Wucheng Loess is earlier than that of Nihewan fauna. The *Prosiphneus intermedius* fossil in the Red Clay layer found here is typical one of Late Pliocene. Therefore, in view of both biostratigraphy and petrostratigraphy, Luochuan loess can be subdivided into Wucheng, Lishi and Malan Loesses, belonging to the Early, Middle and Late Pleistocene, respectively.

3.1.3 Geological-Climatic Events and Loess Sequences

The magnetostratigraphic and lithostratigraphic framework of the Luochuan loess-paleosol sequence has made it possible to further subdivide the loess strata.

The climate changes during the Quaternary in the Luochuan area are recorded in alternating layers of loess and paleosol. The climatostratigraphic units are based on cycles from warm-humid to dry-cold environments, recorded by the multiple occurrences of thick loess layers reflecting prolonged dry-cold dust accumulation periods. As few studies have been done on Wucheng Loess and the lower part of Lishi Loess, their stratigraphic division is considered preliminary. Detailed studies of the Malan Loess and the upper part of Lishi Loess are available. For the time being, the loess strata have been divided, somewhat arbitrarily, into ten complete climato-stratigraphic groups ($A_1 - A_{10}$) and two incomplete groups (A_0, A_{11}) (Fig. 3.4).

Each paleosol or loess layer can be considered a single climatostratigraphic unit which has recorded a warm-humid or dry-cold climatic event. As for the Wucheng Loess, where there are not enough data, the individual loesses and paleosol layers are grouped into assemblages.

3.1.4 Polycyclic Geological-Climatic Events and Brunhes Chronostratigraphy in Northern China

Everywhere in the Loess Plateau, loess-paleosol sequences reflect the multicyclic nature of the climatic change. So do other sediments of the Brunhes chron formed since 0.73 m.y., including lake sediments, cave deposits and alternating sediments of land and sea (Fig. 3.5). In Fig. 3.5a comparison is made between the geological events of various Quaternary sediments in northern China, and the oxygen isotope records (Shackleton and Opdyke 1973) and glacial cycles (Kukla 1977). The odd-numbered stages of oxygen isotope are comparable to warm-humid climate stages reflected by the formations of paleosol, fine-grained travertine layer and marine beds, while even-numbered stages correlate with loess, limestone breccia and coarse-grained layers, as well as river/lake sediments alternating with marine beds. Thus the Luochuan loess-paleosol sequence can serve

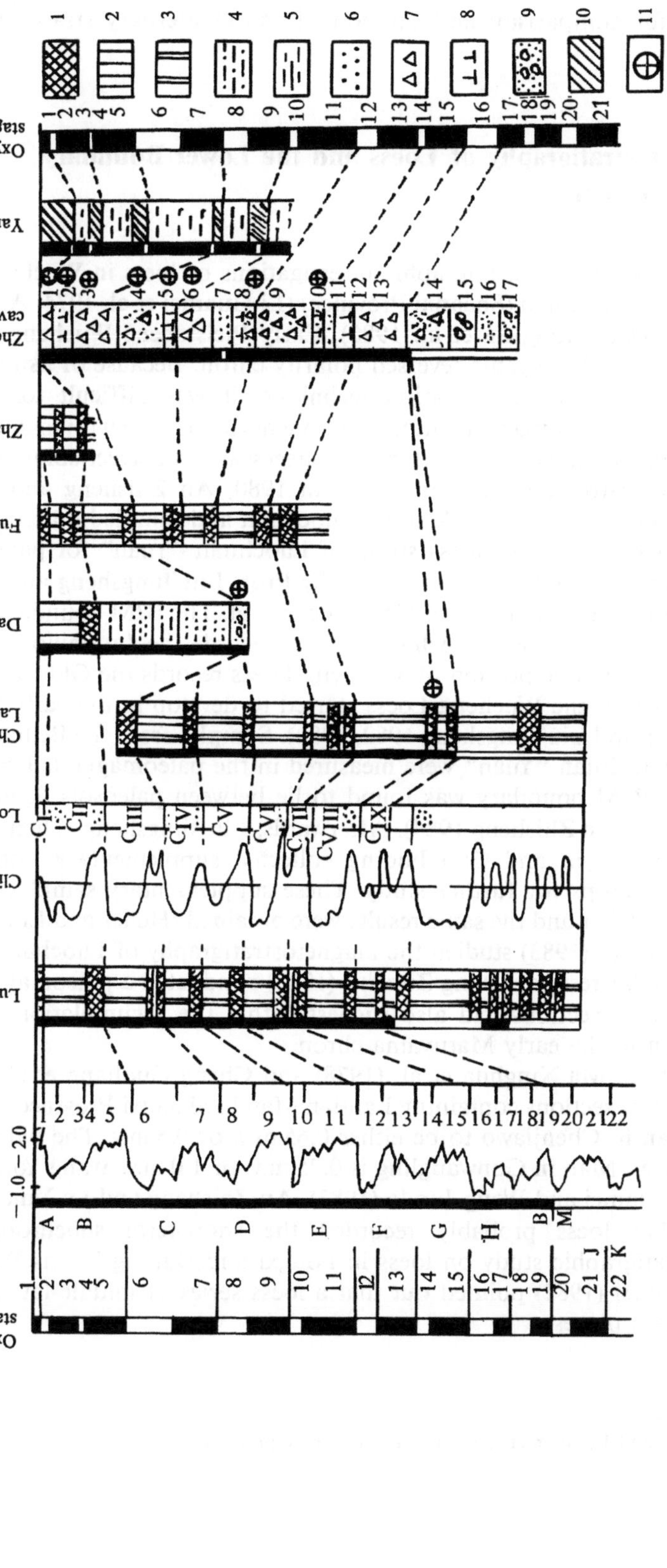

Fig. 3.5. Comparison between paleoclimatic records of land/sea in northern China and deep-sea deposits since 0.73 m.y. (Data partly from Shackleton and Opdyke 1973; Ma Xinghua 1978; Tang Keli 1981; Liu Zechun 1982; Zhao Songling and Chin Yunshan 1982; Wang Yongyan et al. 1983; An Zhisheng and Lu Yanchou 1984). *1* Paleosol; *2* Malan loess; *3* Lishi loess; *4* sandy clay; *5* clay; *6* sand or ash; *7* breccia; *8* travertine; *9* sandy gravel; *10* marine bed; *11* fossil man

as a basis for comparison and division of the Quaternary strata in northern China.

3.2 Magnetostratigraphy of Loess and the Lower Boundary of the Pleistocene

In preliminary magnetostratigraphic investigations of loess in Wucheng (Shanxi Province) and Luochuan (Shaanxi Province) (Li Huamei et al. 1974; An Zhisheng et al. 1977; Wang Yongyan et al. 1980), Early Pleistocene Wucheng Loess was placed within the Matuyama reversed polarity chron. Because of using the alternating field (AF) demagnetization technique, it was difficult to completely remove the secondary magnetization components associated with high coherecivity ferromagnetic minerals. Therefore, it is necessary to correct some of the published magnetostratigraphic data. Early in 1980, An Zhisheng and Liu Chun sampled from Qun 22 core in Yaotun, Luochuan and sampled outcrops near the B/M boundary within the loess strata of Luochuan "Yuan" for paleomagnetic analysis[4]. These samples were measured by Prof. Liu Tungsheng and Dr. Heller with the help of Ge Tongming and W. Gruber et al. The B/M boundary is located in the middle of paleosol S_8, while $L_{10}-S_{11}$ records the Jaramillo normal subchron, and the middle portion of Wucheng Loess records the Olduvai subchron. It is obvious that the Wucheng Loess started to develop in the early Matuyama chron (Heller and Liu Tungsheng 1982, 1984). Samples from the Beihanzhai loess section of Luochuan "Yuan" were measured in the paleomagnetic laboratory of A.N.U. The B/M boundary was found to lie between paleosols S_8 and S_7 (Liu Tungsheng and An Zhisheng 1984). Afterwards, Liu Tungsheng, Chen Mingyang, Bater, Yuan Baoyin and Gao Fuqing collected supplementary samples from Luochuan outcrops for further study. These supplementary samples were measured by Dr. Heller and the same results were obtained (Heller and Liu Tungsheng 1984). Torii et al. (1983) studied the magnetostratigraphy of Luochuan loess and obtained similar results. Zhang Zonghu (1983) conducted a paleomagnetic study on several loess sections and also suggested that the accumulation of loess in China began in the early Matuyama chron.

In addition, Ma Xinghua et al. (1978) and Cheng Guoliang et al. (1978) investigated loess sections containing Lantian Man in Shaanxi Province. They dated Lantian Man in Chenjiawo to be either 0.65 m.y. or 0.5 m.y. The paleomagnetic age of Lantian Man in Gonwangling is 0.75 m.y. or 1.0 m.y. Wang Yongyan et al. (1980), Li Huamei and Wang Junda (1982), An Zhisheng and Lu Yanchou (1984) reported that loess probably recorded the short-term subchron. From a magnetostratigraphic study on loess in Longxi and Datong basins Wang Junda and Li Huamei (1982) pointed out that a loess series of middle-late Matuyama exists in these places.

[4] Wang Junda and Li Huamei took part in the sample processing.

3.2.1 Magnetostratigraphy of Luochuan Loess

The most detailed magnetostratigraphic study of loess in China is on the Luochuan section (Figs. 3.6 and 3.7). Measurements have been made on samples from the Qun 22 core in Yaotun, and the Heimugou and the Beihanzhai sections, using the cryogenic magnetometer and thermal demagnetization in order to revise earlier results. A reliable polarity column for the Luochuan section has been derived and assigned a minimum stage of 2.4 m.y. for the beginning of loess sedimentation. On the basis of Mankinen's magnetic chronological table (Mankinen and Dalrympleo 1979), some important polarity boundaries have been determined: Olduvai subchron (1.87 – 1.67 m.y.), Jaramillo subchron (0.97 to 0.90 m.y.) and Brunhes chron for 0.73 m.y. The measurements on samples from the Qun 22 core in Yaotun and supplementary samples from the Qinjiazhai section in Heimugou provide the most systematic results of Luochuan magnetostratigraphic investigations so far.

In Qun 22 core, the B/M boundary is located at a depth of 53.05 m: the Jaramillo subchron at 67.3 – 72.3 m; Olduvai subchron at 107.4 – 113 m; M/Ga

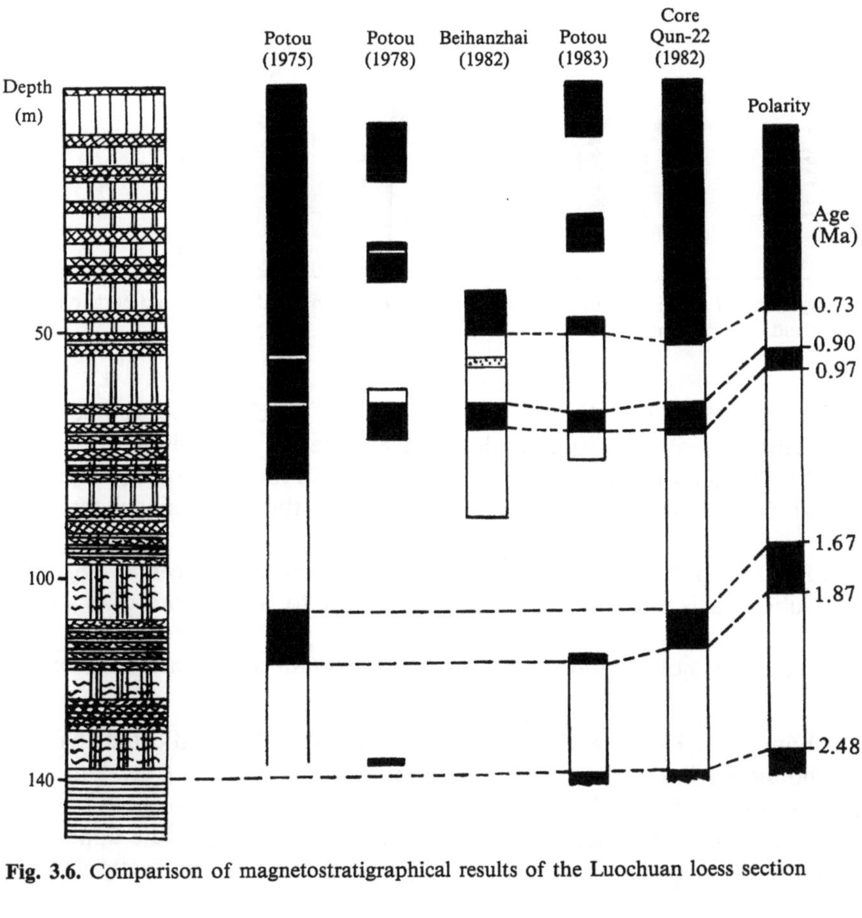

Fig. 3.6. Comparison of magnetostratigraphical results of the Luochuan loess section

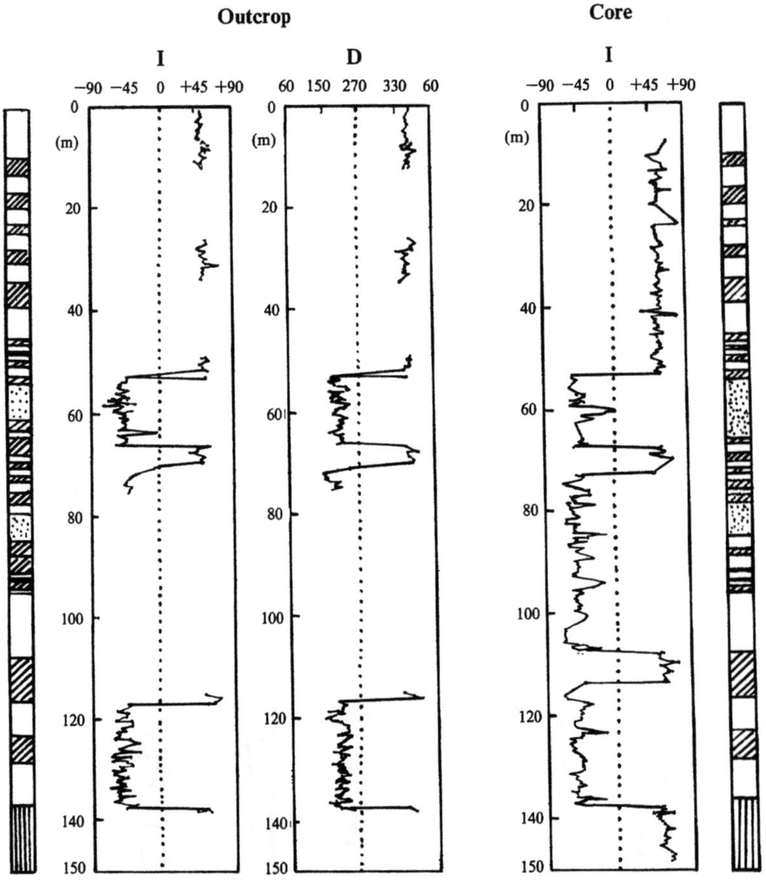

Fig. 3.7. Comparison of magnetostratigraphical data of Luochuan natural outcrop and borehole core. (After Heller and Liu Tungsheng 1984)

boundary at 137.75 m, that is about 2 m below the Wucheng Loess/Red Clay boundary. Three major polarity boundaries established in the Heimugou section are: the B/M boundary is located at a depth of 52.3 m in the section, the Jaramillo subchron at 66.25 − 70.5 m, the lower boundary of the Olduvai subchron at 116.25 m.

In the Beihanzhai section, the B/M boundary is located between $S_7 - S_8$. The Jaramillo subchron is placed from the bottom of S_9 to the bottom of S_{10} (Fig. 3.8).

The positions of polarity boundaries in the Luochuan section are summarized as follows:

1. M/Ga boundary. Heller and Liu Tungsheng (1982, 1984) indicated that the M/Ga boundary lies at the upper portion of the Pliocene Red Clay, 2 m away from Wucheng Loess/Red Clay boundary.
2. Olduvai subchron. The Olduvai subchron is recorded at the middle-upper portion of the W_{S-2} assemblage in Wucheng Loess. Recent examination of the

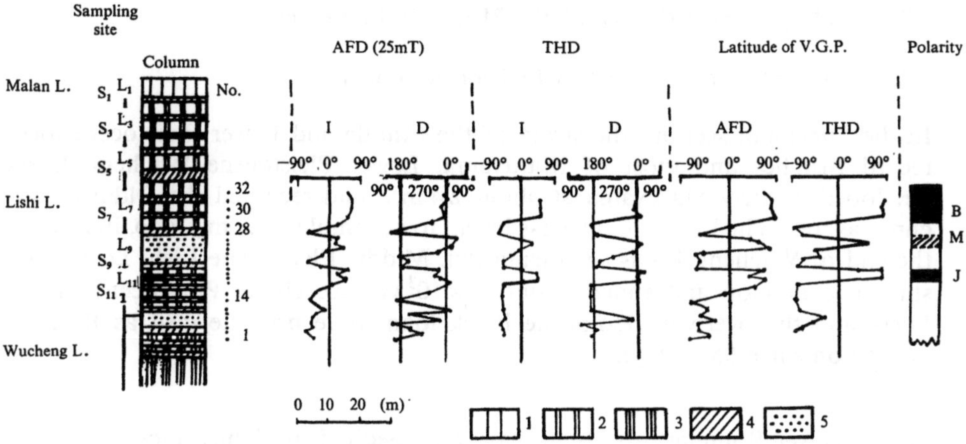

Fig. 3.8. Comprehensive magnetostratigraphical section of Beihanzhai, Luochuan. *1* Malan loess; *2* Lishi loess; *3* Wucheng loess; *4* paleosol; *5* sandy loess

Heimugou section showed that the lower boundary of the Olduvai subchron is near the bottom of the assemblage.

3. Jaramillo subchron. In the Qun 22 core and Heimugou section, the lower boundary of Jaramillo lies at the bottom of paleosol S_{11}. It may be suitable to put the lower boundary of Jaramillo at the bottom of paleosol S_{11}. The upper boundary of the Jaramillo subchron is at the upper portion of loess L_{10} or at the bottom of paleosol S_9.
4. B/M boundary. It is located at the middle of S_8 in the borehole, at the top of S_8 in the Heimugou section, and at the middle-upper portion of S_8 in the Beihanzhai section. The depth discrepancies of three boundaries are within 1–2 m, which make us consider that the B/M boundary recorded in loess-paleosol sequence may be located at the portion from S_8 to L_8. The differences between the stratigraphic horizon and B/M boundary in different sections not far apart from one another may be related to the insufficient identification of strata, inaccurate magnetic examination of imperfect records of the B/M transformation in different sections probably due to the local difference in composition of paleosol and loess, especially the movement of ferruginous components. The movement of an iron component might exist in strongly weathered loess L_8 and well-developed paleosol S_8 in Luochuan section. Consequently the polarity from the lower part may reflect the polarity of the iron component moving downward from the upper part. It might have occurred either at the product at the beginning of Brunhes or at the end of Matuyama. Based on the above discussion, the author suggests that the B/M boundary in the Luochuan loess section be located at the portion from the top of S_8 to the middle-upper part of L_8.

3.2.2 Geological Significance of the Magnetic Boundary

3.2.2.1 Magnetostratigraphy and Pleistocene Loess

In the Luochuan section, the strata of the Middle and Lower Pleistocene loess record Brunhes and most of Matuyama. Lower Pleistocene Wucheng Loess developed in early Matuyama at about 2.4 m.y. and recorded the Olduvai subchron as well. The lower sandy loess began to accumulate 1.2 m.y. ago, indicating the end of Wucheng Loess sedimentation. Middle Pleistocene Lishi Loess then started to develop and ended about 95×10^3 yr ago. Upper Pleistocene Malan Loess and the overlying Holocene black loam have been defined as Brunhes deposition since 95×10^3 yr.

3.2.2.2 Lower Boundary of Quaternary and Geological Significance of the M/Ga Boundary

Prosiphneus intermedius of Late Pliocene was found above the M/Ga boundary which is situated at the top of the Pliocene Red Clay in the Luochuan section. The deposition of Red Clay indicates a warm climate which continued to early Matuyama. The loess started to accumulate in early Matuyama, 2.4 m.y. ago, characterized by a dry cold climate. *Myospalax omegodon* was found at the bottom of Wucheng Loess, so the M/Ga boundary or the bottom of Wucheng Loess may mark the evolution from the most advanced *Prosiphneus* stage to primitive *Myospalax* stage.

The datum levels if biological evolution and climatic change reflected by the M/Ga boundary or loess/Red Clay boundary of the Luochuan section are also recorded in Plio-Pleistocene strata in northern China. In the Chengqiang section of the Nihewan basin, Hebei Province, the layer containing *Equus sanmeniensis* is dated at about 2.4 m.y. B.P.[5], which may be the first horizon of *Equus* in northern China. In Danangou section of Nihewan near Chengqiang section, a mammalian fauna horizon of Dongyaozitou was found with an age of about 2.5 m.y.[6], based on the comparison of sections containing transitional fauna from the Pliocene to the Pleistocene, such as *Hipparion* cf. *houfenese*, *Zygolophodon* sp., *Palaeotragus*, *Antilospira*, *Proboscidipparion sinenses*, *Lynx variabilis*, *Nyctereutes* cf. *sinensis*, *Coelodonta antiquitatis*, *Paracamelus* sp., *Gazella sinensis* (Tang Yingjun and Ji Hongxiang, 1983). This fauna may corre-

Fig. 3.9. A correlation of magnetostratigraphical sections from the Loess Plateau to coastal plain in North China. (Part of data from Liu Chun 1982; Cao Zhaoyuan et al. 1983; Li Huamei and Wang Junda 1982; An Zhisheng and Lu Yanchou, 1984; Li Huamei et al. 1977; Qian Fang 1983; Zhao Songling and Chin Yunshan 1982). *1* Basalt; *2* Malan loess; *3* Lishi loess; *4* Wucheng loess; *5* Red clay; *6* paleosol or weathering layer; *7* sandy clay; *8* sand or ash; *9* sand and gravel; *10* marlite or calcareous sandy layer; *11* calcareous nodule; *12* baked layer; *13* clay; *14* breccia; *15* travertine; *16* marine bed

[5] Based on fossil bed by Wu Zirong et al. (1980) and estimation of magnetostratigraphical results by Li Huamei and Wang Junda (1982).
[6] Based on the estimation of the author.

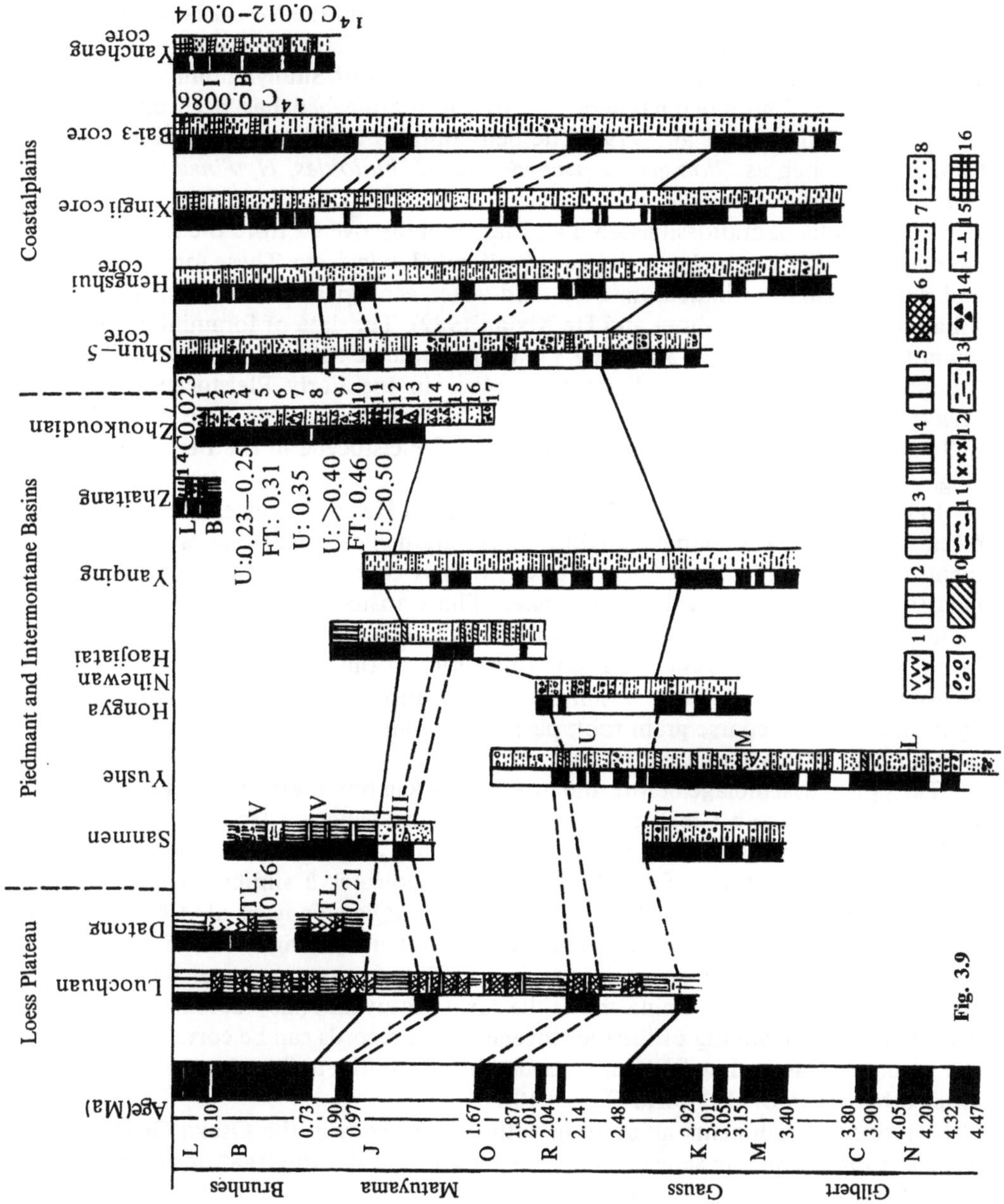
Loess Plateau
Piedmant and Intermontane Basins
Coastalplains
Age(Ma)
Luochuan
Datong
Sanmen
Yushe
Hongya
Nihewan
Haojiatai
Yanjing
Zhaitang
Zhoukoudian
Shun-5 core
Hengshui core
Xingji core
Bai-3 core
Yancheng core
Brunhes
Matuyama
Gauss
Gilbert
L
B
J
O
R
K
M
C
N
0.10
0.73
0.90
0.97
1.67
1.87
2.01
2.04
2.14
2.48
2.92
3.01
3.05
3.15
3.40
3.80
3.90
4.05
4.20
4.32
4.47
TL: 0.16
TL: 0.21
U:0.23–0.25
FT: 0.31
U: 0.35
U: >0.40
FT: 0.46
U: >0.50
14C0.023
14C 0.0086
14C 0.012–0.014

Fig. 3.9

spond to early-middle Villafranchian. The period around the M/Ga boundary is characterized by alternating Pliocene and the early Pleistocene fauna.

A marine bed lies at a depth of 428 m in core S–5 of Shunyi County, northeast of Beijing. The paleomagnetic age of the marine bed has been dated at 2.3 m.y. (An Zhisheng et al. 1979). The bed contains benthonic and planktonic foraminifera, such as *Globigerina pachyderma*, *G. bulloides*, *Hyalinea balthica*, *Buccella frigida*, *Elphidiella arctica* etc.

A calcareous nannofossil assemblage has been discovered from the same bed, characterized by *Coccolithus pelagicus*, *Emiliania huxleyi*, etc. These foraminifera and nannofossils are indicators of a cold-water ecological environment of the subarctic zone (Wang Naiwen and He Xixian 1982). The data of foraminifera and nannofossils from core S–5 dated 2.3 m.y. ago indicate that an extensive Beijing transgression with normal salinity took place in the Early Pleistocene. Since *Hyalinea baltica* was found in the marine bed at the bottom here, the time (2.3–2.4 m.y.) can be recognized as the onset of Pleistocene in the Beijing Plain (Fig. 3.9).

In the Hongya sections of the Nihewan basin in Hebei Province, the sediment of early Matuyama is characterized by coarse-grained calcareous lacustrine sediment, while the sediment of late Gauss chron is characterized by fine-grained lacustrine sediment formed in freshwater. The transformation of depositional facies also reflects the climatic change from humid to dry conditions in Nihewan basin before or after M/Ga boundary. In a series of the Plio-Pleistocene fluvio-lacustrine sediments in Yushe basin, Shanxi Province, the variation of sediments from fine grain to coarse grain took place before and after the M/Ga boundary as well (Cao Zhaoyuan et al. 1983).

The pollen assemblage before and after the M/Ga boundary within the Bai–3 borehole in the North China plain shows the transformation from a coniferous and broad-leaf mixed forest to steppe vegetation (Li Huamei and Wang Junda 1983). The pollen study of the Yanqing basin indicates a similar variation of vegetation before and after the M/Ga boundary (Qian Fang et al. 1983). The climatic deterioration in early Matuyama indicated by magnetostratigraphy , fossils and depositional facies in the Loess Plateau, intermontane basin and coastal plain in northern China shows the early Matuyama (2.3–2.4 m.y.) may be regarded as the beginning of the Pleistocene. These records can be correlated with the records of several sedimentary sequences of continental and marine facies from other parts of the world (Fig. 3.10).

Continental sediments alternating with marine beds in the Osaka Group of Japan record a climatic series of the Plio-Pleistocene showing that during an interval of 2.2–2.3 m.y. the climate changed from warm to cold (Ithihara and Kamei 1982). The analysis of depositional facies for a core from Lake George in Australia showed that the M/Ga boundary was consistent with the transition from a large dry playa to a deep-water lake (Singh et al. 1981). The oxygon isotope record of the deep sea core V28–179 shows that the scale of glaciations increased about 2.5 m.y. ago (Shackleton and Opdyke 1977). The intense cold climate prevailed about 2.5 m.y. ago in the Black Sea (Hsu and Giovanoli 1979). Apparently, an important level of climate change before and after 2.4 m.y. seems to mark the beginning of the Pleistocene. The event of climatic deterioration which

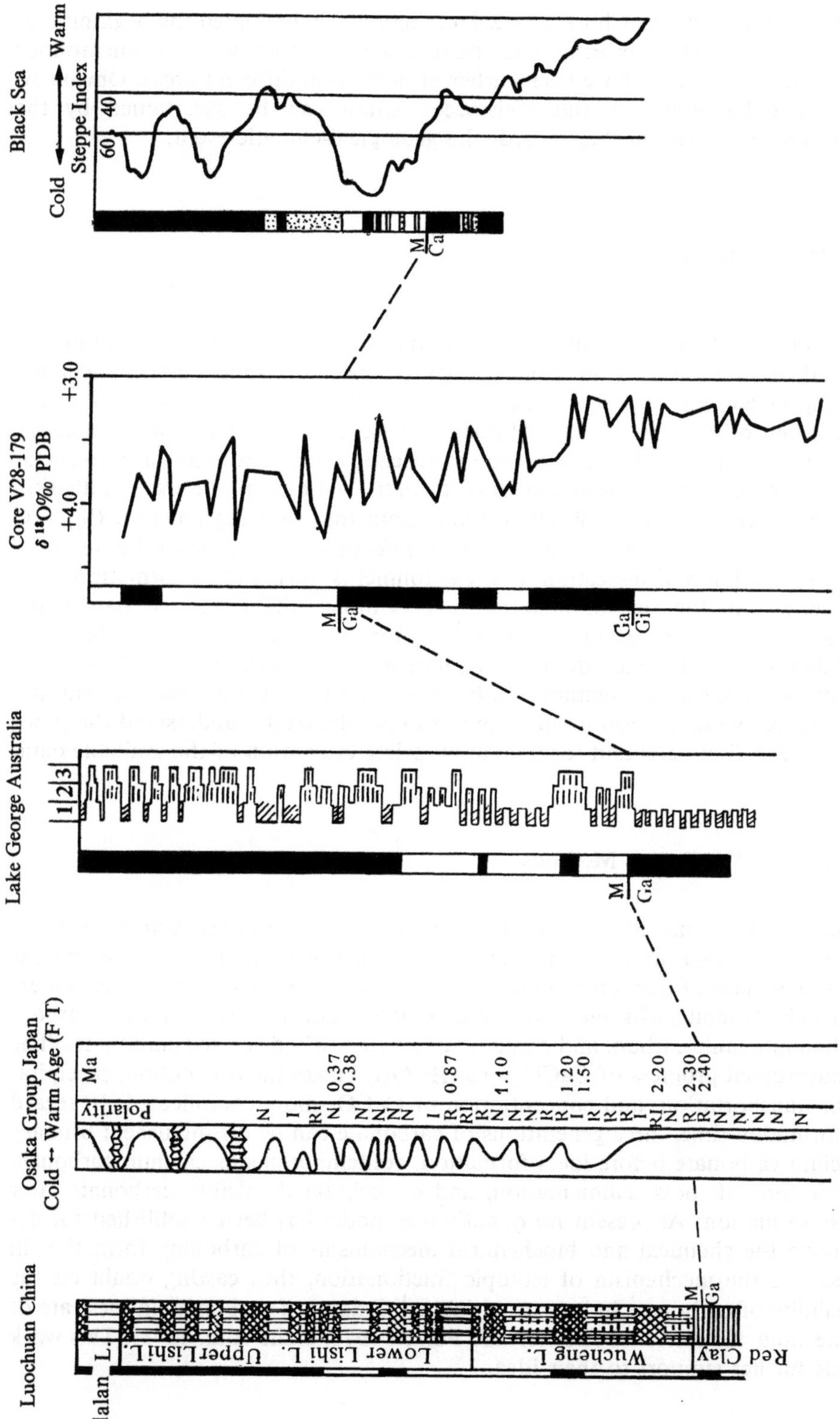

Fig. 3.10. A comparison of the Plio-Pleistocene climatic records on sea and land. *1* Dry lake or ephemeral fan-like deposition; *2* intermediate deep lake or permanent lake; *3* deep lake

took place in northern China 2.3 – 2.4 m.y. ago also represented the beginning of the Pleistocene. The important climatic deterioration event was not synchronous everywhere, and might have been earlier or delayed in different areas. Obviously, the lower boundary of the Pleistocene strata can be ascertained by the stratigraphical position that records the geological-climatic event.

3.3 ^{14}C Dating of Loess

The available data is presented on ^{14}C dating of loess samples collected in vast loess-distributed areas from Longxi, Gansu Province in the west to Zhaitang, Beijing in the east. Major results include: (1) ^{14}C dating of samples from the three layers of black loam in the Holocene loess; (2) confirmation of previous age data from the upper Malan Loess and justification of the extensive occurrence of the Malan Loess, as deduced previously from the sedimentary layers; and division of loess accumulated in the last 0.1 m.y. from the overlying paleosol (S_1). ^{14}C dating of loess can thus provide extensive information on geomorphic features, rates of erosion and dissection, and environments during loess formation.

Because of the lack of material in loess more suitable for ^{14}C dating, the author used calcium carbonate, or minor organic matter present in paleosol as ^{14}C dating material, and adopted several different methods. For this reason, the results are more or less tentative. A brief account of sampling environment and sample treatment is given in an attempt to help the reader understand the problems under discussion and form an appropriate evaluation of the available data.

3.3.1 Loess ^{14}C Dating Methods

Generally, there are four types of C-bearing material in loess suitable for ^{14}C dating: (1) C-bearing materials, such as wood, bone, etc. in a site of ancient cultural remains; (2) organic matter in soil; (3) various forms of carbonate in loess and soil; (4) snail shells and fossil bones in loess. Loess is low in organic carbon, high in inorganic carbon, and contains about 10 – 20% dispersed carbonates. The uninterrupted progress of $CaCO_3 \rightleftharpoons Ca(HCO_3)_2$ makes the dissolution, precipitation, transportation, and concentration of $CaCO_3$ more complicated. In regard to formation time, three generations of carbonate can be distinguished: primary calcium carbonate before loess formation, contemporaneous calcium carbonate at the time of loess sedimentation, and reprecipitated calcium carbonate after loess formation. At present no quantitative model has been established for describing the chemical and biochemical mechanisms of carbonate formation in loess and the mechanism of isotopic fractionation, thus casting doubt on the reliability of carbonate in loess as viriable. ^{14}C dates of carbonate in loess are of some value for reference (Bowler and Polach 1971; Ruhe et al. 1971). This work lends further support to their idea.

Table 3.1. ^{14}C dates of loess from the Heimugou section at Potou, Luochuan

Sampling depth			Age	
Sample No.	(m)	Material	(yr B.P)	$^{13}C(CO_3^{2-})$
Ge-518-1	0.3–0.5	Organic matter in soil	2760±180	
Ge-518-2	0.6–0.8		7650±275	−3.61 (CO_3^{2-})
Ge-518-3	1.3–1.5	Carbonate in loess	11530±460	
Ge-518-5	2.3–2.5		13540±530	
Ge-518-7	3.3–3.5		18830±590	
Ge-518-9	4.3–4.5		19020±640	−5.65
Ge-518-11	5.3–5.5		15000±570	−5.61
Ge-518-12	5.8–6.0		14100±570	
Ge-518-13	6.3–6.5		21450±830	−5.60
Ge-912-1	1	Organic matter	7690±200	
Ge-912		in soil	8100±300	
	1–4	Snail shell	a 9400±280	
Luochuan			b 10440±260	−6.3
	4–6	Snail shell	11300±260	

3.3.2 Results of Loess ^{14}C Dates

1. Luochuan Loess Section, Shaanxi Province. This loess section is located at Heimugou near Potou, Luochuan. The uppermost part of the section is a 0.8 m thick residual black loam, below this is an 8 (0.8–8.8) m thick Malan Loess, 8.8–11.2 m is a layer of cinnamonic paleosol (S_1) and further downwards is the Lishi Loess (L_2). ^{14}C dating of samples less than 5 m depth were collected from the top of the "Yuan" downwards, while samples at depths greater than 5 m were collected from paleosol S_1 upwards. Results of ^{14}C dating of loess are listed in Table 3.1. The "Chinese Sugar Carbon" is adopted as the modern carbon standard, ^{14}C radioactivity is 1.362±0.003 times as great as that of the international NBC oxalic acid standard (Qiu Shihua et al. 1983).

2. The Longxi Loess Section, Gansu Province. In the Longxi basin, loess is distributed mainly on the piedmont hills and high terraces. Total thickness of loess here is less than that of the Luochuan "Yuan". ^{14}C dating samples were collected from the Nuanquangou loess section at Wenfengzhen. Total thickness of loess in this section is 80 m (Wen Qizhong et al. 1982). Within 5 m from the top downwards there are two layers of black loam (i.e., from 2.5 to 3.3 m and from 3.8 to 4.5 m) intercalated in loess. The loess section and ^{14}C data are plotted in Fig. 3.11.

3. Dingxi Loess Section, Gansu Province. Samples for ^{14}C dating were collected from the top of the Anjiapu loess section at Dingxi. Under the cultivated layer is a 0.8 m thick black loam. The results of ^{14}C dating of loess and carbonate in the soil are listed in Table 3.2.

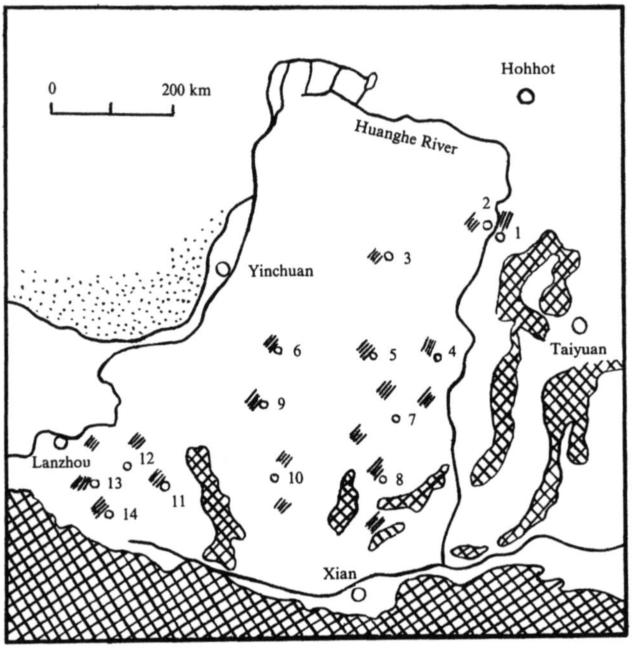

Fig. 3.11. Sketch map showing the distribution of black loam in the middle reaches of the Huanghe River. *1* Baode; *2* Fugu; *3* Yulin; *4* Suide; *5* Jingbian; *6* Dingbian; *7* Yanan; *8* Luochuan; *9* Huanxian; *10* Xifeng; *11* Jingning; *12* Huining; *13* Dingxi; *14* Longxi

4. Other Loess Sections. ^{14}C ages of the selected samples of loessic soil from the middle reaches of the Huanghe River are listed in Table 3.3.

From the above results of ^{14}C ages, we come to the following preliminary conclusions:

1. Samples from the top of the Malan Loess are dated at 10000–20000 yr B.P.
2. Black loam was developed mainly in the Holocene. In accordance with the ^{14}C ages on organic matter in it, three epochs can be distinguished: the Early Holocene (11000–8000) yr B.P.), represented by the lower black loam in the Longxi section; the early stage of the Middle Holocene (8000–5000 yr B.P.),

Table 3.2. ^{14}C dates of loess from Anjiapu, Dingxi

Sample No.	Sampling depth (cm)	^{14}C age on inorganic carbon (yr B.P.)
Ge-242	20–45	4860 ± 200
Ge-244	70–100	6800 ± 250
Ge-245	100–135	9400 ± 250
Ge-247	165–180	11200 ± 300
Ge-249	185–210	15700 ± 500

Table 3.3. ^{14}C dates of black loam from middle reaches of the Huanghe River

No.	Locality	Depth (m)	Age (yr B.P.)	
			t_2	t_1
GC-119	Kangjialing, Xiynag, Shanxi	0.5		8990±220
GC-240	Xiwuling, Heshui, Gansu	0.8		9970±300
GC-860	Niuzhangou, Jingning, Gansu	3.0	3162±160	
GC-861	Baicao "Yuan", Huining, Gansu	1.0	5166±220	
GC-862	Liuhegou, Lanzhou, Gansu	1.0	4620±130	
GC-863	Wujiadun "Yuan", Huanxian, Gansu	1.0	4170±135	
GC-864	Yujiagou, Huanxian, Gansu	4.0	4820±130	
GC-465	Chengdong "Yuan", Huanxian, Gansu	1.5	1935±125	
GC-466	Leijiaxian, Xifeng, Gansu	1.1	2600±140	7778±222
GC-867	Xilicun, Ningxian, Gansu	1.0	3170±190	

t_1: organic matter; t_2: $CaCO_3$.

represented by the upper black loam in Longxi section; the late stage of the Middle Holocene (5000–2000 yr B.P), represented mainly by residual black loam distributed in the middle reaches of the Huanghe River.

3. As deduced from the ^{14}C dates and cumulative thickness of loess, the rate of loess accumulation since the Late Pleistocene is about 0.3–0.4 mm/yr, much higher than that of the older loess.

3.3.3 Discussion

1. ^{14}C Dating. ^{14}C age data represent radiometric values. The error is not the difference between the true age and the measured one. The source of errors and the magnitude of variation must be fully understood. Methods of correction commonly used are ^{14}C-half life correction, tree-ring age correction and isotopic fractionation correction. Proper corrections can improve the reliability of ^{14}C ages. However, uncorrected age data may also be used only to establish the time scale for specimens, collected according to the arrangement of geological position. In this work $^{13}C/^{12}C$ ratios are determined only in carbonates from the Luochuan and Longxi loess sections without any other correction.

2. ^{14}C Dating of Loess. One of the compositional characteristics of loessic material is the low content of organic matter and the high carbonate content. Specimens for ^{14}C dating are difficult to obtain. Carbonates are specimens easy to collect for ^{14}C dating. Carbonates in loess are predominately calcite. The dissolution and precipitation of calcite depend on climate, especially rainfall, which exerts direct influence on the partial pressure of CO_2 (f_{CO_2} in soil). The equilibrium equation for calcium carbonate is given as follows:

$$CaCO_3(solid) + CO_2(gas) + H_2O = Ca^{2+}(solution) + 2HCO_3^-(solution) .$$

CO_2 is derived from the exchange reservoirs of carbon (e.g., the atmosphere, hydrosphere and biosphere), and in the resultant HCO_3^-, 50% carbon is derived from exchangeable carbon, and 50% from solid $CaCO_3$ which is probably ^{14}C-free dead carbon. Obviously, the more times the dissolution and precipitation of $CaCO_3$ occur, the higher the proportion of exchangeable carbon will be. If the main body of loess carbonate is formed during the process of loess formation, this type of loess specimen is considered suitable for ^{14}C dating. According to experimental results in this study, carbonate can be used for ^{14}C dating. Its ^{14}C ages are positively correlated with sampling depth, indicating the important role of exchangeable carbon in the formation of carbonate in loess, especially in the sediments of arid areas. The value of using carbonate for ^{14}C dating needs further study. Additionally, carbonate in loess may provide a new approach to the study of geochemical processes of loess formation.

^{14}C dating of organic matter in soil also involves a very complicated procedure. ^{14}C ages of soil are always difficult to explain due to the lack of distinct formation ages and initial ^{14}C concentration of the soil samples. After soil is buried, the decay of root systems, the precipitation of soluble organic substances and the disturbance of soil layers by animals continues. Under such circumstances, soil may even be influenced by the modern root system, resulting in ages which are too young. Therefore, choosing well-buried soil (e.g., the Longxi loess section), collecting samples from the lower part of a soil section (e.g., sample GC−518−2) where humus was preserved undisturbed, and using insoluble organic matter which undergoes change represent effective measures to improve the reliability of ^{14}C dating results.

3.4 Loess Sedimentation Sequence in the Last 10000 Years

Pleistocene loess and modern loess (modern atmospheric dust falls) have been studied by many scientists, but reports of systematic study on Holocene loess are few. Early in the 1960s, Liu Tungsheng noted that Holocene loess was developed in the middle reaches of the Huanghe River, and observed buried black loam developed in some localities of Shandong Province, inferring the overlying loess as Holocene (Liu Tungsheng et al. 1965). Since the 1970s, the occurrence of aeolian Holocene loess has been further confirmed in northern China with the application of radiocarbon dating technique in loess research. In recent years, typical sites of Holocene loess in northern China have been reported (Zheng Honghan 1984). Thus more attention has increasingly been given to the problems of Holocene loess. One major reason is that Holocene loess and pedogenesis have an important bearing on the development of human economics and modern environment. Holocene loess is the product formed by interaction of the atmosphere and the lithosphere in recent geological history.

3.4.1 Distribution of Holocene Loess

Holocene loess, which is composed commonly of upper loess and lower black loam – a good marker for stratigraphic subdivision – can be distinguished from

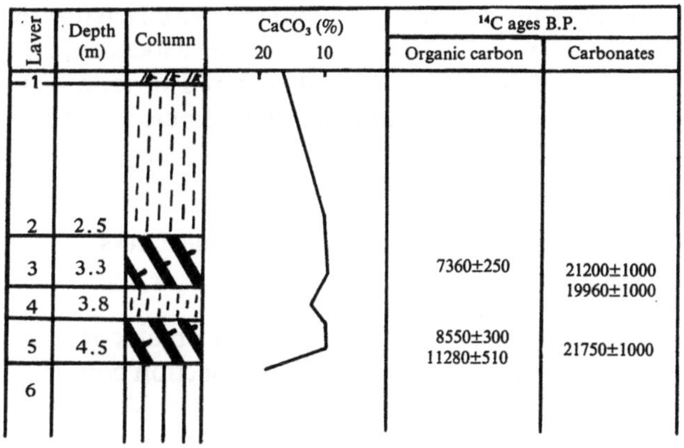

Fig. 3.12. The Nuanquangou loess section at Longxi and its ^{14}C ages

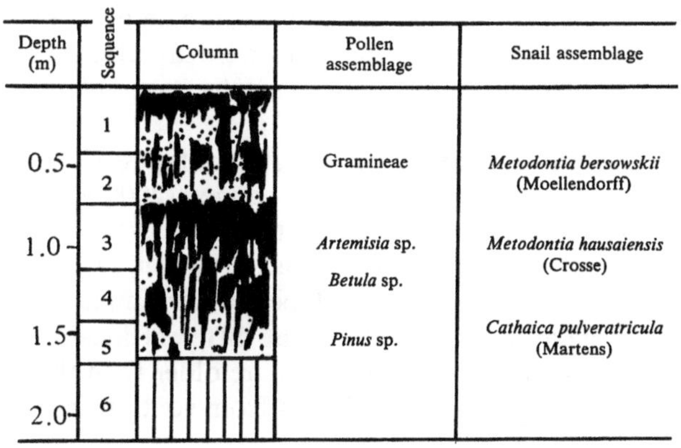

Fig. 3.13. Black loam section at Renjiahutong, Luochuan. *1* Modern soil; *2* Transition horizon; *3* argillic horizon with black color; *4* argollic horizon with greyish-black color; *5* carbonate illuvium; *6* Malan loess

the underlying Malan Loess by the basal boundary of black loam. Black loam is widespread in the areas of the central Loess Plateau and the eastern foothills and plains within the loess deposition belt of northern China. However, no black loam is developed in the area of western inland basins of western Gansu Province, Qinghai Province and the Uygur Autonomous Region of Xinjiang. Black loam is well developed near Yidu county at foothills in the central and southeast areas of Shandong Province; in Wuzhai, Pianguan and Baode counties of Shanxi Province and in Yulin, Suide, Jingbian counties of Shaanxi Province (Liu Tungsheng et al. 1965). In recent years, the author and his coworkers have also found black loam or similar soil exposed in Huanxian, Qingyang, Xifeng, Jingning, Huining, Dingxi and Lanzhou of Gansu Province (Fig. 3.11).

Regarding the occurrence of black loam in strata, two types are present: one is relict soil observed in the Luochuan section (Fig. 3.13), and the other is buried and overlain by loess about 1 m thick. Two or more layers of black loam can be recognized in some sections, showing slight difference in thickness and soil maturity. The Nuanquangou section at Longxi is a typical example (Fig. 3.12).

Holocene loess, including the underlying black loam, is best developed and preserved on levelling geomorphic positions, including loess "Yuan", ancient terraces on the edges of intermontane basins, and gently dividing ridges covered with loess or flat tops of loess hills. The geomorphic units with Holocene loess are different in altitude, and even the same geomorphic type also shows variations in altitude from place to place. Holocene loess overlying the Malan Loess and other strata runs parallel with basal relief.

3.4.2 Age of Black Loam and Stratigraphic Division of Holocene Loess in the Middle Reaches of the Huanghe River

Generally, the argillic horizon of black loam contains about 1% organic matter. Samples of black loam collected from the top of the Luochuan section comprise 0.35–1.23% organic matter (Dai Yingsheng et al. 1980) and hence are ideal materials for radiocarbon dating and soil age research. Table 3.4 indicates the organic matter ages extracted from samples by Zhou Mingfu. These age data are more reliable than those yielded by inorganic carbonate; the latter would be older due to the influence of older carbonate. In addition, the samples were all taken from buried soil free from the influence of modern plant roots, so the data obtained represent ages of various pedogenetic stages.

The ages in Table 3.4 are arranged in accordance with different types of landform from south to north. Ages generally fall into three ranges (2000–3000, 4600–7400 and 8100–9900 yr. B.P. respectively), showing that the development of black loam involves three different stages of the Holocene. However, layers of black loam at similar stratigraphic levels which were developed under similar conditions in terms of parent rock, vegetation and landform at different localities, show variations in age that indicate samples from the south are older than those from the north, suggesting that the climatic belt was shifted from south to north during pedogenesis.

Table 3.4. Radiocabon ages of black loam in the middle reaches of Huanghe River

Geomorphic position	Sampling locality	^{14}C age (yr B.P.)
Loess "Yuan"	Chengdong "Yuan", east of Huanxian	1935±130
	Leijiaxian, about 40 km north of Xifeng	2600±140
	Xilicun, about 25 km south of Xifeng	3169±190
	Baicao "Yuan", about 120 km north of Huining	5166±220
	Potou, Luochuan	8100±200
Ancient terrace on the edges of basin	Liugouhe, east of Lanzhou	4620±130
	Anjiapu, Dingxi[a]	6800±250
	Nuanquangou, Longxi[b] (upper layer)	7360±250
	(lower layer)	8550±300
	Yuijiagou, about 10 km south of Huanxian	4820±130
	Majiang, Qishan[c]	5120±200
Dividing ridge and loess hill	Niuzhangou, Jingning	3162±108
	Ziwuling[a] (layer A)	6820±300
	(layer B)	9900±500

[a] Qiao Yulou et al. (1979). [b] Zheng Honghan et al. (1982). [c] Zheng Honghan et al. (1979, unpubl.).

Black loam generally contains more pollen and spores as compared with loess. Pollen assemblages found in black loam samples from Luochuan, Xifeng and Longxi indicate that layers of black loam are developed in an arid grassland or forest-grassland environment, similar to the ecological environment recorded by fossil snails in the Luochuan section. Some differences are noticed in pollen assemblages in samples from different localities and stratigraphic levels. For example, the lower layer of black loam in the Longxi section contains primarily herb pollen, such as *Artemisia* sp., Polygonaceae and Chenopodiaceae, but no woody plant pollen, reflecting an arid-cool climatic condition. In contract, the upper layer has considerable woody plant pollen, such as *Betula* sp., *Carpinus* sp. and *Pinus* sp., though herb pollen is dominant. It is evident that this upper layer was developed under a warm-moist climatic condition. Pollen and spore assemblages in black loam developed in the ancient terraces on the edges of intermontane basins are mostly similar to those identified in the samples from the upper layer of black loam in the Longxi section. Similar assemblages are recognized in the samples of black loam from Xilicun, south of Xifeng "Yuan", Leijiaxian, north of Xifeng "Yuan", and Chengdong "Yuan" – a broken loess "Yuan" extending further northwards. Herb pollen is predominated by *Artemisia* sp. and Rosaceae with considerable woody plant pollen, such as *Betula* sp. It seems reasonable to assume that black loam in these three localities was developed under similar wet-cool bioclimatic conditions during pedogenesis. The pollen percentages from different localities are shown in Table 3.5.

Table 3.5. The contents of spore-pollens in black loam from different localities

Time	Locality	Pollen percentage			
		Tree		Herb	
Late Holocene	Chengdong "Yuan"	*Betula* sp.		*Artemisia* sp.	77.9
				Polygonaceae	2.0
	Leijiaxian	*Betula* sp.	25.0	Chenopodiaceae	26.0
		Pinus sp.	37.5		
	Xilicun			*Artemisia* sp.	76.2
				Polygonaceae	12.7
	Niuzangou	*Pinus* sp.		Polygonaceae	68.5
				Chenopodiaceae	10.5
Middle Holocene	Liugouhe	Ulmaceae	28.6	Polygonaceae	6.1
		Larix sp.	28.5	*Artemisia* sp.	29.4
		Quercus sp.	14.3		
	Baicao "Yuan"	*Pincus* sp.		Polygonaceae	45.7
				Gramineae	5.7
	Longxi (upper layer)	*Betula* sp.		*Artemisia* sp.	61.9
				Gramineae	19.1
Early Holocene	Longxi (lower layer)			*Artemisia* sp.	34.8
				Gramineae	52.2
	Luochuan	*Betula* sp.	22.3	*Artemisia* sp.	10.4
		Pinus sp.	11.1	Gramineae	18.6

Percentage of tree and herb were counted seperately. Analyset: Lin Shaomeng.

Table 3.6. Tentative scheme of stratigraphic subdivision of Holocene loess

Time	Formation	Sediment type	Thickness (m)	Locality	Pollen assemblage	^{14}C age (yr B.P.)
Late Holocene	Xifeng	loess (L_{h-3})	0.8 – 1.0	Xilicun Leijiaxian	*Betula* sp. *Artemisia* sp. Polygonaceae	
		black loam (S_{0-3})	0.8 – 1.0	Chengdong "Yuan" Niuzhangou	Rosaceae	2000 – 3000
Middle Holocene	Longxi	loess (L_{h-2})	2.0 – 2.5	Baican "Yuan" Liugouhe Yuijagou	*Quercus* sp. *Betula* sp. *Carpinus* sp.	
		black loam (S_{o-2})	0.4 – 1.2	Anjipapu Nuanquangou	*Pinus* sp. Ulmaceae	4600 – 7400
Early Holocene	Luochuan	loess (L_{h-1})	0.0 – 0.9	Potou Nuanquangou	*Pinus* sp. *Artemisia* sp. Chenopodiaceae	
		black loam (S_{o-1})	0.7 – 1.0		Polygonaceae	8100 – 9900

The alternative occurrence of loess and black loam, radiocarbon age of black loam in three different ranges and marked differences in pollen assemblages at different stratigraphic levels indicate that the existence of the Holocene loess underlain by black loam in the middle reaches of the Huanghe River is a proof of loess accumulation, pedogenesis and environmental evolution in this region in the last 10000 yr. Based on the data, a tentative scheme of stratigraphic subdivision of the Holocene loess is proposed (Table 3.6).

It is evident from Table 3.7 that the climatic history and ages of climatic events are in good agreement with those recorded in Holocene sediments of southern Liaoning (Laboratory of Quaternary Palynology and Laboratory of Radiocarbone, 1978) and the North China Plain (Chen Fangji 1980), indicating a good parallel of climatic conditions in this latitudinal zone (Table 3.7 also indicates that a younger black loam may exist in the middle reaches of the Huanghe River, i.e., a relatively warm stage occurred between the so-called second cold stage proposed by Zhu Kezhen (1500 – 1600 yr B.P.) and the Little Ice Stage (100 – 500 yr B.P.) (Zhu Kezhen 1973; Hecht et al. 1979). Perhaps some sections with only one black loam recorded the superimposed pedogenesis during this period (1500 – 500 yr B.P.).

3.4.3 Particle Size and Mineral Composition of Holocene Loess

In the middle reaches of the Huanghe River, particle size of the Holocene loess and the underlying black loam is composed mainly of silt-sized particles,

Table 3.7. Holocene climatostratigraphy of northern China

Age (B.P.)	N. Europe (Blytt-Sernander)	S. Liaoning (Institute of Geochem. 1977)	Beijing (Chen Fangji 1980)	N. China (Zhu Kezhen 1973)	Huanghe (Zheng Honghan 1984)	
		Zhuanghe II		Little glacial		
			Liubintun		?	
	Sub-Atlantic	Zhuanghe I				
2000				2nd cold stage		L_{h-3}
					Xifeng	S_{o-3}
4000		Dagushan II				
	Sub-Boreal		Yingezhuang	1st cold stage		L_{h-2}
					Longxi	
6000				warm stage (Yangshao Culture)		S_{o-2}
	Atlantic	Dagushan I	Yinjiahe			
8000					Luo-chuan	L_{h-1}
	Boreal	Pulandian	Changgou			S_{o-1}
10000	Pre-Boreal					
	Arctic					

0.075 – 0.002 mm in diameter (comprising about 80 – 85% of the total composition) with some fine sandy and clayey materials. These characteristics of particle size are generally similar to those Malan Loess and modern loess (dust fall) (Liu Tungsheng et al. 1981), indicating the continuation from Pleistocene loess through Holocene loess to modern dust fall. The general tendency of reginal variation of particle size is that the samples from the west and north of Liupan Mt. contain large amount of coarser particles. Further southward, the particle size tends to decrease gradually. Concerning mineral composition Holocene loess contains more than 30 kinds of heavy minerals in particles of 0.25 – 0.01 mm in diameter predominated by hornblende, epidote and opaque minerals (accounting for about 80%); biotite, garnet, tourmaline, titanite and zircon rank next; the most stable minerals are rarely found (about 2 – 4%). Clay minerals are illite with varying amounts of kaolinite, chlorite, montmorillonite and vermiculite. In some samples halloysite and mixed-layer minerals were found (Zheng Honghan 1984). In the Tuomur Peak area of the Tianshan Mts. in the Uygur Autonomous Region of Xinjiang, according to the results of particle size analysis for samples taken from loess sections at 2500 m camp site and Xiata electric station in the lower reaches of the Qiongtailan glacial valley, Holocene loess contains 0.73–7.83% of fine sand, 11.31–16.58% of microsand (0.1–0.005 mm in diameter) and 9.44–21.63% of clay particles (less than 0.005 mm in diameter). The difference between the characteristics of this particle size composition and the middle reaches of the Huanghe River is that coarser particles are predominant in the loess of Tuomur Peak, with more of those 0.1 and 0.1–0.5 mm in diameter. Heavy mineral composition of Holocene loess in the area of Tuomur Peak shows close similarity with that of the middle reaches of the Huanghe River predominated by hornblende, epidote and opaque minerals. Additionally, there is a difference in sharpness of mineral particles from place to place. For example, most of the heavy mineral particles from loess sample of 2500 m camp site section show angular or subangular sharpness. Rounded particles have been found rarely, whereas those from the section by the Ili-Urmuqi highway, heavy mineral particle shows subangular or sub-rounded sharpness, and sometimes rounded particles can be found as well (Liu Tungsheng et al. 1985). Loess of Tuomur Peak is clearly distributed in the lower reaches of a valley glacier with obvious aeolian occurrence, 5250±200 yr B.P. in age. Its particle size, heavy mineral composition and sharpness show clearly regional characteristics. The loess of Tuomur Peak distributed in glaciated area is another type of aeolian loess in China; its material was produced by glacial grinding. Such loess differs from the loess of the middle reaches of the Huanghe River, where constituent materials have a wider convenance.

3.5 Discussion of Loess-Paleosol Sequence and of Geological Events in the Past 2.4 Million Years

The Luochuan loess-paleosol sequence has recorded a number of geological events that have happened in the past 2.4 m.y., thus making it possible to reconstruct the history of the Quaternary climates in China, and in the Northern Hemisphere and to understand the future climate trends.

Magnetostratigraphic studies have confirmed that the loess in China has been deposited over the last 2.4 m.y. Several important time lines have been recognized in the Luochuan series, including those of 1.87–1.67 m.y. (Olduvai subchron), 0.97–0.90 m.y. (Jaramillo subchron) and 0.73 m.y. (Brunhes/Matuyama). Investigation of loess in China has entered a new stage in which a time scale has been made available. Investigators have realized that the loess in China is of utmost importance for the study of the world's Quaternary.

3.5.1 Time Sequence of Luochuan Loess-Paleosols

The time scale of the Luochuan loess-paleosol sequence is determined by paleomagnetic boundaries, thermoluminescence and ^{14}C dating, and is given in Table 3.8.

The main characteristics of loess deposits is that they are homogeneous and contain multiple buried paleosols in different stages of development. The loess-

Table 3.8. Ages of Luochuan strategraphic section

Stratum	Paleomagnetism ($\times 10^3$ yr)	Thermoluminescence ($\times 10^3$ yr)	^{14}C ($\times 10^3$ yr)
S_0			7.69±0.2 8.10±0.3
L_1		19±2 14±4 103±1	11.53±0.5 13.54±0.5 18.83±0.6 19.02±0.6
L_2		148±16	
L_5		357±31	
L_8	730	696±6	
L_9			
	900		
L_{10}		876±95	
S_{11}			
	970		
L_{12}			
W_{L-1}			
	1670		
W_{S-2}			
	1870		
W_{L-2}			
W_{L-3}			
	2480		
Red clay			

paleosol sequence is more or less continuous in time but different in composition and in development. As for the ages, one might ask: "Is there a way to establish ages of the geologic events recorded by individual stratigraphical units, i.e., by individual loess and paleosol layers (or their assemblages), when there are no detailed chronological data available?" The answer is "yes", and the need to solve such a problem is urgent. As the geological characteristics of the Luochuan section may well represent the conditions of the whole Loess Plateau, it is possible to take the Luochuan section as a model or even a stratotype for this purpose, and as a strating point to explain the history of environmental changes in the whole Loess Plateau.

With reference to other ^{14}C and thermoluminescence ages, the magnetostratigraphic boundaries and ^{14}C ages of organic matter in black loam are employed as a basis for the establishment of the time sequence of Luochuan loess-paleosol sequence.

Table 3.9. Time sequence of Luochuan loess paleosols assemblage

Sequence	Age ($\times 10^3$ yr)	Sequence	Age ($\times 10^3$ yr)	Sequence	Age ($\times 10^3$ yr)
S_0	10	S_6	660	S_{12}	
L_1		L_7		L_{13}	
S_1	140	S_7	720	S_{13}	
			730		
L_2		L_8		L_{14}	
S_2	250	S_8	770	S_{14}	1090
L_3		L_9		L_{15}	
S_3	330	S_9	900	W_{S-1}	1480
L_4		L_{10}		W_{L-1}	1670
S_4	410	S_{10}		W_{S-2}	1870
L_5		L_{11}		W_{L-2}	
S_5	560	S_{11}	970	W_{S-3}	2220
L_6		L_{12}		W_{L-3}	2400

An important problem concerning the establishment of the sequence (geochronological table) of the Luochuan section is how to calculate the age of individual loess and paleosol layers from the position of polarity reversals. If based merely on the time range delimited by the reversals, accurate ages could not be obtained on the basis of depth through extrapolation and interpolation. From the discussion on the loess and paleosol formation, it is known that loess was formed when the rate of dust accumulation was greater than that of soil formation. During the soil forming process, the dust accumulation did not stop. But the rate was slower than that of loess formation. Therefore, it is obvious that the loess and paleosol of the same thickness should span different time intervals with the former spanning shorter than the latter.

The actual amount deposited and the rate of accumulation should be taken into account when calculating the stratigraphic ages of loess and paleosol. The time required for the accumulation of the loess and paleosol layers have been calculated (see Chap. 6, Sect. II). Because loess has a higher depositional rate than paleosol, significant errors may result if ages of individual loess and paleosol layers are calculated based merely on the deposition amount. Accordingly, most calculations were made for individual loess-paleosol assemblages (Table 3.9), in order to minimize errors in calculated ages caused by the differences in the depositional rates. However, this assumes that the depositional rate is constant for a given loess-paleosol assemblage within an individual age interval determined by paleomagnetic boundaries.

The depositional amount-based dating method, although not precise, appears preferable to the depth-based method. Thus, it may serve as a basis for the establishment of the time sequence of the loess strata.

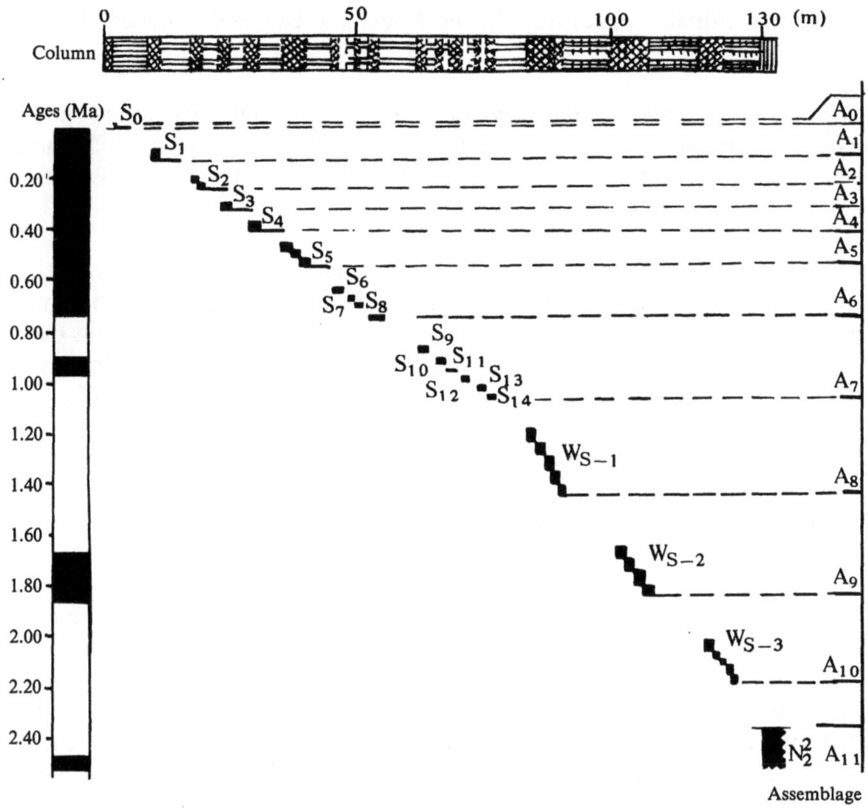

Fig. 3.14. Time sequence of Luochuan paleosols and their assemblages

Based on the depositioned amounts and with reference to the paleomagnetic boundary ages, we calculated the starting ages of each paleosol or paleosol assemblage shown in Table 3.9. It should be pointed out, however, that the data in Table 3.9 are only approximate.

3.5.2 Geological Events over the Past 2.4 Million Years

The Luochuan series has recorded multiple loess/paleosol alternations over the past 2.4 m.y. which are related to environmental or climatic change.

1. Climatic-Deterioration 2.4 m.y. Ago. The Wucheng Loess/Red Clay boundary of 2.4 m.y. marks the beginning of the loess-dust accumulation. Records of fossils and sedimentary facies in northern China indicate that this boundary can be considered as the start of the Quaternary period in northern China, when the climate deteriorated drastically from warm-humid to dry-cold (see Section II of Chapter 3).

2. Climatic-Optimum Event of 1.8 m.y. Ago. The well-developed paleosol assemblage (W_{S-2} developed in the middle of Wucheng Loess 1.87 − 1.67 m.y., recording the optimum climatic conditions in the Loess Plateau and northern China around 1.8 m.y.

3. Dry-Cold Climatic Events of 0.80 and 1.15 m.y. Ago. Two coarse-grained sandy loess deposits occur extensively in the Plateau, reflecting severe dry-cold climatic conditions and harsh steppe–desert steppe environment at 0.8 m.y. (L_9) and 1.15 m.y. (L_{15}).

4. The Optimum Climatic Event of 0.5 m.y. Ago. The paleosol S_5 in the Luochuan section which developed about 0.46–0.56 m.y., is a composite of three soil layers, reflecting the optimum climatic event and the climatic oscillations during this period. This event is comparable in age to the fine-grained travertine layer in cave deposits of Locality 1, Zhoukoudian, where the fossil Peking Man (*Homo erectus pekinensis*) was discovered (fission track age 0.47 m.y., Guo Shilun et al. 1980) and where relatively warm climatic conditions prevailed. About 0.5 m.y., marine transgressions occurred at the coastal areas of North China Plain reflecting the optimum climatic event in northern China.

5. Climatic and Geological Events Since 0.14 m.y. The portion of paleosol S_1 – Malan Loess L_1 – the Holocene black loam S_0 (Fig. 3.15) has recorded several climatic and geologic events that have occurred in the past 0.14 m.y. The paleosol S_1 is well developed and indicates a warm-humid soil-forming period between 140 and 90×10^3 yr. The Malan Loess developed about $95-10 \times 10^3$ yr and can be subdivided into three portions – upper and lower coarse-grained layers (L_{1-3}, L_{1-1}) and middle fine-grained layer (L_{1-2}) (An Zhisheng and Lu Yanchou 1984).

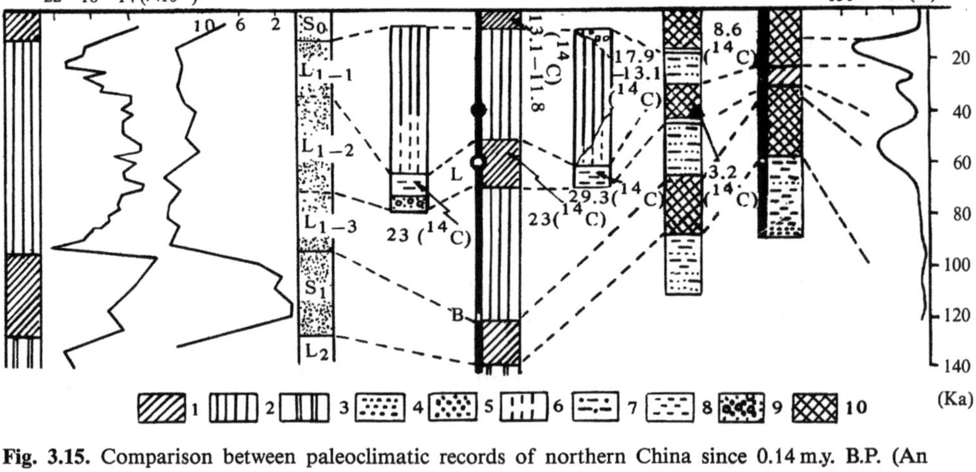

Fig. 3.15. Comparison between paleoclimatic records of northern China since 0.14 m.y. B.P. (An Zhisheng and Lu Yanchou 1984). *1* Paleosol; *2* Malan loess; *3* Lishi loess; *4* fine-grained layer; *5* coarse-grained layer; *6* loess-like silt; *7* sandy clay; *8* clay; *9* sand and gravel; *10* marine bed

The L_{1-2} fine-grained layer is relatively low in $CaCO_3$, but contains broad-leaved tree pollen such as *Alnus*, *Betula*, *Quercus* and Ulmaceae, indicating a moderately cool and humid climatic period as contrasted to the dry-cold climatic events reflected by the coarse-grained loess layers. The black loam, developed since 10000 yr B.P., suggests a warm-humid climatic event of post-glacial period. The climatic events which occurred in the Loess Plateau since 140×10^3 B.P. have also been recorded in mountainous basins on the plains as well as in land/sea sediments (Fig. 3.15). Five climatic events occurred in northern China during that period, i.e., a warm-humid one between 140 and 95×10^3 yr B.P., a dry-cold one at $95-50 \times 10^3$ yr B.P., a moderately cool one at $50-25 \times 10^3$ yr B.P., a dry-cold one at $25-10 \times 10^3$ yr B.P. and a warm-humid one since 10×10^3 yr B.P.

During the warm-humid climatic periods, fine-grained deposits or soil developed in northern China, travertine deposits were formed in piedmont caves and transgressions took place along the east coast. In contrast, the advent of dry-cold climatic events led to the increase of dust accumulation, deposition of coarse-grained loess or loess-like silt, and the development of river/lake sedimentary beds in the coastal areas due to the fall of the sea level.

3.5.3 Discussion on the Age of Early Human Fossils in Northern China

Fission track and U-series datings have yielded the age of the fossil Peking Man (*Homo erectus pekinensis*) discovered in cave deposits of Location 1, Zhoukoudian, northern China, to be 0.5–0.2 m.y. (Guo Shilun et al. 1980; Zhao Shusen et al. 1982).

In the past, the fossil men found in Chenjiawo and Gongwangling, Lantinan County, Shaanxi Province were designated as Lantian Man (*Homo erectus Lantianensis*; Wu Rukang 1964, 1966). Ma Xinghua et al. (1978) reported the paleomagnetic age of the Man unearthed in Chenjiawo to be 0.65 m.y., and the Lantian Man at Gongwangling may date back to 0.75–0.8 m.y. Cheng Guoliang et al. (1978), however, considered that the ages of Lantian Man from the locations mentioned above are 0.5 m.y. and 0.75 m.y., respectively. As the locations where the Chenjiawo and Gongwangling Lantian men were discovered were within loess sections, it is possible to make a comparison with the Luochuan section. The ages of Lantian Man can be determined from the ages of relevant stratigraphical positions in the Luochuan section. According to An Zhisheng and Sun Jianzhong, the paleosol layer containing Chenjiawo Lantian Man is comparable to the paleosol S_6 of the Luochuan section, which was formed about 0.65 m.y. B.P. Similarly, the sandy loess containing Gongwangling Lantian Man seems comparable to the lower sandy loess L_{15} of the Luochuan section, suggesting the fossil man to be 1.15 m.y. old.

Having compared the stratigraphic positions where early human fossils were found in China with the paleoclimatic cycles of loess/deep-sea sediments, Liu Tungsheng and Ding Menglin (1984) stated that the Lantian Man layers may date back 0.5–0.59 m.y. and 0.73–0.8 m.y., respectively.

In short, the fossil-man layers occurring in northern China may be roughly attributed to three periods, i.e., 0.73–0.8 m.y. or 1.15 m.y. (Gongwangling), 0.5–0.59 m.y. or 0.65 m.y. (Chenjiawo) and 0.5–0.2 m.y. B.P. (Zhoukoudian).

CHAPTER 4

Fossils in Loess and Its Ecological Environment

4.1 Fossil Mammals with Special Evidence of the Evolution of *Myospalax*

The mammalian fossils including Insectivora, Lagomorpha, Rodentia, Carnivora and Artiodactyla collected from Luochuan during field investigation from 1980 to 1982 can be preliminarily listed as follows:

Scaptochirus cf. *moschatus* M.-Edwards, *Ochotonoides complicidens* (Boule et Teilhard), *Ochotona dauurica* Pallas, *Allocricetus ehiki* Schaub, *Kowalskia yananica* sp. nov., *Cricetinus varians* Zdansky, *Meriones* sp., *Prosiphneus intermedius* Teilhard et Young, *P.* cf. *intermedius* Teilhard et Young, *Myospalax arvicolinus* (Nehring), *M. omegodon* (Teilhard et Young), *M. chaoyatseni* (Teilhard et Young), *M. hsuchiapinensis* (Teilhard et Young), *M. fontanieri* (M.-Edwards), *M. chanchenensis* (Teilhard et Young), *Microtus brandtioides* Young, *Apodemus agrarius* (Pallas), *Ursus* sp., *Lynx shansius* Teilhard, *Gazella* sp. (Fig. 4.1).

Among the rodents the genera *Allocricetus, Cricetinus* and *Kowalskia* are known in loess beds, and *Kowalskia* is found in the Pleistocene for the first time in fossil record.

4.1.1 Description of New Species and Its Importance to the Age of Loess

Kowalskia yananica sp. nov. (1981 *Cricetulus* sp., Xue Xiangxu, p. 35, pl. II Fig. 7, Youhe, Shaanxi Province).

Holotype: An almost complete skull with M^{1-3} tooth rows (Qv 10005).
Type Locality: Jiuyangou, Yanan County, Shaanxi Province.
Horizon: Upper part of Wucheng Loess.

Diagnosis: The most advanced species of *Kowalskia* with the skull having a narrower interorbital space and a more slender rostrum than that of *Kowalskia yinanensis* Zheng and *Cricetinus varians* Zdansky. On M^1 two anterocones are separated by an opening forward cavity and the crests linking the main cusps are shortened as in *C. varians*. Mesolophs on M^{1-2} are long and reach to the teeth, but absent on M^3 as in *K. yinanensis*.

Description: The skull is well preserved except the nasal file. The nostrils expanding slightly at their distal ends are about 14 mm long. The rostrum, having no

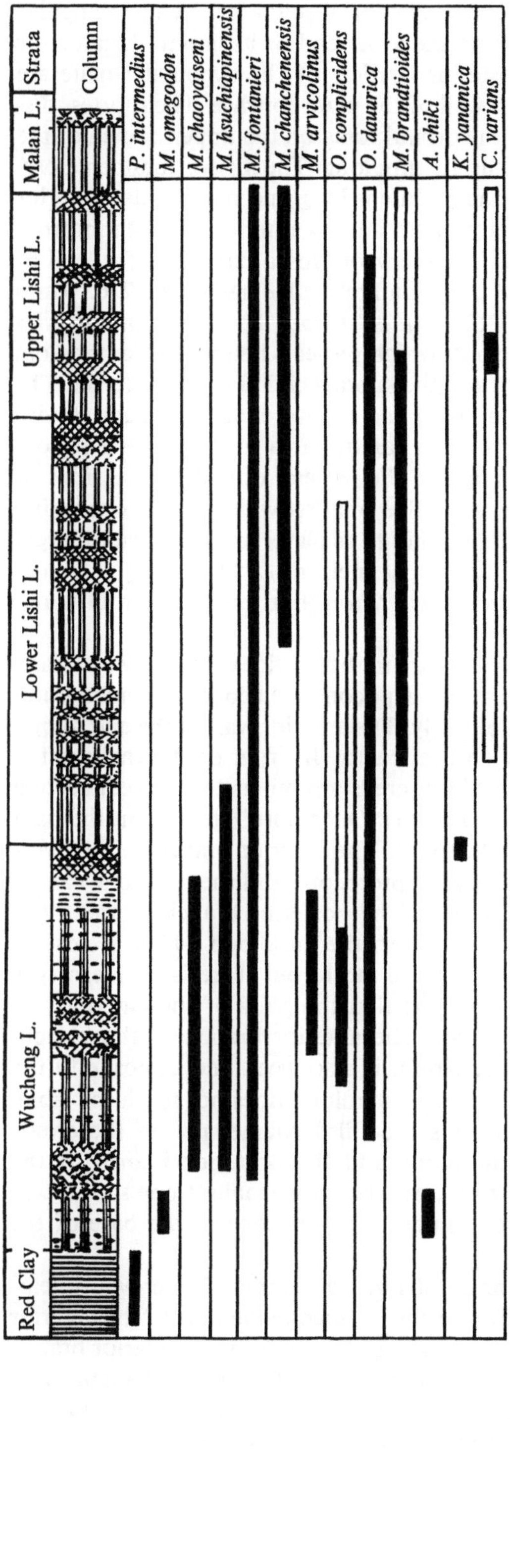

Fig. 4.1. The distribution and evolution of fossil mammals in the loess strata, Luochuan

swelling at the upper incisor roots, is slender. The infraorbital foramen is triangular from the outer view and 4.2 mm long from the lower end of its gap to the anterior edge of M^1. The anterior and the posterior borders of the zygomatic plate are nearly vertical downwards and the outer wall of the plate concave. The shortest distance from the infraorbital foramen to the orbital fossa on the upper branch of the anterior root of the zygomatic arch is only 1.5 mm long. The incisor foramen crossed by the premaxilla-maxilla suture in the middle is long and narrow, its posterior end about 1 mm from the anterior margin of M^1 and its anterior end 2.4 mm from the posterior margin of the incisor. The interpterygoid fossa extends to the level of M^3. The palate keeps 1.3 mm behind M^3. The audital bullae having a 7.7/6.10 mm diameter as big as that of the recent *"Cricetulus" triton*. The occipital condyles project on the general plane of the occiput, which projects behind the lambdoidal crest. The occipital crest is more robust and the supraoccipital process clearer. The interparietal (3.9 mm long and 10.2 mm wide) and its two ends just intersect with the temporal crests which expand and form a sagittal area in the parietal region. The frontal-parietal suture is a semi-circle and the parietal extends to the front of the temporal-frontal suture on the outer side. The smallest width of the interorbital space is 4.8 mm. The frontal-nasal suture reaches to the level of the anterior margin of the orbital fossa. The curvature of the upper incisor (1.78 mm long, 1.24 mm wide) is as strong as in the recent *"Cricetulus" triton*, but the incisor root is not swelling on the rostrum.

M^1 (2.45 mm long, 1.40 mm front and 1.50 mm back wide): The crown slightly narrow anteriorly, and its largest width at the hypocone-metacone. The main outer and inner cusps are arranged almost alternately. The anterior wall of the anteroloph is straight and separated anteriorly by an oval cavity. In front of the cavity exists a small cingulum cusp and at the back of it there is a semi-circle ridge connecting with the forearm of the protocone. An anterior "Quersporn" goes from the base of the labial anterocone to the edge of the tooth. The anterior branch of the "Vorjochkante" is so weak that the pit between the protocone and the paracone can be formed only when the tooth is worn heavily. The posterior branch of the "Vorjochkante" is stronger and connects directly with the mesoloph which is close to the metacone, and reaches to the margin of the tooth, but there is no mesostyle. The crests linking the main cusps are so short that the pairs of the anterior and the posterior cusps are close to each other. The anterior branch of the "Nachjochkante", which connects with the mesoloph, is also weak and opposite to the posterior branch of the "Vorjochkante". The posteroloph does not reach the labial side of the tooth, but encircles and forms the third enamel pit on the crown together with the back wall of the metacone and the posterior branch of the "Nachjochkante". The tooth has four roots, of which the smallest one supporting the protocone is separated from the one under the hypocone at the base of the crown.

M^2 (1.89 mm long, 1.57 mm front and 1.40 mm back wide): The crown is nearly square in shape and its largest width is at the protocone-paracone. The anterior cingulum is weaker on the lingual side than on the labial side. The posterior branch of the "Nachjochkante" is quite weak, but the two pits between the protocone and the paracone, the hypocone and the metacone, can be formed. The mesoloph reaches to the edge of the tooth, but the mesostyle does not exist. Four roots.

M^3 (1.80 mm long, 1.35 mm front and 1.20 mm back wide): The tooth is less reduced. The anterior cingulum on the lingual side is rather low. The paracone is more robust and the metacone more reduced than that on M^2. There is a supplementary crest extending to the bottom of the outer sinus on the back wall of the metacone. The posterior branch of the "Nachjochkante" is so weak that the sinus between the metacone and the posteroloph often open post-externally. There is a small enamel lake between the labial side of the metacone and the cingulum. The mesoloph absent. Three roots.

Comparison: The skull studied here is compared directly with the skulls of *Kowalskia yinanensis* Zheng from the latest Pliocene of Shandong Province, *Allocricetus ehiki* Schaub of Early Pleistocene at Locality 18 of Zhoukoudian, *Cricetinus varians* Zdansky of the Middle Pleistocene of Zhoukoudian district and the recent "*Cricetulus*" *triton* of Asia. *K. yananica* differs from *A. ehiki* in the skull in having a narrower nasal, a more projecting occiput and quite large audital bulla. The skull of *K. yananica* is similar to that of *K. yinanensis* in the lengths of the nasal and the diastema, the position of the nasalfrontal suture situated at the posterior edge of the infraorbital foramen, the size of the interparietal, the width of the rostrum near the incisor and the space between the occipital condyles, the incisor foramen closing to the anterior end of M^1, the concave wall of zygomatic plate and the smaller infraorbital foramen. The skull is also closer to that of *C. varians* and *"C". triton* in the premaxilla-maxilla suture standing ventrally at 1/3 of the incisor foramen anteriorly, the length of the frontal along their suture line, the largest width of the sagittal area, the diameter of the audital bulla, the shortest distance between the infraorbital foramen and the orbital fossa, the smallest width of the zygomatic plate and of the palate between the two M^1. While some characteristics, such as the lengths of the skull and the upper tooth row, the widths of the rostrum before the infraorbital foramen and the interorbital space and the distance from anterior edge of the incisor to anterior edge of the pterygopalatina fossa, are different from them.

As far as the structure of the upper molars is concerned, the new species seems to be a transitional form from *K. yinanensis* to *C. varians.* They have four roots on M^1 and the absence of the mesoloph on M^3 in common. The new species, on the one hand, is in agreement with *K. yinanensis* on the flattened anterior wall of the anterocone, and anterior "Quersporn" and the long mesoloph reaching to the edge of tooth on M^{1-2}, on the other, more similar to *C. varians* in the two tubercles of the anterocone separated from the front, double "Nachjochkante" on M^{1-2}, the shortened crests linking the main cusps and the branch of the "Nachjochkante" moving toward the labial side. All the above-mentioned facts enable us to distinguish the *K. yananica* from those species of the genus *Kowalskia* known in Europe and Asia.

Discussion: Many species of *Kowalskia* have been found in Eurasia. The Vallesian *Kowalskia* sp. from Vosendorf of the Vienna Basin in Europe is considered as the oldest representative, the Ruscinian *K. intermedia* from Ivanovce, Czechoslovakia the youngest one. In China, Baodean (equal to *Turonian* of Europe) *K. gansunica* from Tianzu of Gansu Province is known as the oldest one and *K. yinanensis* from Yinan of Shandong Province is comparable with *K. intermedia* in age. The new species, *K. yananica*, from the Wucheng Loess shows that *Kowalskia* could continue to the Early Pleistocene.

V. Fahlbusch (1969) thought that *Kowalskia* must have been the ancestor of one or more of the Pleistocene and recent small and middle hamsters (*Allocricetus, Cricetinus, Cricetulus, Mesocricetus*). The fact the *Kowalskia* coexists with the *Cricetus* in Poland and the *Allocricetus* in Czechoslovakia shows that the former is not the forerunner of the latter. Meanwhile, the skulls from Shandong and Shaanxi Provinces indicate that *Kowalskia* is also not the ancestor of the recent *Mesocricetus* whose skull is very similar to that of *Cricetus*. The structure of the molars of *Cricetulus* is closer to that of the Early Pleistocene *Allocricetus* and the Pliocene *Sinocricetus* (or *Nannocricetus*). From the above analysis we have sufficient reason to infer that *Cricetinus* is probably the direct descendant of *Kowalskia*.

Now, we agree with the opinion that the genus *Kowalskia* originates from the Miocene *Democricetodon gaillardi* group which has not been found in China yet.

Though some species of Pleistocene cricetids have been found in many places in China, we cannot utilize them to make a stratigraphic subdivision. However, the materials obtained from the loess district make a superficial understanding possible about the distribution of these genera and species stratigraphically: both *Allocricetus ehiki* and *Kowalskia yananica* sp. nov. appear only in the Early Pleistocene; both *Allocricetus teilhardi* and *Cricetinus varians* seem to be found only in the early Middle Pleistocene. Because A. *ehiki* exists either in the lower part of the Wucheng Loess bed or in Locality 18 of Zhoukoudian, the faunas can be regarded as contemporaneous and comparable with those of Villany, Kadzielnia and Kamyk of eastern Europe in age. The fossil myospalacini are most abundant in the Luochuan district. In order to get a more definite concept, we use some new methods and parameters. It has been found that the mode connecting the temporal crest line with the parietal-temporal suture line is different in various types: the two lines are always intersected in *Myospalax tingi* group and parallel to each other in *M. fontanieri* and *M. arvicolinus* groups. Each species of the three groups has its intrinsic features, for example, the temporal crest of *M. tingi* converges forward more gently, so that the junction of the temporal crest and the parietal-temporal crest suture is near the midpoint of the latter; but in *M. chaoyatseni* of the same group the junction at one third of the suture anteriorly for the crest converges slowly posteriorly, then turning abruptly anteriorly and forming an obvious turning point before intersecting it with the suture.

At the same time the parameters E_1, E_2, A and B of M^1 of the rooted myospalacini, which represent the variation of the labial enamel-free area, are also important for determining the height of the crown, the more primitive the species is the smaller these parameters will be. Judged from these parameters only, *Prosiphneus intermedius*, found in the Red Clay of Luochuan, should be one of the most advanced species of *Prosiphneus*.

The synthetic section drawing shows clearly the distribution of various fossils collected from different localities of Luochuan district. According to the mammalian complex this section can be divided into the following mammalian fossil zones from the bottom to the top:

1. Final *Prosiphneus* zone: *Prosiphneus intermedius* is a good indicator marking a transitional position from the rooted *Prosiphneus* to the rootless Myospalax, and indicates that the bed containing this animal is of the "Jingle" period, Late Pliocene in age.

2. *Myospalax tingi* group zone: The members of this group are dominant in variety and number, but that of the *Myospalax fontanieri* group is rather dull. Such an evolutionary stage is essentially in agreement with the common Wucheng Loess, Early Pleistocene in age, and shows a drier and colder climatic environment than the above-mentioned one.

a) *Myospalax omegodon* subzone: It occupies the lowest part of the zone in the Luochuan section and may be comparable with the so-called Zone A of the Reddish Clay, of the early Early Pleistocene.

b) *M. chaoyatseni-M. hsuchiapinensis* subzone: The fauna consists of *M. chaoyatseni, M. hsuchiapinensis, M. arvicolinus, M. fontanieri, Kowalskia yananica* sp. nov. and *Ochotonoides complicidens*. The faunal complex shows that this subzone should be in agreement with Zone B of the Reddish Clay and the Nihewan Bed in the strict sense, which is of the middle-late Early Pleistocene. The dominant cold- and dry enduring elements found in North China from this period indicate that there was then a colder and drier climatic condition.

3. *M. fontanieri* group zone: The fauna contains *M. hsuchiapinensis, M. fontanieri, M. chanchenensis, Microtus brandtioides, Ochotona dauurica, Crincetinus varians* and *Megaloceros luochuanensis*. In accordance with the materials recorded in North China, *Ochotonoides* could survive to the middle and *Cricetinus* variants traced to the lower of the zone. Both *Myospalax minor* and *M.* cf. *cansus* are also found in this stage. This zone belongs to the Lishi Loess and is comparable with the so-called Zone C of the Reddish Clay, and of the Middle Pleistocene in age. Some hydrophilous and thermophilic elements, such as *Palaeoloxodon* and *Bubalus* (from the Dale Man Locality), *Hystrix, Elephas, Megamacaca, Ailuropoda, Stegodon, Dicerorhinus, Megatapirus, Tapirus* and *Capricornis sumtraensis* (from the Locality of Lantian Man), show that the climatic change of this stage was quite rapid. This hypothesis can be also borne out by the alternate deposition of the greyish-yellow loess and the dark reddish fossil soil.

4.2 Fossil Snails and Their Environment

Snails are very common invertebrates in loess areas and consequently their fossils are frequently found in loess layers. Their appearance in loess is connected with an environment rich in calcium carbonate. Different combinations of snails are present in the loess layers. Ecologically, these combinations of fossil snails reflect various natural environments. Therefore, snails can be regarded as a kind of "indicator animal" for the paleoenvironment in which the loess was formed.

The fossil land mollusks in China appeared in the Late Cretaceous period, and became most abundant in the Cenozoic era. In the loess layers of the Quaternary period, the number of both individuals and variety of species is plentiful. Although a large number of these are now extinct, the majority belong to existing species. E. V. Martens (1879), V. Hilber (1882) and A. Andraea (1913) have been engaged in the study of the Quaternary fossil snails in loess layers of the northern and northwestern parts of China. They have identified and described the genera and species concerned, but no details about their locality, horizon and geological period have been given. Ping Chi (1929, 1930, 1931) studied a number of the Cenozoic freshwater and land fossil mollusks of the areas mentioned above. King (1926) reported on the Pleistocene fossil snails, including bivalves, freshwater and land gastropoda from the Sanmen system. S. C. Hsu (1936) recorded gastropoda fossils from Xiashu layer, near Nanjing of Jiangsu Province. Yen Tengchien (1936, 1938, 1943) systematically summarized the study on nonmarine fossil mollusks of China and described 72 species of fossil snails, five of which are Tertiary species and the rest are the Quaternary species. Yen's research work has laid a foundation for the study of fossil snails of the Quaternary period.

After the founding of the People's Republic of China, Yu Wen (1963) wrote a book entitled On Gastropoda Fossils of China, in which she described 722 species and varieties of gastropoda fossils reported before 1962 belonging to 267 genera, 114 families and 44 super-families respectively. Among them, 58 species and varieties contained fossil snails. Li Yuntong (1966) reported 33 species and one new species of the Cenozoic mollusk fossils discovered in the areas of Lantian and Weinan, Shaanxi Province. In addition, 21 species (including one new species) of fossils are also found in the loess layers of the middle and upper Pleistocene. Li's paper discussed the formation of the Loess Plateau and paleoclimate. Pei Wenzhong and Zhou Mingzhen (1958), Chia Lanpo and Wei Chi (1976) recorded the species of fossil snails of the Quaternary period. Chen Deniu et al. (1982) recorded 12 species of fossil snails in the loess layer of the Heimugou, Luochuan. They divided them into assemblages and discussed their ecological significance. Other records on the fossil snails of the Quaternary period can also be seen in the early regional paleontological reports in China.

According to the fossil snails contained in the loess layers in North America and Central Europe, V. Lozek (1964) and G. J. Kukla (1977) divided them into four groups and established four climate patterns. This shows that the fossil snails in the loess layers are good recorders of the changes in the paleoclimate and paleogeography. A. Piechocki (1977) discussed the significance of mollusk fossils in both the environment and stratigraphy and compared the fossil snails of the Quaternary period with the habitat of the corresponding existing species. He divided them into ten ecological types, and established the paleoclimatic and paleogeographic environments.

From 1980 to 1981, at Heimugou near Potou, Luochuan, 2178 individuals belonging to five families, eight genera and seventeen species were collected by the authors from 85 loess layers in accordance with the stratigraphic sequence. Among them *Cathaica* and *Metodontia* predominated, with *Pupilla* and *Vallonia* subordinate. The individual numbers of different species in each layer are shown in Fig. 4.2. They can be divided into a group of *Pupilla aeoli* and *Cathaica* and a group of *Cyclophorus* and *Metodontia*, according to the character of the community of the fossil snails. The number of fossil snails in the loess layers has been

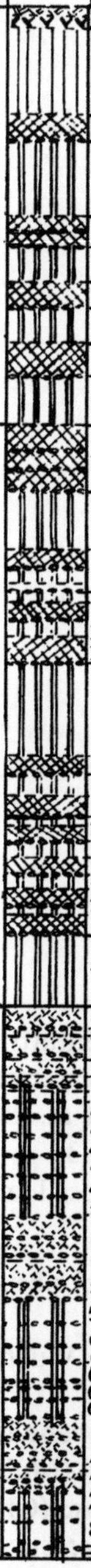

Strata	Column	Layer	C. pulvera trix	P. aeoli	V. costata	V. tenera	C. fasciola	C. lutuosa	C. pulveratricula	C. shikauensis	C. subrugos	P. futtereri	M. beresowskii	M. hausaiensis	M. yantaiensis	G. areasiana	S. erythrophana	C. teilhardi	B. similaris	Total
Malan L.			35						19			1		30	4				2	91
			26				4	12	2			4		29						77
			22	7	1	3	7	4	7			6		87						144
			16		3	2	8	1	13				4	167				3		217
		2	19	6		5	19		44					129			4			226
Upper Lishi L.		3	158	12	2	2	3		11					59	34					281
		4	52	7	6	7			4					70	10		1			157
		5	38	6	3	4							3	15	11					80
		6	13	4									3	11	2	5	3			41
		7	19	5									1	5	2					32
		8	7	12					2					20	11					52
		9	24	3					2					86				2		117
Lower Lishi L.		10	10	4										90	35			2		141
		11	6	1										9						16
		12	4										3	97	2					106
		13	5											13						18
		16	4											10						14
		17	4											21						25
		18	8											19						27
		19	4											14						18
		20	5											4						9
		21																		
		22	2							10				6						18
		23	14											15						29
		24	5							8				6						19
		25	4							10				2						16
		26	2											5						7
		27	4											4						8
		28	4											3						7
		29																		
Wucheng L.		30																		
		31																		
		33	2							4				4						10
		35	6							5				10						21
		37									7			8						15
			5							4				4						
		41	2											2						4
		43	2							7				2						11
		45	2											5						
		47																		
		49									6			2						8
		51												2						2
		53	2							1										3
		55									4									4
		57																		
		59												4						4
		61												3						3
		63																		
		65																		
		67	6							1				7						14
		69	2							6				2						10
		71																		
		73												4						4
		75												5						5
		77							3											7
		79	2																	2
		81	2											3						5
		83																		

Fig. 4.2. Distribution of fossil snails in section of Luochuan

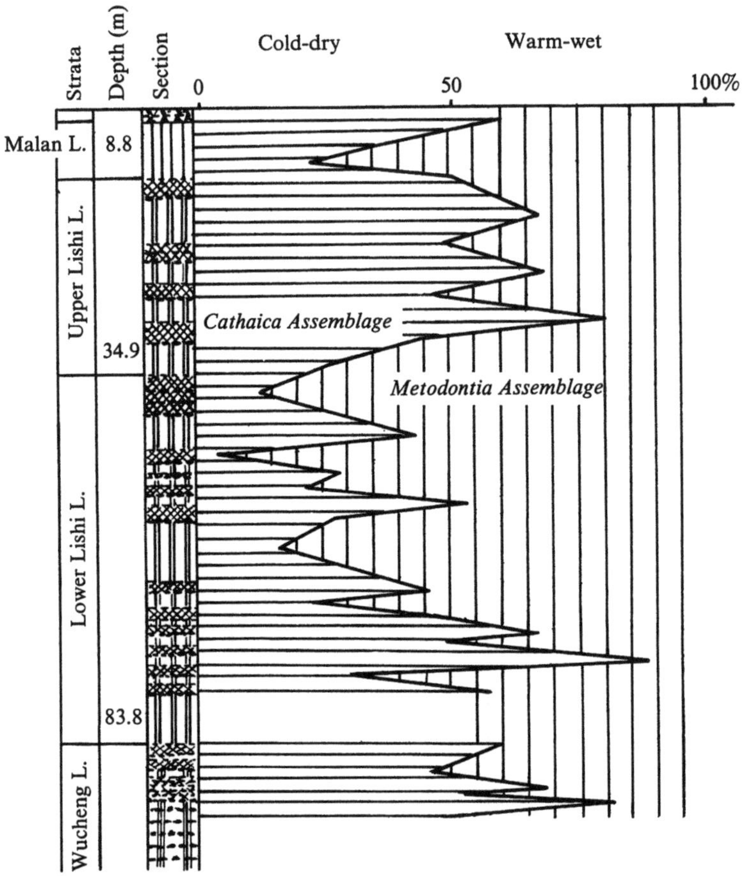

Fig. 4.3. Environmental change reflected by fossil snails of Luochuan loess

calculated. Taking the group of *Pupilla aeoli* and *Cathaica* as a representative of an arid environment and the group of *Cyclophorus* and *Metodontia* a wet environment, a variation curve of cold-warm environment (Chen Deniu et al. 1982) represented by the two groups of fossil snails is thus shown in Fig. 4.3.

On the basis of the above work, the distribution of 17 species of fossil snails that are most frequently encountered can be divided into four types:

1. A series of *Cathaica pulveratrix* and *Metodontia hausaiensis* (Crosse) occur in approximately 85 sampled layers of the loess.
2. *Cathaica shikouensis* Yen, *C. subrugos* (Deshayes) and appear in Wucheng Loess.
3. Twelve species and genera occurred in the Lishi Loess and the Holocene loess since 0.73 m.y. They are *Pupilla aeoli* (Hilber), *Vallonia costata* (Muller), *Vallonia tenera* (Reinhard), *Cathaica fasciola* (Draparnaud), *C. lutuosa* (Crosse et Deshayes), *C. Pulveratricula* (Mortens), *C. teilhardi* Yen, *Platypetasus futterei* (Andreae), *Metodontia beresowskii* Moillendorf, *M. yantaiensis* (Crosse

et Debaeux), *Ganesella areassiana* (Crosse et Debaeux) and *Succinea erythrophana* Ancey.
4. Among the above twelve species, eight appeared only from 0.5 – 0.6 m.y. to the present, and are found above the fifth paleosol. They are *Vallonia costata* (Muller), *Vallonia tenera* (Reinhard), *Cathaica fasciola* (Draparnaud), *Cathaica teilhardi* Yen, *Platypetasus futtereri* (Andreae), *Ganesella areasiana* (Crosse et Debaeux), *Succinea erythrophana* Ancey and *Bradybaena similaris* (Ferusac).

The following preliminary conclusions can be drawn from the stratigraphic distribution of about 5000 individuals collected in 1979 and 1980 – 1981.

a) There is a considerable amount of *Cathaica pulveratrix* (Martens) and *Metodontia hausaiensis* (Crosse) throughout the 85 sampled layers. They appeared continuously over the last 2.4 m.y., demonstrating that they had the ability to adapt themselves to the loess environment. Therefore they can be regarded as the predominate species of fossil snails in loess-paleosol sequence.

b) Over the past 0.73 m.y. (Brunhes chron) numerous new species have appeared, suggesting an epoch of differentiation and evolution of snails. However, only *Cathaica pulveratrix* (Mortens) and *Metodontia hausaiensis* (Crosse), were abundant in this period. This may be related to the dry environment of the Brunhes chron (see Sect. 3 in this chapter and Sect. 3 in Chap. VI). Such a condition may be suitable for species differentiation, but may not be advantageous to individual reproduction. It is not clear whether the appearance and extinction of some species are connected with the change of environment.

c) It is worth noting that many species developed rapidly above the fifth paleosol (S_5) in the recent 0.5 – 0.6 m.y. This period was climatically favorable for forming soil, and the development of species.

The S_5 period was a climatic optimum not only in northern China, but also in the rest of the world.

d) Fewer species of snail in Wucheng Loess appeared before 1.1 m.y., which appears to contradict the conclusion that the climate was wetter and warmer at that time, as evidenced from spore-pollen, deposition and geochemistry. This is possibly related to poor preservation of the species due to the depositional environment.

From the analysis of fossil snails (Fig. 4.3), we can see that *Metodontia hausaiensis* (Crosse) predominated in 2.4 – 0.9 m.y. (Matuyama chron), with *Cathaica pulveratrix* (Martens) and *C. shikauensis* Yen subordinate. The former reflected a warm, wet grassland, while the latter represented a dry, cold desert grassland.

In the accumulating layers, namely the lower part of Lishi Loess, 0.9 – 0.6 m.y. *Metodontia hausaiensis* (Cross) still predominates, with *Cathaica pulveratrix* (Martens) subordinate. *Metodontia beresowskii* (Moellendorff) and *Metodontia yantaiensis* (Crosse et Debaux) have also been found. The small rodent fossils in this period are relict species, such as *Myospalax chaoyatseni* and *Myospalax arvicolinus*.

In the upper part of Lishi Loess, *Metodontia hausaiensis* predominates. In the lower part of the fifth paleosol 0.56 – 0.46 m.y., *Cathaica pulveratrix* (Martens) ac-

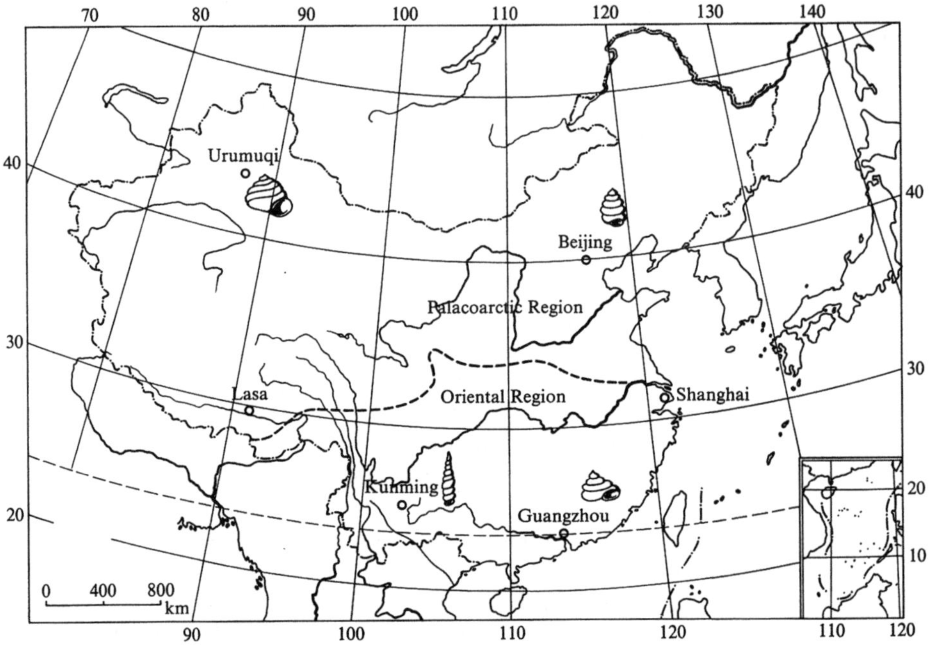

Fig. 4.4. Difference of distribution between the existing species of snails in the Oriental Region and that in the Paleoarctic Region

counts for only 10%. The fossil of *Megaloceros* cf. *pachyosteus*, a representative of northern forest grassland, has been found in this layer. This period progressed from the warm and humid environment of forest grassland to the semi-humid and semi-arid grassland.

In the Malan Loess of the Late Pleistocene (0.1 – 0.01 m.y.). *Metodontia hausaiensis* (Crosse) tended to decrease in number, while the combinations of drought-enduring and cold-resistant *Pupilla aeoli* (Hilber), *Vallonia costata* (Muller), *Cathaica fasciola* (Draparnaud) and *Cathaica pulveratrix* (Martens) gradually increased. Some semi-arid and humidiphilus species became extinct, such as *Succinea erythrophana* Ancey. Spore-pollen analysis shows that small shrubs and herbs predominated in the vegetation of this period. The character of the grassland gradually changed from humid to dry conditions.

In the Holocene, the loess containing black loam was deposited on the Loess Plateau, dated at 8500 – 10000 yr B.P. by ^{14}C dating (Zheng Honghan 1984). The fossil snails appearing in this period are basically similar to the existing species, in which the eurytropic species, *Bradybaena ravida* (Benson) for example, is distributed from the tropical and subtropical zones to temperate zone. The climatic environment reflected by them is similar to the present one.

The ability to migrate and the range of activity of snails are restricted. The burial sites of snail shells in the eolian loess are the same as the sites of their habitat.

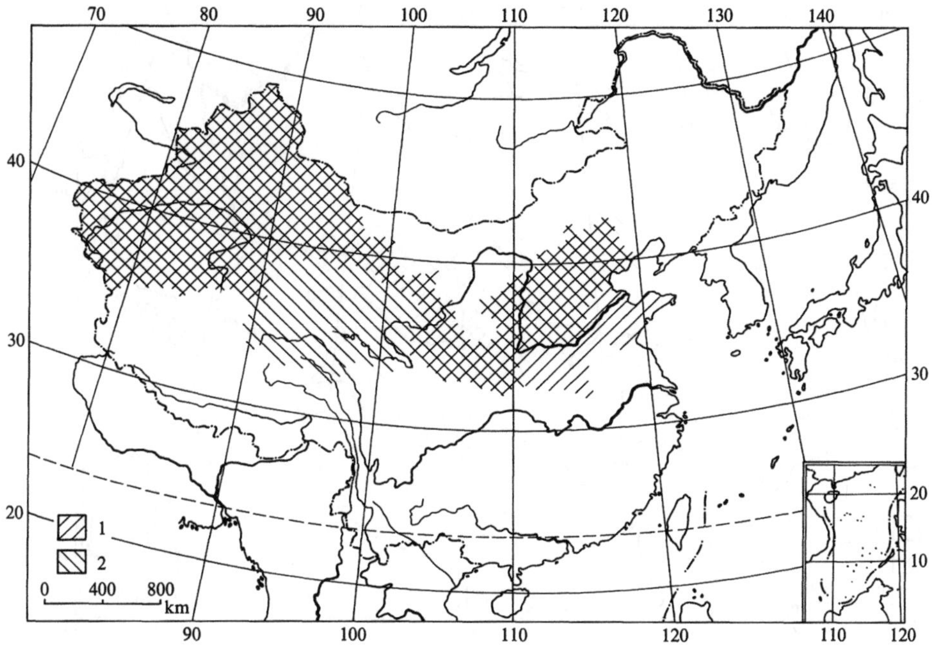

Fig. 4.5. Distribution of *Cathaica*, a type of aridphilus. *1* Distribution of the existing species; *2* distribution of fossil species

Up to now, seven families, 13 genera and 35 species have been found in the loess layers of China. There are two kinds of species: One if found of living in a warm and wet environment and is distributed in the Oriental Region, such as *Cyclophorus martensianus, C. punctatus, Diplommatina pupinella, Hemiphaedusa heudeana, Vitreapygmaen, Macrochlamys amdonan, Kaliella lamprocvstis, Xestina chrysorapha,* and *Succinea*. The other prefers living in an arid and cold environment and is distributed in the Paleoarctic Region, such as *Pupilla aeoli* (Hilber), *Vallonia costata* (Muller), and *Cathaica*. Some species, such as *Metodontia hausaiensis* (Crosse), are distributed in the semi-humid and semi-arid region. In addition, some species, such as *Cyclophorus*, *Diplommatina* and *Hemiphaedusa*, mainly dwell in the southern areas of the Changjiang River, but are not found now in the northern areas of the Huanghe River.

Species in the Oriental Region are completely different from those in the Paleoarctic Region. *Cyclophorus* and *Hemiphaedusa* represent the former region, while *Pupilla aeoli* and *Vallonia costata* represent the later region. Take *Cathaica* as a representative for aridphilus (Fig. 4.5), *Metodontia* for thermo-humidiphilus (Fig. 4.6), *Cyclophorus* for thermo-humidiphilus (Fig. 4.7) and *Succinea* for semi-arid and wetphilus (Fig. 4.8), from which we can observe that the distribution of the four species is quite different. From figures showing the distribution sites of the corresponding species of fossils found over the past 2.4 m.y., we will find the changes of their distribution and infer the ecological changes for this period.

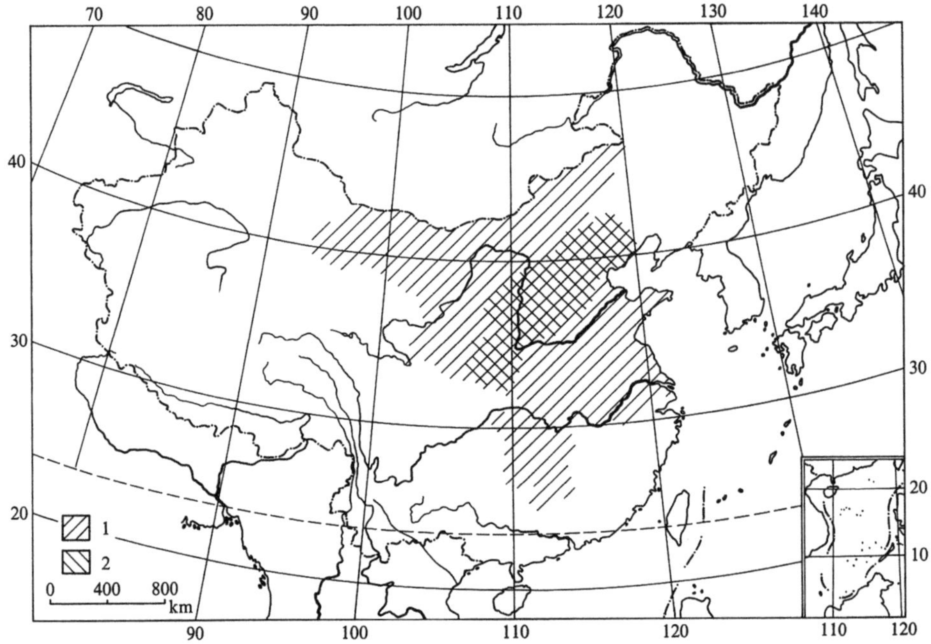

Fig. 4.6. Distribution of *Metodontia*, a type of thermohumidiphilus. *1* Distribution of the existing species; *2* distribution of fossil species

Cathaica is an animal community found in the Paleoarctic Region, and is mainly distributed in the loess region to the north of the Changjiang River. When the present distribution of *Cathaica* is compared to the mapped loess distribution, we find that both are identical. There is no difference between the distribution of fossil *Cathaica* and that of the existing species (except the case in Qinghai Province due to lack of data). This shows that, at least from the beginning of the Pleistocene Period, i.e., 2.4 m.y., *Cathaica* had developed and dwelled steadily on the loess, so it became the most adaptable genus to the ecological system of loess.

The distribution of the existing species of *Metodontia* lies mainly in central China, i.e., from Inner Mongolia, Shanxi and Shaanxi Provinces to Jiangus and Hunan Provinces. *Metodontia* is not distributed in the subtropical zone as Guangdong Province and Zhuang Autonomous Region of Guangxi. *Metodontia* belongs to an intermediate type of environment, and is mainly distributed in the Loess Plateau. As far as the data reveal, the fossil *Metodontia* is distributed also to the east of the Loess Plateau.

The common character of fossil snail distribution represented by *Cathaica* and *Metodontia* is that their range is smaller than that of their present-day distribution.

The existing species *Succinea* present in the Loess Plateau, the Uygur Autonomous Region of Xinjang, Sichuan and Guangdong Provinces, is more widely distributed and extends to the east and the south compared to the fossil range.

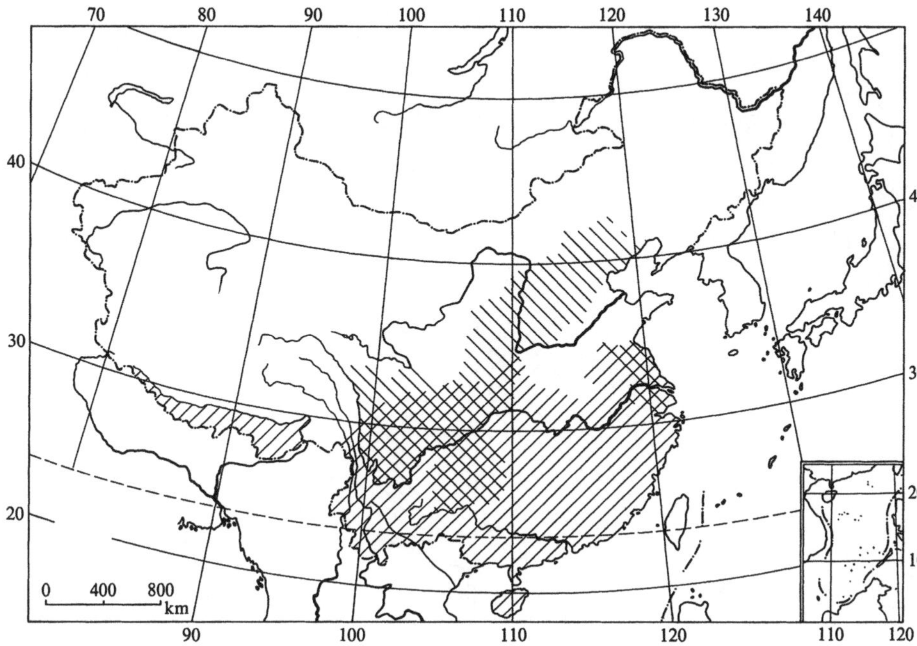

Fig. 4.7. *Cyclophorus, Macrochlamys* and *Hemiphaedusa, Vitrea*, stypes of thermo-humidiphilus, *1* Distribution of the existing species; *2* distribution of fossil species

Another point of interest is the limited distribution area of fossil *Succinea*, which indicates that the species has gradually moved from its original habitat and migrated to a warmer and wetter environment. The tendency to migrate is more obvious for *Cyclophorus, Macrochlamys, Hemiphaedusa*, and *Vitrea*, which are fond of a warm and wet environment. The existing species are mainly found in the Oriental Region distributed along the Changjiang River, in Guangdong and Yunnan Provinces, and Zhuang Autonomous Region of Guangxi. The fossil species has been found as far north as the Paleoarctic Region. Such fossil species are found in the Loess Plateau and northwestern part of Sichuan Province, but no existing species can be found there at present. This suggests that this type of snail has gradually migrated to the south, adapting themselves to a warmer and wetter environment. As fossil forms have been found to the north of the habitat of the existing species, this means that the environment of the Loess Plateau during the formation of paleosol was like that of their present wetter and warmer habitat.

To summarize, *Cathaica* and *Metodontia*, (especially *Cathaica*), are animals in the loess ecological environment either at present or in the Quaternary period. But the distributions of *Succinea*, that is semi-arid and thermophilus and *Cyclophorus*, that is thermo-humidiphilus, are quite different in the recent Loess Plateau from the geological period.

Comparison of the range of distribution between fossil and present-day species provides important information on the paleoenvironment and hence

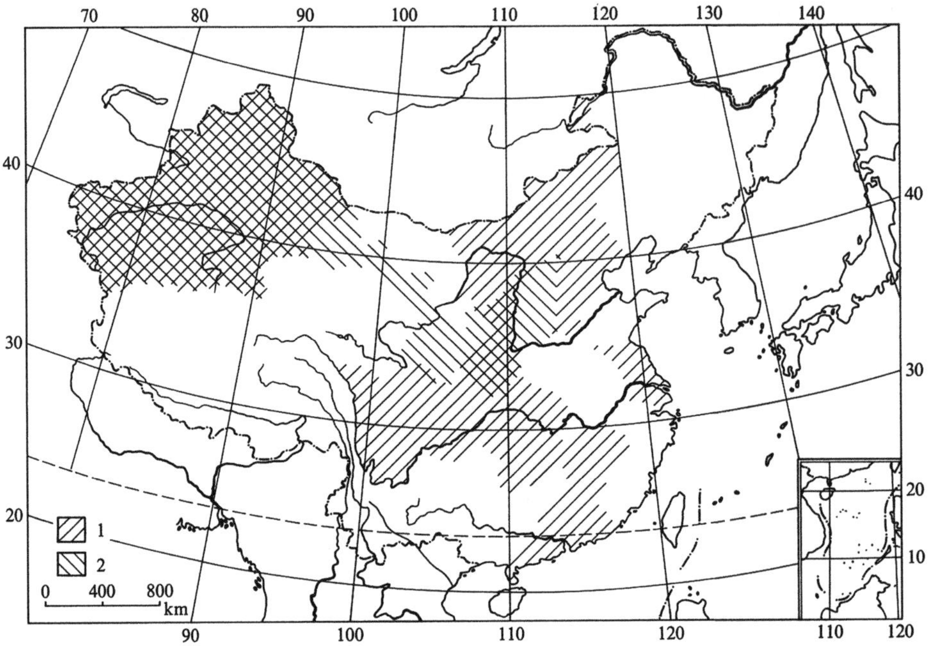

Fig. 4.8. Distribution of *Succinea*, a type of semi-arid and humidiphilus. *1* Distribution of the existing species; *2* distribution of fossil species

Table 4.1. Comparison of meteorological data between Luochuan and Hanzhong (1961 to 1970)

	Annual average of air temperature (°C)	Annual average of absolute humidity	Annual average of relative humidity (%)	Annual average of precipitation (mm)	Annual average of surface temperature (°C)
Luochuan	9.1	9.0	6.4	661.3	12.2
Hanzhong	14.3	14.5	8.0	903.9	16.9

paleoclimate. Some species, such as *Hemiphaedusa*, are mainly distributed to the south of Hanzhong, Shaanxi Province (33°04′N, 107°02′E; 508.3 m above sea level), but these fossil species also have been found in Luochuan (35°49′N, 109°30′E, 1158.3 m about sea level) and Lantian, Shaanxi Province. According to meteorological data, the surface air temperature, humidity and precipitation of Luochuan are lower or less than those of Hanzhong, which is located to the south of the Qinling, Mountains. In terms of fauna, Hanzhong belongs to the Oriental Region, while Luochuan belongs to the Paleoarctic Region. If the annual mean temperature and humidity of the northernmost limits of the present distribution area represent those of the ancient Luochuan where *Hemiphedusa* lived, then the annual mean temperature of the ancient Luochuan was 5.2 °C higher, the annual precipitation was 242.6 mm higher and the surface temperature that influenced the life of snail was 4.7 °C higher than that of the present. This marked difference of temperature and precipitation is an important factor in determining the history of loess formation.

4.3 Spores, Pollen and Paleovegetation

Palynological research on loess of China began in the 1950s, and the article on loess sporo-pollen analyses was in 1960 (Zhou Kunshu et al. 1960). Later, a series of articles on palynological analysis of loess were published (Liu Tungsheng and Chang Tsunghu 1962; Guo Xudong et al. 1981; Xia Yumei et al. 1983; Wen Qizhong et al. 1982; Zhou Kunshu et al. 1982, 1984). The results of these studies show that the loess of China formed in an environment of semi-arid steppe or semi-humid forest and steppe mainly under dry-cold conditions.

Luochuan "Yuan", a typical loess basin with an elevation of about 1100 m and with loess over 100 m thick, is surrounded by the Huanglong, Laoshan and Ziwuling Mountains. The Luohe River cuts through the basin, eventually joining the Huanghe River (Fig. 4.9).

This region is situated at the northern end of the warm temperate zone with an average annual temperature of about 10 °C, frost-free period of 160–180 days, ⩾10 °C accumulated temperature of 3000–3600 °C, precipitation of 550 to 650 mm and evaporation of 1600–2000 mm a year. The soil in this region is brown and grey-browns soil, with deciduous broad-leaved forest.

The hills in this region are covered with well-preserved secondary forest composed mainly of *Quercus liotungensis, Pinus tabulaeformis*. In the border between "Yuan" and the hills lies to forest and steppe. Luochuan "Yuan" has been reclaimed as an annual wheat and maize area, with the original vegetation almost destroyed. Meadow and steppe mixed with deciduous bush vegetation are dominant in the gullies and wasteland.

4.3.1 Pollen Analysis of Loess

In the pollen analysis of the Luochuan "Yuan", five samples respectively from Wucheng Loess, the lower part of Lishi Loess, the fifth paleosol, the upper part of Lishi Loess and Malan Loess in the Nianzigou profile and Qingchi, were analyzed. Three different methods, conventional, washing and HF methods, were used. After comparison (see Table 4.2) the conventional method was selected.

4.3.1.1 Surface Soil

Spore and pollen samples were taken and analyzed from the surface soil of the Huanglong Mountains which is covered by a deciduous broadleaf forest, Tundi

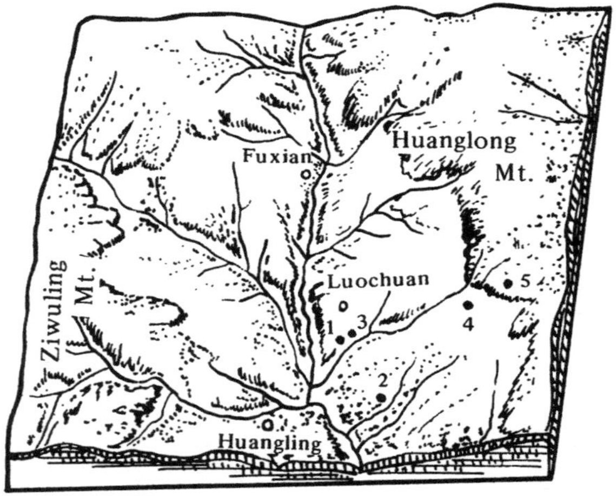

Fig. 4.9. Luochuon "Yuan" and Sampling locations of spores and pollen. *1* Potou village; *2* Xiatun village; *3* Guzui village; *4* Tundi village; 5 Huanglong Mt

Fig. 4.10. Geological profile, vegetation and analysis results of spores and pollen from the surface soil of Luohe River-Potou village-Tundi Village-Huanglong Mountain in Luochuan "Yuan", Shaanxi. *1.* Sandstone, shale of Triassic; *2.* Malan Loess; *3.* Lishi Loess, Wucheng Loess and red clay; *4. Pinus tabulaeformis*; *5.* broad-leaved tree; *6.* small leaf tree; *7.* shrub; *8.* herb. *a* Gramineae; *b Artemisia; c Compositae; d* Chenopodiaceae; *e Polygonum; f Ephedra; g Rhus; h Niteraia; i Pinus; j Betula; k Alnus; l Corylus; m Salix; n Quercus; o Elaegnus; p Selaginilla*

▼

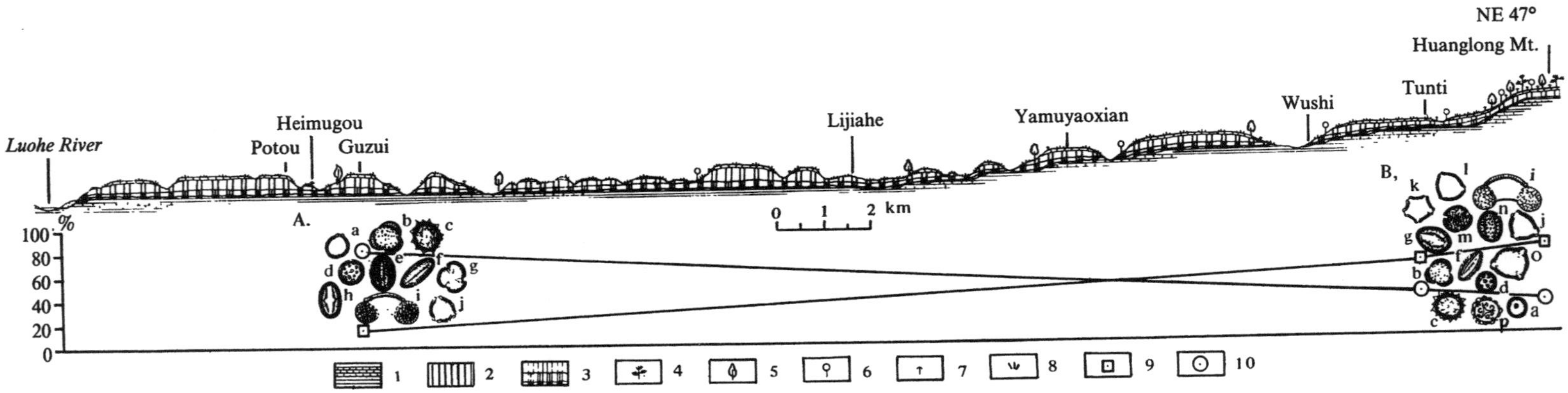

Table 4.2. Experimental results of sporo-pollen analysis methods of loess

Sample	Conventional	Number (grain) HF	Washing
Wucheng Loess	81	4	3
Lower Lishi Loess	275	5	2
Fifth bed of lossil soil	300	7	2
Upper Lishi Loess	298	17	2
Malan Loess	208	127	1
Amount	1162	160	10

village in front of Huanglong Mountains vegetated by forest and steppe and the Guzui village in the central area of Luochuan "Yuan" which is mainly meadow and steppe (Figs. 4.9 and 4.10).

Abundant spores and pollen grains are found in the above samples. In the Huanglong Mountains, tree pollen amounts to 73% of the total amount, herb pollen 26.2% and fern spores 0.6%. In the Tundi village, tree pollen amounts to 62.6% of the total amount, herb pollen 35.9% and fern spores 1.3%. In the Guzui village, tree pollen occupied 15.7% of the total amount, herb pollen 82.6%, fern spores 1.6%. These indicate that the palynological assemblages of the surface soil is in accord with the general vegetation of the locality.

4.3.1.2 Loess Profiles

1. Heimugou Profile. Ninety samples were systematically analyzed from the profile located in the central area of Luochuan "Yuan" during June-July, 1980 (Table 4.3 and Fig. 4.11).

Based on the result of spore-pollen analysis of the Heimugou profile given above, six spore and pollen zones can be classified in ascending order:

Zone I. The zone is the section of Wucheng Loess, dated at 2.4–2.1 m. y., 130–120 m deep. It contains only a few spores and pollen which are the common elements of deciduous boadleaf forest, meadow and steppe of warm temperate zone with the exception of two pollen grains of *Fagus* which grow in subtropical zone.

Zone II. The elements of the subtropical zone such as *Fagus, Symplocos, Carya, Hicriopteris,* etc. are found here, in addition to the common species of deciduous broadleaf forest, meadow and steppe of the locality. This zone is dated at 2.1–1.48 m.y., 122–92 m deep. This section contains abundant spore and pollen grains and more components of subtropical plants.

Zone III. This is the section of Wucheng Loess–Lishi Loess, dated at 1.48–1.05 m.y., 92–75 m deep, with few spore and pollen grains.

Zone IV. This zone is located at a depth of 75–56 m of Lishi Loess, and dated to 1.05–0.92 m.y. Many spore and pollen grains are found, similar to those of present-day deciduous broadleaf forest, meadow and steppe.

Table 4.3. Results of pollen analysis of the loess profile of Heimugou at the Potou village of Luochuan, Shaanxi

	Number (grain)				Number (grain)		
	M.L.	L.L.	W.L.		M.L.	L.L.	W.L.
Pollen of tress	88	248	209	*Humulus*	13	6	2
Pinaceae			1	Cruciferae	2	6	1
Pinus	5	16	38	Polygonaceae	12	8	2
Taxodiaceae		2		Umbelliferae		3	
Cupressaceae	3	1		*Aster*			2
Betula	11	28	38	*Artemisia*	98	101	138
Alnus	40	148	33	Compositae	20	23	2
Corylus		4	8	*Gentiana*			1
Carpinus	2	8	5	Cyperaceae	12	3	1
Quercus	7	11	58	Geraniaceae		1	
Fagus		2	7	*Typha*	10		
Tilia		4		Rosaceae		2	
Moraceae	6	3		Labiatae			3
Morus			1	Amaranthaceae		1	
Broussonetia	6	3		Berberidaceae	1		
Carya			2	Leguminosae	1		
Pterocarya		1	5	Chenopodiaceae	11	12	5
Ulmaceae	5	4	1	Gramineae	38	73	39
Celtis		1		Fern spores	22	47	68
Acer		2		*Lycopodium*			3
Oleaceae		2	1	*Selaginella*		6	2
Rhus		1	1	*Athyrium*	1		
Symplocos		1	1	*Osmunda*		2	
Nitraria			1	*Dryopteris*		1	1
Liquidambar	1	2		Gleicheniaceae			2
Simaroubaceae			1	*Hicropteris*			3
Ailanthus altissima			3	*Gonocormus*		7	27
Poncirus trifoliata		1		Aspleniaceae			1
Salix	2	3	4	Polypodiaceae	5	6	6
Pollen of brush and herb	220	243	201	*Pteris*	1		
				Pteridium	4	13	10
Ephedra			2	*Microlipia*		1	3
Ranunculaceae		2		*Adiantum*	3		
Sanguisorba	1	2		*Davalliaceae*			1
Rubiaceae	1			Filicales	9	6	2
Urticaceae			1				

M.L. Malan Loess; *L.L.* Lishi Loess; *W.L.* Wucheng Loess.

One pollen grain of *Symplocos*, which is an element of the subtropical zone, is noted.

Zone V. Only few spore and pollen grains are found at a depth of 56–30 m in Lishi Loess and at an age of 0.92–0.41 m.y.

Zone VI. This zone extends from a depth of 30 m to the surface, and is dated at 0.41 m.y. Species of deciduous broadleaf forest, meadow and steppe are represented as well as three pollen grains of *Liquidambar* which is a

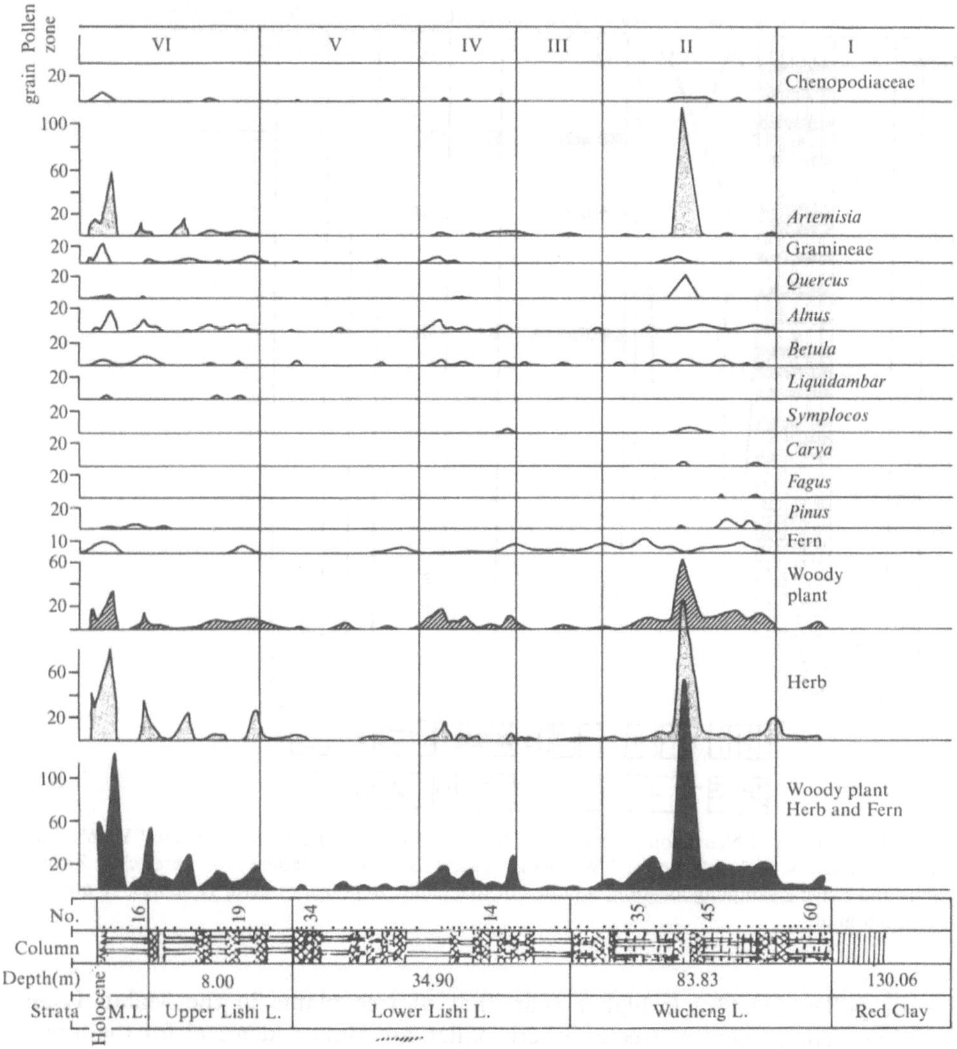

Fig. 4.11. Diagram of Heimugou profile at the Potou village of Luochuan

component found in the southern part of warm temperate zone and subtropical zone.

2. Nianzigou Profile. This profile is located near the Xiatun village of Qingchi, about 15 km southeast of Luochuan City, south of the central area in Luochuan "Yuan". Five samples were taken and analyzed presenting Wucheng Loess, the lower part of Lishi Loess, the fifth paleosol, the upper part of Lishi Loess and Malan Loess (see. Fig. 4.12).

It can be seen from Fig. 4.12 that the amount of tree pollen in Wucheng Loess, the lower part of Lishi Loess and the fifth paleosol is between 66.6% – 73%, while herbaceous pollen is only between 19 – 24%. Conversely the amount of herba-

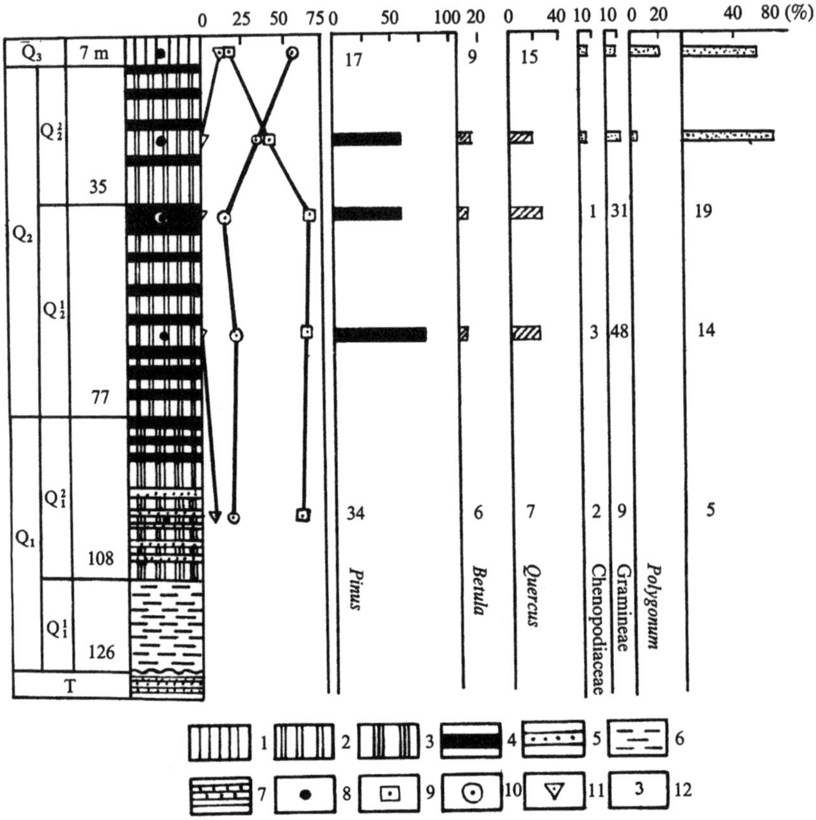

Fig. 4.12. Diagram of Nianzigou profile at the Xiatun village of Qingchi in Luochuan "Yuan", Shaanxi. *1* Malan Loess; *2* Lishi Loess; *3* Wucheng Loess; *4* fossil soil; *5* nodule bed; *6* grey clay; *7* sandstone and shale; *8* sampling location; *9* pollen of tree plant; *10* pollen of herb; *11* spores; *12* number of pollen grain

ceous pollen (63.4%) is higher than that of tree plants in the Malan Loess (20.6%). The percent of tree and herb pollen are almost identical for the upper part of Lishi Loess, amounting to 49.3% and 45.3% respectively.

4.3.2 Spores and Pollen Assemblage, Paleovegetation and Paleoenvironment of the Luochuan Region

The sporo-pollen analysis based on the Heimugou profile gives a clear succession of the Quaternary vegetation. Zones I, III and V contain only a few spores and pollen, indicating that at that time the vegetation was not luxuriant, and an adverse steppe environment with a dry-cold climate existed. But Zones II, IV and VI contain a greater quantity of spores and pollen, suggesting a more dense forest and steppe vegetation with a temperate and moist climate. Based on spores of *Dryopteris* and *Micriolepia*, which prefer a moist environment, and some spores and pollen of subtropical plants such as *Fagus, Carya, Symplocos* and *Hicriopte-*

ris, which do not grow at the present location, but are only found south of Qinling Mountains and the Huaihe River, we can suggest that at that time the region was forested, the climate was optimal, and the Quaternary vegetation had a transitional character. The amount of spores and pollen in Zone VI is variable, reflecting an unstable environment.

The B/M boundary in L_8 is in the lower part of Zone V, reflecting a dry-cold climate. The Jaramillo subchron of $L_{10}-S_{11}$ located in Zone IV and the Olduvai subchron of $W_{S\text{-}2}$ in Zone II are all normal polarity subchrons of the Matuyama reversed chron. At that time the climate was relatively temperate and moist.

Past palynological results for Shanxi loess as well as the present results indicate an increase in the percent of herb pollen, especially *Artemisia*. This indicates that the loess environment has a dry character.

It is possible that the rapid uplift of the Qinghai-Xizang Plateau and the high mountains resulting from the uplift of the Qinling Mountains during the Quaternary blocked the Indian Ocean Monsoon to northern Xizang and Southeast Monsoon to northwestern China. Also, the thickening of loess accumulation, the deepening of valleys due to loess erosion and the lowering of the erosion basis of the river caused the lowering of underground water, resulting in less water in loess covering valleys. The desiccation of the environment was accelerated by human activities.

Pollen analysis of the Nianzigou profile shows that the assemblage of the loess under the fifth paleosol is dominated by tree pollen and differs from the vegetation of the surface soil. The reason is that in the early period of the Quaternary a small lake developed. This is known from the sediments of lake facies of 18 m. The humidity in the microclimatic environment was high, favorable for trees to grow. In the later development of loess after the late Middle Pleistocene, this microclimatic condition disappeared and was replaced with a great thickening of loess, the degeneration of forest and the increase of herbaceous plants.

CHAPTER 5

Composition of Loess

5.1 Grain-Size Distribution

Since the 1930s, a great deal of research on the grain-size distribution of loess has been accomplished. The characteristics of grain-size distribution are known to be crucial to the study of loess. Since the founding of the People's Republic of China in 1949, several authors, including Xiong Yi and Wen Qixiao (1953), Liu Tungsheng and Chang Tsunghu (1962), and Liu Tungsheng et al. (1965, 1966), have made granulometric studies of loess.

A series of analysis of the Heimugou loess section in Luochuan, Shaanxi Province, has been made by the Quaternary Research Laboratory of the Institute of Geology, Academia Sinica since 1962. The following is a description of the features of the grain-size distribution of loess and paleosol of various geological periods for the Heimugou section and a discussion on some related problems.

5.1.1 A General Description of the Features of Grain-Size Distribution

The loess and paleosol sequences of various geological times are developed in the Heimugou section, 130 m thickness. This section is a representative of the typical loess sections on the Loess Plateau. In order to make a comparative analysis of the general features of the grain-size distribution of loess and paleosol, the analytical results of 109 samples, not published before, are shown in Fig. 5.1.

From Fig. 5.1 we can see that the general appearance of the grain-size distribution is similar for most samples. Almost no grains greater than 0.25 mm in diameter can be found, and grains between 0.25 and 0.1 mm are also very few (less than 1% for an overwhelming majority of the samples). The percentage of 0.1 – 0.05 mm fraction is generally between 3% – 10%.

The content of the 0.05 – 0.01 mm fraction is very high, often between 50% and 60% for Malan Loess and Lishi Loess, between 35% and 45% for Wucheng Loess, and between 40% and 50% for paleosol. The variation in the content of this fraction provides an informative clue to understanding the sedimentary cycle and climatic fluctuation. The content of the 0.01 – 0.005 mm fraction does not change very much and is generally between 10% – 15%.

Both Malan and Lishi Loess show a relatively low percentage of < 0.005 mm fraction, often around 20% – 35%, whereas Wucheng Loess and paleosols have a higher content of this grain-size fraction about 35% – 45%. The percentage of

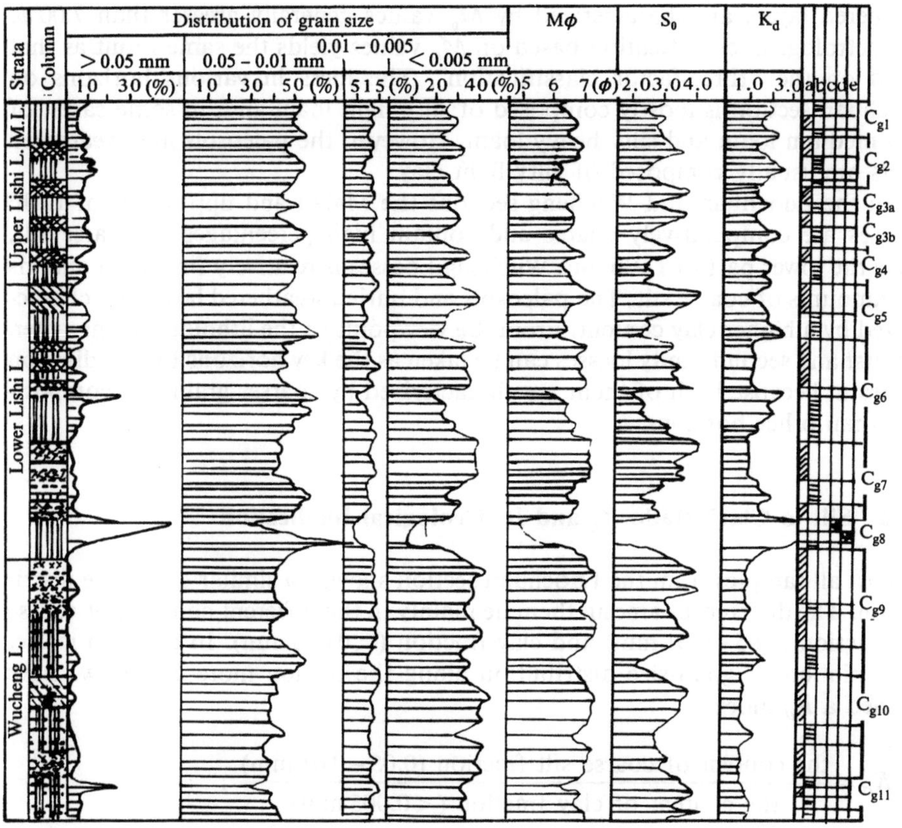

Fig. 5.1. Diagram showing the grain-size distribution, the variation in grain-size parameters and the grain-size cycles for Heimugou section. Luochuan, Shaanxi Province

this fraction shows a clear negative correlation with that of the 0.01 – 0.005 mm fraction.

The percentage of the < 0.001 mm fraction is similar to that of the < 0.005 mm fraction, generally accounting for 10% – 20%.

Statistical measures, such as the mean (M_ϕ), standard deviation (σ_ϕ), skewness ($S_{k\phi}$) and kurtosis (K_ϕ), can be used to depict the characteristics of the grain-size distribution in various detrital sediments (Krumbien 1938). Using these statistical parameters, Liu Tungsheng et al. (1966) described the grain-size distribution of loess, secondary loess and paleosol. In loessial sediments a linear positive correlation exists between M_ϕ and σ_ϕ and a linear negative correlation between M_ϕ and $S_{k\phi}$ as well as M_ϕ and K_ϕ. Because the latter are high-order moments associated with large uncertainties, the characteristics of grain-size distribution will be depicted only by the low-order moment values of M_ϕ, which can be compared with sorting coefficient S_0 and M_ϕ values obtained by the use of quartile measurements (see Fig. 5.1).

The M_ϕ values of the Luochuan section fall in the range of 5.17 – 7.54 ϕ. The majority of samples have M_ϕ values between 6.30 – 7.00 ϕ; but paleosols and weathered layers are characterized by M_ϕ values generally greater than 7.00 ϕ. The petrological classification based on M_ϕ values yields the same result as that based on the "three-factor classification". The loess in various horizons on Luochuan section is mostly composed of silty light loam and, in some cases, of silty medium loam and silty heavy loam. However, the paleosols and weathered layers are usually composed of silty light clay.

In the Luochuan and Wucheng sections the Malan and upper part of Lishi Loesses are comparatively coarse and contain high percentages of coarse silt, while the lower part of Lishi and Wucheng Loesses is relatively fine and contains low amounts of coarse silt. The paleosols and buried weathered layers are characterized by a higher clay content. From the viewpoint of the lithological character, the first and second sandy loesses can be taken as the key horizons for the division of strata, because both of them are characterized by coarser grain-size composition than other horizons.

5.1.2 "Silt to Clay" Ratio K_d and Its Geological Implication

Almost all samples from the Luochuan section showed a similar grain-size distribution. The differences among them lie mainly in the different contents of coarse silt fraction (0.05 – 0.01 mm) and clay fraction (<0.005 mm). In order to reflect the variation of grain-size distribution along the section more clearly we have defined a K_d index:

$$K_d = \frac{\text{the content of coarse silt fraction (0.05}-\text{0.01 mm)}}{\text{the content of clay fraction (<0.005 mm)}} .$$

K_d values magnify intuitively the difference in grain-size distribution among various samples of the lithologically similar loessial sediments. From Fig. 5.1, we can see that the K_d values vary greatly from horizon to horizon. The physical meaning of K_d values for aeolian dust accumulations and paleosols and weathered layers developed on loess can be understood in the following two respects.

First, according to the experimental wind tunnel results by R.A. Bagnold (1941), the grains greater than 0.3 mm can only be transported by jumping. The grains in the fraction of 0.3 – 0.07 mm are also very difficult to transport in suspension, whereas the grains less than 0.005 mm cannot be drifted by wind. The dust grain experiments by L. Moldvay (1962) showed that the grains of 0.05 – 0.01 mm drift easily and thus are called the "basic fraction" of the aeolian dust, while the grains between 0.01 – 0.005 mm drift and disperse poorly, and are called the "additional fraction". The comparison between the grain-size distribution in loess from various localities in China and the above experimental results favors the explanation that loess is an aeolian deposit. Loess in China contains almost no grains greater than 0.25 mm, and also very few grains greater than 0.05 mm or 0.0625 mm. The average fraction of loess are grains between 0.05 – 0.01 mm. In this paper this fraction is called the basic fraction of aeolian dust – "loess fraction".

The percentage of the <0.005 mm fraction is called the "attached fraction". The grains in this fraction are mainly composed of the material that came from the source region but are attached to the surfaces of the coarse silt grains (Han Jiamao 1982). The attached clay particle material has been observed on the surfaces of the quartz grains of the silt fraction under scanning electron microscope. The above characteristics in grain-size distribution suggest aeolian transportation of the grain portion of loess material.

K_d value is a measure of the ratio of "basic fraction" versus "attached fraction" of aeolian deposit. The production rates of both fractions depend on whether the source regions were humid or arid. If other conditions are the same, an arid and cold environment is favorable for producing silt dust, but not clay particles, whereas the warm and humid environment produces the opposite. Thus, K_d index indicates the climatic conditions: the greater K_d values mean a drier and colder environment.

Secondly, part of the clay and of the Fe-Mn oxides is concentrated along the fissures of the argillic horizons in the cinnamon type paleosols, indicating the migration and concentration of the colloidal particles of clay (An Zhisheng and Wei Lanying 1979, 1980). In this sense, the clay particle fraction of <0.005 mm is a secondary fraction which is susceptible of change and its content increases with the development of weathering and soil-forming processes. The silt-dust fraction of $0.05-0.01$ mm is mainly composed of quartz and feldspars, and has changed inconsiderably during the whole formation process of paleosols. Thus, from the point of the weathering and soil-forming processes, K_d values indicate the extent of reformation of loess after deposition by the weathering and soil-forming processes, and also reflect the biological and climatic intensity at that time.

M_ϕ (the mean) of loessial deposits may represent the basic characteristics of the grain-size distribution, and thus can be considered as an index of the formation of deposits. Based on the statistics of 109 samples from the Luochuan section, K_d and M_ϕ are related to each other, showing a very pronounced linear negative correlation ($r = -0.773$). According to Lu Yạnchou and An Zhisheng (1979) and Lu Yanchou (1981), the $CaCO_3$ content can be taken as another index of the environment during the formation of loess-paleosols. A pronounced positive correlation of $r = 0.794$ between K_d values and $CaCO_3$ content exists.

In summary, the physical meaning of K_d value can be described as follows: (1) It represents the quantitative ratio between "basic fraction" and "attached fraction" of the aeolian dust deposits; (2) it reflects the quantitative relationship between the weathered dust and the secondary fractions. K_d values are controlled not only by the humidity and the biological and climatic conditions in the source regions, but also by the climatic environment and the soil-forming processes in deposition regions. Therefore, K_d value can be taken as an index of the environment of loess and paleosol and the subdivision of the lithologic-stratigraphic units of loess sections.

5.1.3 Grain-Size Cycles and the Lithologic-Stratigraphic Association

The regular fluctuation in grain size from coarse to fine or from fine to coarse in a sedimentary section is defined as a grain-size cycle. From Fig. 5.1, one can see certain differences between M_ϕ and S_0 in different horizons and a cyclic or rhythmic variation in M_ϕ and S_0 values. In the horizons above the second sandy loess a good correspondence can be observed between the grain-size cycles and the loess-paleosol sequence. Though there is only a slight difference in grain-size distribution in Wucheng Loess, the statistics suggest that the loess strata have also incorporated several coarser horizons, providing evidence for the existence of grain-size cycles.

The basic principle that the authors follow for the division of grain-size cycles is to take a sequence from coarse to fine as a cycle. Eleven cycles can be distinguished in the Luochuan section, designated respectively as C_{g1}, $C_{g2} \dots C_{g11}$. Each of C_{g1}, C_{g2}, C_{g4} and C_{g5} is built up by one layer of coarse loess and several layers of paleosols as well as some thin layers of finer loess. These cycles reflect, in fact, the process that started with a great rate of deposition and culminated in a small one, followed by the weathering and soil-forming process. The lower part of each cycle is often composed of silty medium loam or silty heavy loam, while

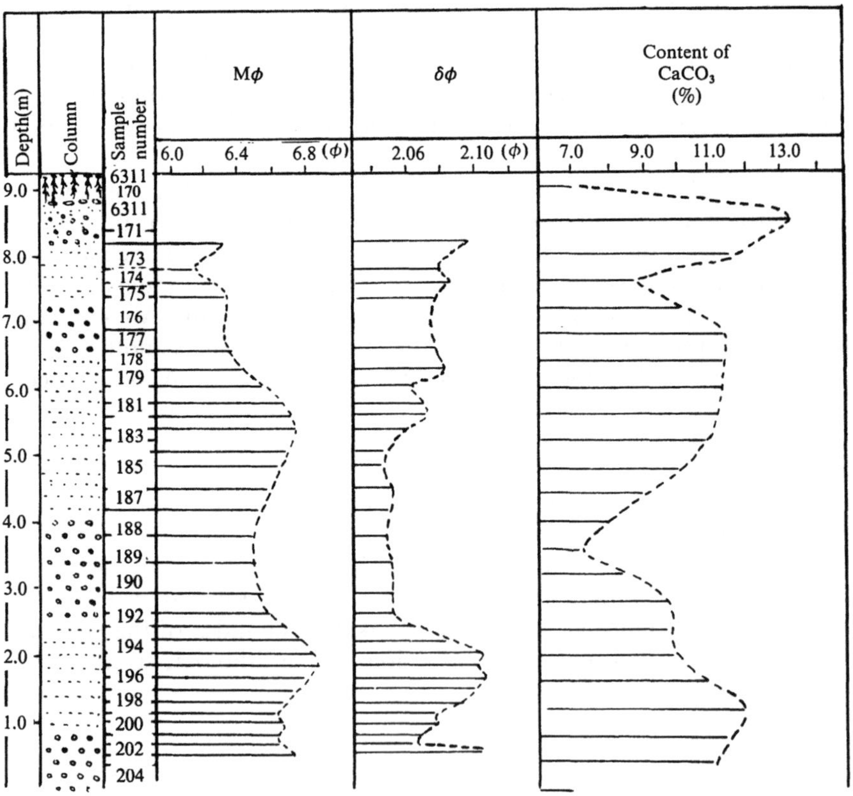

Fig. 5.2. Division of the grain-size of the loess L_9, Luochuan

the upper part is usually composed of silty light clay or silty heavy loam. Therefore, the grain-size cycles can be considered lithologic-stratigraphic associations. However, the implication of grain-size cycles is different from that of climatic cycles represented by paleosol and the overlying loess (see Sect. III, Chap. 6 in detail).

In addition, even within a thicker horizon of loess there still exists some regular variation in grain size, called "grain-size rhythm", implying a minor pulsative fluctuation in climate in an extensive environment. Taking Malan Loess L_1 and the upper sandy loess L_9 as examples, there is a synchronous variation in grain-size and in $CaCO_3$ content: the higher $CaCO_3$ contents coincide with coarser grains, while the lower $CaCO_3$ contents coincide with finer grains. The cyclic variation in K_d values of the horizons above the second sandy loess compared with the cyclic variation in $CaCO_3$ content. This synchronism can be seen more distinctly in Fig. 5.2.

5.1.4 Formation of Loess in Light of the Grain Size Distribution

In the extensive area from the Gansu Province through the Shandong Province, which is as much as 2000 km in an east-west direction and as wide as 500 km from north to south, the loess developed on various geomorphological units shows approximately the same grain-size distribution. The major fraction of loess is 0.05 – 0.01 mm, which accounts for about 45% – 60%; the next is <0.005 mm fraction, accounting for about 15% – 25%. There are almost no grains greater than 0.25 mm. This fact favors the origin of loess as an aeolin deposit in arid and semi-arid areas. For example, the Danube Basin in Hungary (Bognar 1980), the Altay mountainous region in the USSR (Kriger 1965), Kansas and Nebraska in USA (Swineford and Frye 1955) and eastern Argentina (Teruggi 1957) all have loess and secondary loess which are comparable in grain-size distribution with loess, especially the sandy loess in China.

According to Liu Tungsheng et al. (1965), who made an extensive investigation of the grain-size of loess in the middle reaches of the Huanghe River and in the Qaidam Basin, Qinghai Province, the loess in the middle reaches of the Hanghe River can be divided into three zones according to the average M_ϕ. The southern boundaries of these three zones correspond approximately to the Baiyu Mountain, Huanglong and Qinling Mountains respectively (Fig. 5.3).

I. Sandy loess zone: $\bar{M}_\phi = 4.939 \pm 0.203\ \phi$. In this zone, the loess grains are comparatively coarse and belong to the silt clayey sand zone.

II. Loess zone: $\bar{M}_\phi = 5.611 \pm 0.308\phi$. This zone is silt light clay zone.

III. Clayey loess zone: $\bar{M}_\phi = 6.049 \pm 0.211\ \phi$. This zone is mainly composed of silt medium and heavy clay.

From north to south, the grain-size distribution of loess in the middle reaches of the Huanghe River varies markedly, becoming finer from northwest to southeast or from NNW to SEE. On the contrary, no remarkable grain-size variation can be recognized along east-west direction. Thus, the grain-size distribution seems to be affected considerably by Liupan Mountains and Luliang Mountains that stretch approximately in a north-south direction. On both the east and west

Fig. 5.3. Diagram showing the zonation of grain size of loess in the middle reaches of the Huanghe River. *1* Desert; *2* sandy loess zone; *3* loess zone; *4* clayey loess zone; *5* flow direction on 1500 m a.s.l. in January; *6* flow direction on 3000 m a.s.l. in January

sides of these two mountain ranges the grain-size of loess changes considerably, implying that the transportation of most loess material occurs below the altitude of 3000 m. In northern Shaanxi and eastern Gansu the grain size of loess varies mainly in a north-south direction, also suggesting a relation between the transportation-deposition of Malan Loess in the middle reaches of the Huanghe River and the atmospheric movement in altitude of 2000 – 3000 m.

5.2 Heavy Minerals

About 130 m thick loess exposed on both sides of the Heimugou section south of Luochuan City has recorded the evolutionary history of the natural environment over the past 2.4 m.y. The compositional characteristics of heavy minerals

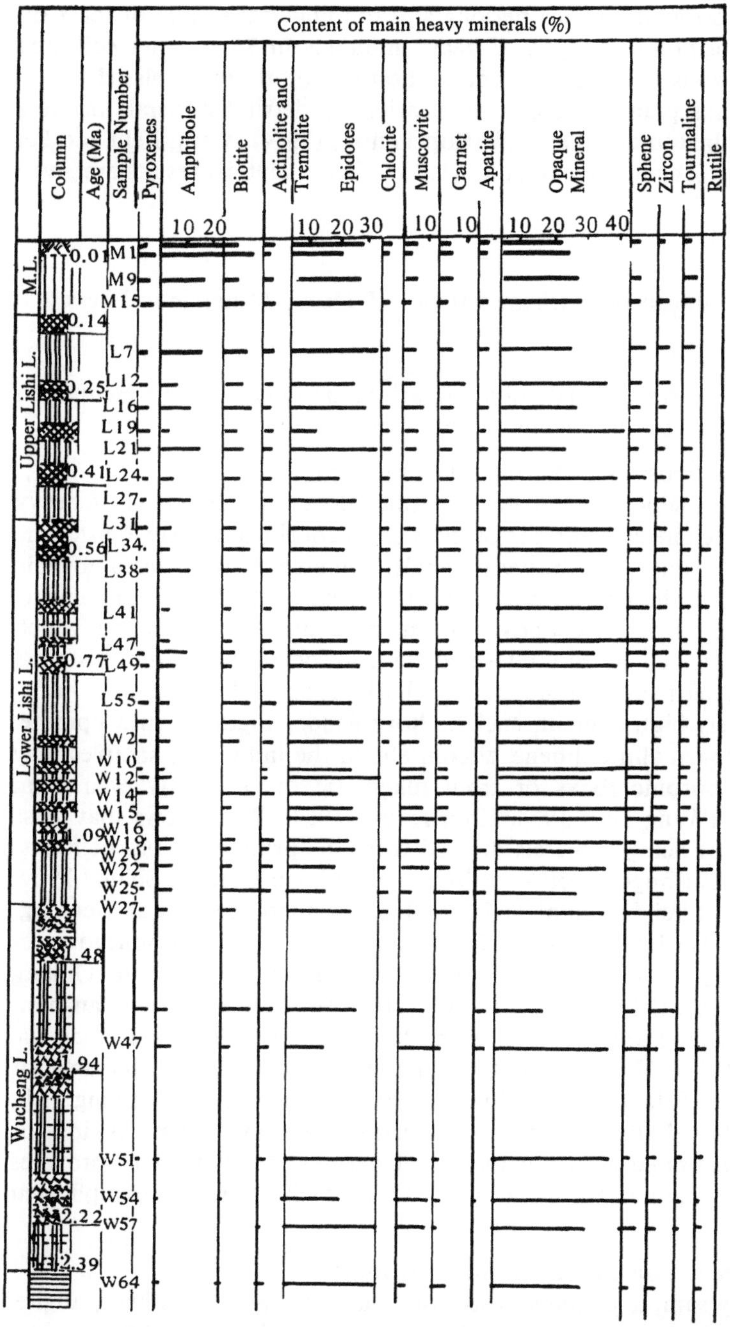

Fig. 5.4. Contents of major heavy minerals in the Heimugou section at Potoucun, Luochuan

in loesses and paleosols of various periods and the material source of loess on the basis of compositional analysis of heavy minerals in 36 samples collected from different horizons in the Heimugou section at Potou are discussed.

The methods of separation and identification of detrital minerals in loess samples were described in detail by E. V. Kopchenova (1954), Wang Kelu and Pei Jingxian (1964), and Zhang Suyuan et al. (1965). Our analytical results are given in Fig. 5.4.

5.2.1 The Main Compositional Characteristics of Heavy Minerals in Loess and Paleosol

Heavy minerals in loesses and paleosols of various periods from the Heimugou section at Potou share the following compositional characteristics:

1. Complexity in composition and diversity in category

It can be seen from Table 5.1 and Fig. 5.4 that more than 30 kinds of heavy minerals are contained in each horizon of loess or paleosol from Early Pleistocene to the Holocene. Among these heavy minerals, some came from metamorphic rocks (e.g. kyanite, anthephyllite, actinolite, tremolite, wollastonite, andalusite, sillimanite, garnet); some were derived from magmatic rocks, among which spodumene, rutile, etc., came from intermediate-acid rocks, and zoisite, enstatite, etc. came from intermediate-basic or ultrabasic rocks. However, the heavy minerals of relatively high content, such as hornblende, augite, biotite, epidote, opaque minerals, tourmaline, sphene, zircon, etc. in the individual samples may come either metamorphic rocks or from magmatic rocks, partly even from sedimentary rocks (Wang Dezi 1974, Kopchenova 1954). This shows that loess contains a variety of heavy minerals derived from the principal types of rocks.

Data from the Liushugou section at Wucheng, Xixian County, Shanxi Province (Wang Kelu and Pei Jingxian 1964), the Nuanquangou section at Wenfeng, Longxi County, Gansu Province (Wen Qizhong et al. 1982), the Xiaohegou section south of Xifeng, Gansu Province (Wang Yongyan et al. 1966), the Gulang area of Gansu Province (Zhang Shuyuan 1965), the Dabaigou section at Yangcun, Henan Province (Zhang Shuyuan 1965), the Malan Loess distributed along the middle reaches of the Huanghe River and the Xiashu loess in the area of Nanjing[7], and the Holocene loess in the Mt. Tourmor area, Xinjiang (Liu Tungsheng et al. 1985) indicate an extensive occurrence of more than 30 kinds of heavy minerals mentioned above in loesses of various localities. This provides evidence suggesting that heavy minerals in the loess of China are very complicated in composition.

2. Similarity in heavy mineral composition of loess

(1) In almost every sample of loess, heavy minerals such as hypersthene, augite, hornblende, anthophyllite, biotite, tremolite, actinolite, epidote, zoisite, clinozoisite, allanite, chlorite, wollastonite, apatite, garnet, muscovite, opaque minerals (including magnetite, limonite, hematite, etc.), sphene, zircon, rutile, tourmaline,

[7] Wei Lanying, 1982, Studies on heavy minerals in the Malan Loess.

Table 5.1. Comparison of average heavy mineral compositions of loesses (L) and paleosols (PS) of different periods in the Luochuan (LC), Longxi (LX) and Wucheng (WC) loess sections

	Holocene loess		Malan loess			Lishi loess								Wucheng loess		
						Upper part				Lower part						
	LX	LC	LX	LC	WC	LX		LC				WC		LX	LC	WC
						PS	L	PS	L	PS	L	PS	L			
Unstable minerals																
Hypersthene	0.21		0.36	0.82	1.27	0.48	0.32	0.71	0.40	0.60	0.82	1.09	1.10	0.30	0.51	1.01
Enstatite	0.43	0.19	0.79	0.58	0.75	0.34	0.56	0.47	0.80	0.38	0.97	2.07	1.05	0.30	0.87	1.37
Augite	1.74	1.95	1.21	2.05	4.85	0.83	1.68	0.99	1.77	0.76	0.82	3.49	5.63	0.31	0.92	3.07
Spodumene				0.20			0.07									
Hornblende	21.07	25.11	21.22	20.04	11.18	8.68	14.94	3.11	13.22	3.01	5.67	12.78	16.03	6.76	2.68	6.57
Riebeckite		0.19			0.37			0.24	0.40	0.22	0.28	0.23	0.34		0.47	0.75
Anthophyllite	0.11	0.58		0.91	0.34	1.58	0.40	0.95	0.30	0.42	0.55	0.68	0.31	0.41	0.23	0.17
Biotite	2.48	4.58	2.34	7.19	0.45	2.30	4.27	5.58	6.63	2.73	6.66	7.06	12.48	4.49	3.09	11.00
Sum	26.03	32.06	25.91	31.35	28.95	13.14	21.48	11.18	22.57	7.55	14.78	29.99	37.94	12.56	5.03	24.00
Relatively stable minerals																
Diopside	0.65	0.19	0.53	0.17	1.44	0.61	0.55	0.98	0.20	0.52	0.39	1.56	2.08	0.43	1.25	2.56
Actinolite	0.43	1.17	1.24	1.20	4.14	0.51	0.57	1.24	0.98	0.58	1.15	2.73	3.99	0.40	0.75	2.82
Tremolite	2.46	1.17	2.89	1.52	2.22	1.79	2.11	0.93	0.82	0.92	1.24	2.16	3.36	1.80	1.34	1.69
Epidote	30.95	27.02	25.21	22.12	6.25	28.78	25.10	10.02	22.35	16.74	14.90	6.38	6.83	29.63	15.95	6.98
Zoisite	2.26	0.19	1.18	1.78	0.41	1.53	1.00	2.76	3.50	3.63	3.38	1.86	2.87	1.64	5.00	1.74
Clinozoisite	0.54		0.20	0.71		0.61	0.53	2.48	2.37	3.04	4.81			0.17	3.41	
Allanite	0.32	0.39	0.55	0.49	0.37	0.17	0.24	0.46	0.58	0.50	0.83	1.36	0.53	0.33	0.23	0.21
Chlorite	0.33	0.78	1.34	1.07	2.76	0.64	0.72	0.70	0.83	0.68	0.85	0.31	0.23	0.51	0.42	0.09
Wollastonite	0.11	0.19	0.45	0.25	0.25	0.22	0.34	0.97	0.47	0.70	0.49	0.52	0.30	0.30	1.23	0.79
Sillimanite	0.22		0.67	0.25	0.62	0.07	0.19	1.27	0.20	0.29	0.58	0.56	0.07		0.23	0.08
Barite	0.33	0.19	0.24	0.20		0.01	0.07	0.24	0.20	0.39	0.45				0.24	

Table 5.1 (continued)

	Holocene loess		Malan loess			Lishi loess								Wucheng loess		
						Upper part				Lower part						
	LX	LC	LX	LC	WC	LX		LC				WC		LX	LC	WC
						PS	L	PS	L	PS	L	PS	L			
Apatite	1.09	1.15	0.75	0.87	0.88	0.85	1.02	0.85	0.71	1.07	1.10	0.75	0.57	0.97	0.88	0.79
Muscovite	1.41	2.29	1.88	2.97	1.94	1.76	2.29	5.92	3.00	2.88	5.99	4.39	5.62	1.58	1.35	7.18
Garnet	2.46	3.70	2.49	3.28	3.68	2.77	3.22	5.24	5.10	6.03	6.57	2.24	1.71	2.66	8.95	5.17
Sum	43.12	38.58	39.58	35.49	24.45	40.01	38.18	33.37	41.40	39.29	39.70	24.81	28.17	40.44	39.47	29.31
Stable minerals																
Opaque minerals	25.94	24.19	28.65	27.90	27.05	40.36	34.47	44.70	28.95	43.10	35.81	27.47	21.13	40.83	39.83	29.62
Sphene	1.93	1.36	1.76	2.08	0.93	1.98	2.09	5.74	3.04	5.74	5.04	0.50	0.34	2.16	6.13	0.70
Kyanite	0.32	0.19		0.20	0.47	0.11	0.26	0.46	0.20	0.46	0.32	0.31	0.39	0.17	0.61	0.53
Sum	28.19	25.74	30.40	30.04	28.46	42.48	29.99	90.71	32.04	49.31	41.21	28.28	23.01	43.17	46.46	30.85
Extremely stable minerals																
Spinel	0.11		0.22		0.52	0.08	0.03			0.23	0.36	0.34	0.13	0.08	5.30	0.13
Zircon	1.09	1.78	1.26	1.59	10.92	2.00	1.20	2.83	2.20	2.84	2.73	10.54	6.33	2.14	4.63	7.56
Rutile	0.21	0.19	0.29	0.60	2.05	0.22	0.37	0.23	0.51	0.57	0.39	2.37	1.91	0.74	0.36	2.35
Tourmaline	0.64	1.17	2.22	0.90	2.44	1.24	0.98	0.46	1.55	1.36	1.14	2.11	2.83	1.21	1.04	5.01
Anatase	0.11	0.19	0.12	0.30		0.41	0.23	0.92	0.26	0.48	0.38	0.31	0.14	0.35	0.37	
Andalusite	0.11			0.20		0.13	0.34	0.55		0.28	0.36			0.06	0.33	
Topaz						0.44	0.03	0.24		0.37	0.35			0.03	1.66	
Brookite										0.22	0.21					
Staurolite								0.22	0.20	0.22	0.30				0.25	
Monazite										0.26	0.23				0.23	
Sum	2.25	3.33	4.10	3.15	15.93	4.18	2.93	4.60	4.56	5.91	4.95	15.67	13.34	4.18	5.74	15.05

Table 5.2. Linear regressions and correlation coefficients for the percentage of the selected heavy minerals

Variable pair	Regression	Correlation coefficient	Degree of freedom	Significance
Amphiboles (X)/unstable minerals (U) (%)	$U = 4.28 + 1.29X$	0.896	35	a
Epidotes (Y)/relatively stable minerals (W) (%)	$W = 21.99 + 0.67Y$	0.702	35	a
Opaque minerals (Z)/stable minerals (Q) (%)	$Q = 0.19 + 1.10Z$	0.950	35	a
Amphiboles (X)/opaque minerals (Z) (%)	$Z = 4.38 - 0.92X$	−0.650	35	a
Amphiboles (X)/epidotes (Y) (%)	$Y = 20.56 - 0.45X$	0.360	35	
Epidotes (Y)/opaque minerals (Z) (%)	$Z = 49.36 - 0.46Y$	−0.383	35	b

[a] $r > r_{0.01}$ $(n-2)$; [b] $r > r_{0.05}$ $(n-2)$; n− −statistical number of samples; confidence $\alpha = 0.01$.

anatase, etc. can be found. These minerals account for 95% of the bulk amount of heavy minerals in loess and paleosol. Other minerals are also recognized in the samples, but only comprise 50% of the total amount.

(2) The dominant heavy minerals are almost the same in various samples of loess and paleosol. Although there are many kinds of heavy minerals in loess, only a few kinds or categories are high or relatively high in content. Each sample of loess or paleosol has a relatively high content of the opaque minerals, epidote and hornblende, Opaque minerals account for 30%−40% or even up to 51% of the total heavy minerals; epidote generally comprises 15%−25%, slightly higher in the Malan Loess and Lishi Loess, but somewhat lower in the Wucheng Loess and hornblende comes to 10%−15% in the Malan and Lishi Loesses, and only 2%−5% in the Wucheng Loess and its paleosols. Since these three kinds of mineral comprise 80% or more of the total heavy minerals, they are referred to as the dominant heavy minerals.

With the exception of the Liushugou section at Wucheng, which contains abundant pyroxene and minor epidote, the analytical results of heavy minerals in the other loess sections are all comparable with those of the Heimugou section at Potou, Luochuan (see Table 5.2). Thus, amphibole, epidote and the opaque minerals are designated respectively to the unstable, relatively stable and stable heavy minerals, in weathering environments (Wang Kelu and Pei Jingxian 1964).

5.2.2. Differences in Composition of Heavy Minerals in Loesses of Different Periods

Although heavy minerals in loesses and paleosols of various periods show many similarities in composition, as mentioned above, some differences are noted, which are reflected mainly in the variation of their contents. The variation in content of heavy minerals possesses certain regularities.

Table 5.1 shows that significant differences in the contents of unstable (amphibole minerals) and stable minerals (opaque minerals) in loesses of various periods, and that some differences are also noted in the content of relatively stable minerals (epidote minerals). With the exception of mica, other heavy minerals exhibit no remarkable regularities.

Except for paleosol, the Malan Loess contains 31.4% unstable minerals in its total heavy minerals, on an average, and the Wucheng Loess contains only 5% on an average, indicating a strong tendency to decreasing. The opposite tendency is observed with stable minerals, as evidenced by an average value of 30.04% for the Malan Loess and an average value of 51.2% for the Wucheng Loess. As the contents or relatively stable minerals of the samples lie scattered, no remarkable regularities could be summarized. It is evident from the above fact that the compositional differences of the heavy minerals in loesses of different periods refer mainly to those of unstable (mainly amphibole minerals) and stable minerals (mainly opaque minerals). Certain differences are also observed for the surface features of heavy mineral grains in different loess. For example, heavy mineral grains in the Malan Loess are generally much fresher, and adherent materials on the grain surface are less than those in the Lishi and Wucheng Loesses. This probably reflects the differences in bioclimatic conditions during the formation of the loess of various times.

More schistose minerals, such as biotite and muscovite are present in the upper and the lower sandy loess layers (L_9 and L_{15}), and loess layer L_{12} than in others. Attention should be paid to such a remarkable increase of schistose minerals, so as to elucidate whether it is caused by reworking of the water during sedimentation or by the action of freeze-thaw in the periglacial environment, or by other processes.

5.2.3 Characteristics of Heavy Mineral Assemblage in Cinnamonic Paleosol

The Heimugou section at Potou consists of a great number of well-preserved paleosols. For example, in the Lishi Loess 14 paleosols occur, most of which are well-preserved pedogenetic sections, constituting a cinnamonic soil sequence composed of carbonate-drab soil, drab soil and brown-drab soil (Section I, Chapter 3; Lu Yanchou and An Zhisheng 1979; An Zhisheng and Wei Lanying 1980). For simplicity, these three kinds are called cinnamonic paleosol. Eleven samples of cinnamonic paleosol collected from horizon AB were analyzed for heavy minerals, and the results of the analysis show that there is no remarkable difference in the type of heavy minerals found in the cinnamonic paleosol, the Malan Loess, the Lishi Loess and the Wucheng Loess, though significant differences are observed in their content. In each of the paleosols, unstable minerals are less if compared with its underlying loess, and stable minerals much greater. It can be seen from Table 5.1 that there is no remarkable difference in the content of relatively stable heavy minerals between the paleosols and the loesses in the Lishi Loess. From the dominant heavy mineral assemblages, the cinnamonic paleosol is characterized by epidote and the opaque minerals, while the Malan Loess and Lishi Loess are generally noted for the assemblages of amphiboles, epidotes and the opaque minerals (or hornblende-epidote assemblages). The differences in the contents of amphibole minerals and opaque minerals between the cinnamonic paleosol and the host loess may reflect, to a great extent, the influence of pedogenesis.

It is evident from the weathering features as observed on the surface of mineral grains that hornblende, pyroxene, biotite, epidote, etc. have undergone strong weathering. Loessic ferromanganesian materials absorbed on the surface of mineral grain and partial conversion of augite into limonite give proof of weathering. The heavy mineral composition of black loam (S_0) is different from that of the cinnamonic paleosol, but similar to that of the Malan Loess. Owing to weak weathering, heavy mineral grains in the black loam have been subjected to little abrasion and thus are relatively clean. Moreover, prismatic grains often exhibit well-preserved angles and joint planes, as well as distinct joint lamellae. The differences in heavy mineral composition between the black loam and the cinnamonic paleosol may reflect different formation environments and/or different material sources.

5.2.4 Discussion

1. Material Source of Loess in China. With regard to the category, assemblage and surface features of detrital minerals in it, loess is quite different from other types of Quaternary deposits. Loess materials can come from vast areas, instead of being the products of small-scale weathering of basement rocks. Silt-dust particles which compose loess are derived not only from products of weathering of adjacent basement rocks, but also, for the most part, from vast Gobi and desert areas, where generally the Mesozoic and the Cenozoic detrital rocks developed. In these areas strong weathering prevailed for a long period of time, producing large amounts of detrital material. These materials were then transported to the upper air by wind and thoroughly mixed there, as is evidenced by the analysis of heavy minerals.

2. Index for Discriminating Different Types of Loess and Paleosol. The principal differences in heavy mineral composition for loesses and paleosols of various times lie in the different content of unstable, relatively stable and stable minerals, with amphiboles, epidotes and opaque minerals as dominant heavy minerals. Among the above-mentioned minerals there exist different linear correlations in their content: amphiboles show remarkable negative correlations with opaque minerals, epidotes have negative, but not too obvious correlations with opaque minerals, and amphiboles are independent of epidotes (Table 5.2). The above-described correlations may provide information on how to distinguish loess from paleosol, and loesses of different periods (Fig. 5.5).

Let the content ratio of amphiboles to epidotes equal, K_1, that of amphiboles to opaque minerals equal K_2, and that of epidotes to opaque minerals equal K_3. All the ratios are listed in Table 5.2 and plotted in Fig. 5.6. Assuming that loesses and paleosols of different periods share almost the same material source, the values of K_1, K_2 and K_3 can be considered to reflect not only the degree of

Table 5.3. Content ratios of the three kinds of heavy minerals – amphiboles, epidotes and opaque minerals

Sequence	Sample No.	K_1	K_2	K_3	F
S_0	M_1	0.94	1.07	1.14	2.01
L_1	ML_{91}				
L_1	M_9	0.60	0.56	0.93	1.16
L_1	M_{15}	0.68	0.65	0.95	1.33
L_2	L_7	0.53	0.57	1.07	1.10
S_2	L_{12}	0.28	0.15	0.53	0.43
L_3	L_{16}	0.43	0.44	1.02	0.87
S_3	L_{19}	0.27	0.05	0.20	0.32
L_4	L_{21}	0.42	0.54	1.27	0.97
S_4	L_{24}	0.19	0.07	0.35	0.28
L_5	L_{27}	0.51	0.36	0.70	0.87
S_5	L_{31}	0.18	0.09	0.50	0.27
S_5	L_{34}	0.17	0.09	0.50	0.26
L_6	L_{38}	0.53	0.45	0.85	0.98
S_6	L_{41}	0.11	0.08	0.73	0.19
L_7	L_{43}				
S_7	L_{45}	0.10	0.04	0.38	0.14
L_8	L_{47}	0.36	0.31	0.86	0.67
S_8	L_{49}	0.23	0.13	0.55	0.36
L_9	L_{55}	0.18	0.13	0.75	0.31
L_9	L_{58}	0.30	0.18	0.60	0.48
S_9	W_2	0.15	0.11	0.71	0.26
S_{10}	W_{10}	0.03	0.20	0.46	0.23
L_{11}	W_{12}	0.09	0.10	1.07	0.19
L_{12}	W_{14}	0.29	0.14	0.49	0.43
L_{13}	W_{16}	0.08	0.05	0.62	0.13
L_{13}	W_{19}	0.28	0.11	0.42	0.39
S_{13}	W_{20}	0.22	0.18	0.81	0.40
L_{14}	W_{22}	0.37	0.15	0.40	0.52
L_{15}	W_{25}	0.50	0.22	0.43	0.72
S_{15}	W_{27}	0.12	0.07	0.57	0.17
	W_{47}	0.47	0.13	0.28	0.60
	W_{51}	0.06	0.04	0.77	0.10
	W_{54}	0.04	0.02	0.39	0.06
	W_{57}	0	0	1.11	0
	W_{64}	0.05	0.05	0.89	0.10

$K_1 = \frac{\text{Amphiboles}}{\text{Epidotes}}$; $K_2 = \frac{\text{Amphiboles}}{\text{Opaque Minerals}}$; $K_3 = \frac{\text{Epidotes}}{\text{Opaque minerals}}$; $F = K_1 + K_2$.

weathering in the source region of loess, but also the degree of pedogenetic reworking of loess during post-accumulation. In accordance with K_1 and K_2, loess can be readily distinguished from paleosol, and variations can be recognized in loesses of different periods. K_3 can sharply reflect the differences among loesses deposited in the last 730000 yr or so, but cannot distinguish any difference among loesses older than 730000 yr. This can be seen clearly in Table 5.4. Field observations are also consistent with the data of Table 5.4. Letting F represent the sum of K_1 and K_2, one can use F as an index to distinguish loess from paleosol, and loesses of different periods. To some extent the F value can also reflect the

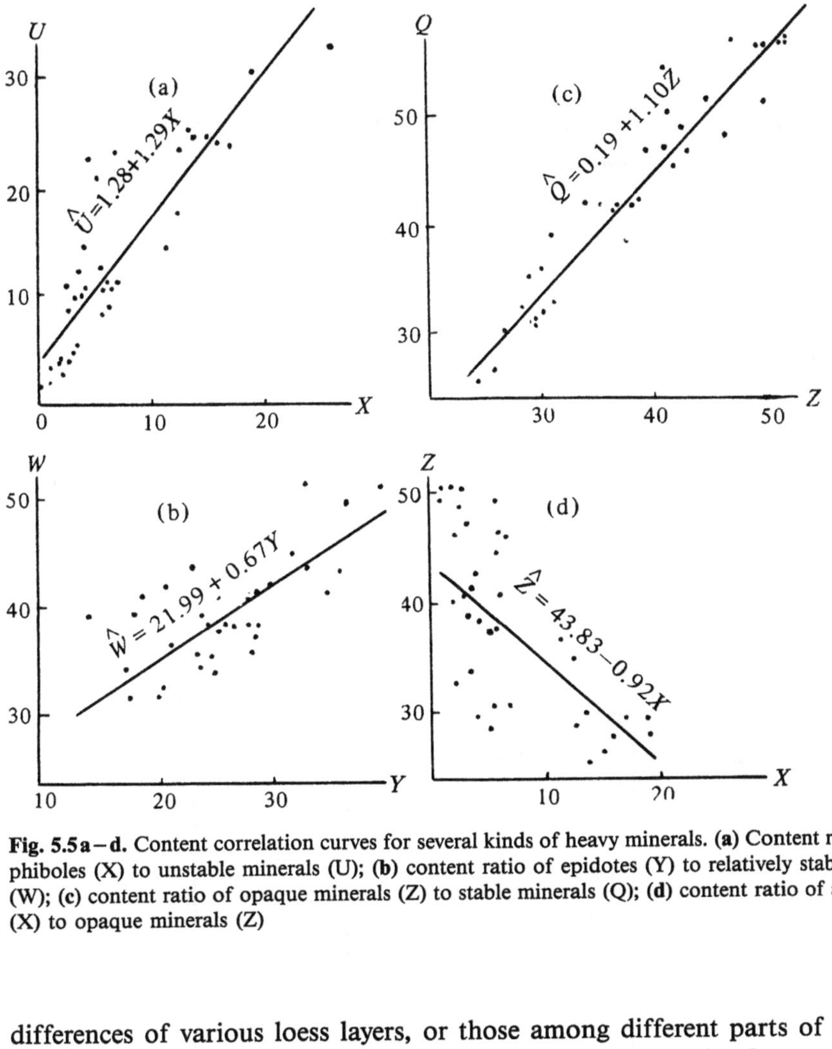

Fig. 5.5 a–d. Content correlation curves for several kinds of heavy minerals. **(a)** Content ratio of amphiboles (X) to unstable minerals (U); **(b)** content ratio of epidotes (Y) to relatively stable minerals (W); **(c)** content ratio of opaque minerals (Z) to stable minerals (Q); **(d)** content ratio of amphiboles (X) to opaque minerals (Z)

differences of various loess layers, or those among different parts of the same loess layer. For example, the upper and lower parts of the Malan Loess are coarse, whereas the middle part is fine; the content of $CaCO_3$ in the upper and lower parts is much higher than that in the middle part; weathering-pedogenesis is relatively prominent in the middle part (Lu Yanchou 1981; An Zhisheng and Lu Yanchou 1984). It can be seen from Table 5.4 that $F = 2.50$ for sample LM91 (the upper part), $F = 1.16$ for sample M9 (the middle part), and $F = 1.33$ for sample M15 (the lower part). That no remarkable differences are found in the lower parts below S_8 may be due to the small magnitude of climatic changes or the long-time intensive pedogenetic reworking.

The first and second sandy loesses (L_9 and L_{15}), and loess L_{12} are the aquifer beds in the Luochuan area, with much lower F values than those of the individual layers in the Malan loess and Lishi Loess. As described previously, this feature may reflect the influence of flows, or mud flows in the periglacial environment during loess formation.

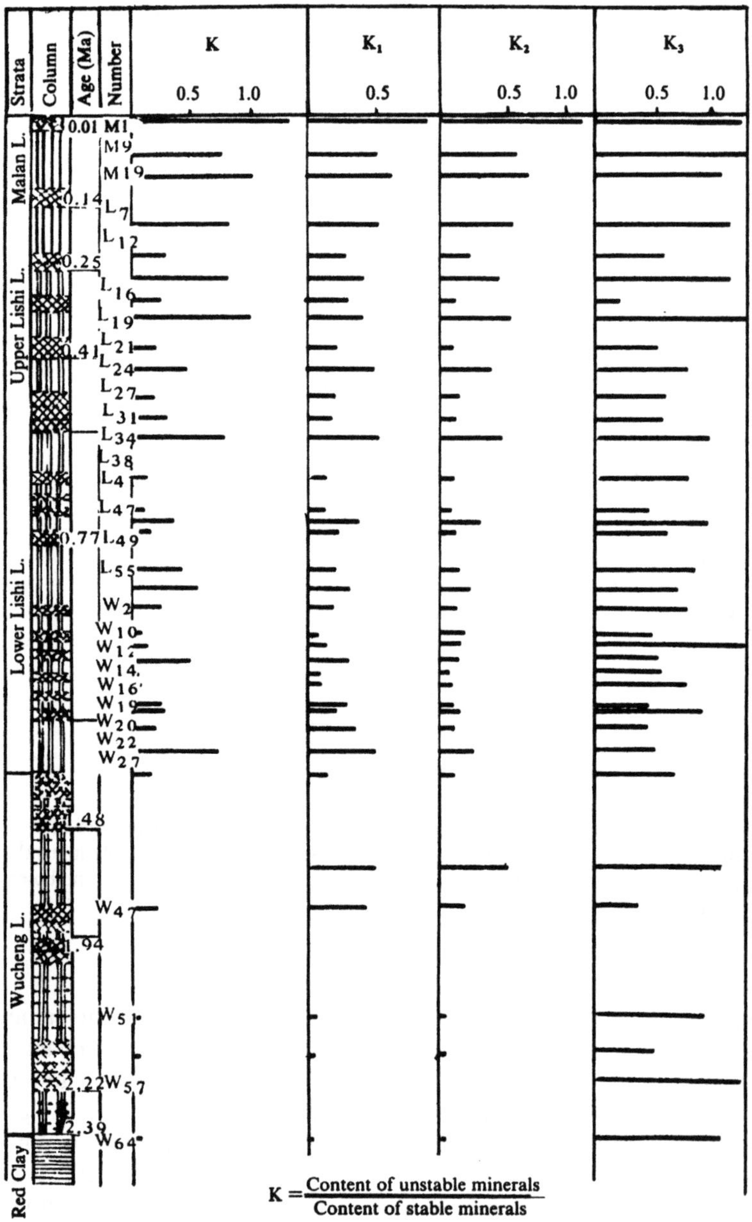

Fig. 5.6. Content ratio distribution of the selected heavy minerals in loess of the Heimugou section, at Potou, Luochuan

Table 5.4. Average value of K_1, K_2, K_3 and F of loesses (L) and paleosol (P) in Luochuan section

Chron			Number of samples	K_1	K_2	K_3	F
Brunhes	Lishi	L	(9)	0.51 ± 0.09	0.50 ± 0.11	0.98 ± 0.17	1.16 ± 0.51
		P	(8)	0.20 ± 0.06	0.09 ± 0.03	0.48 ± 0.15	0.29 ± 0.09
Matuyama	Lishi	L	(9)	0.26 ± 0.13	0.14 ± 0.05	0.60 ± 0.21	0.40 ± 0.18
		P	(4)	0.13 ± 0.07	0.14 ± 0.06	0.62 ± 0.15	0.26 ± 0.09
	Wucheng	L	(6)	0.12 ± 0.06	0.05 ± 0.04	0.67 ± 0.29	0.18 ± 0.19

To sum up, we consider that F value can be used as an index to distinguish loess from paleosol, and loesses of different periods. Also, it can be regarded as a record of bioenvironmental changes, i.e., the greater F value, the more characteristic the loess is of silty sediments deposited in a steppe or desert-steppe environment.

5.3 Clay Mineral

During the various stages of formation, transportation, deposition and diagenesis of loess materials, clay minerals, as a result of the change in physical, chemical and biological conditions, altered their chemical composition and crystal structure. Thus, as early as the 1950s, some researchers initiated qualitative investigations of the colloidal dispersive mineral contained in the loess in China (Sedletsk and Ananev 1953; Zhang Shaozhen 1958). Because the change of clay minerals in species, quantity and structures under the ordinary pressure is closely related to the climatic conditions (temperature, humidity and their rates of change), this can also be taken as one of the indices to the vicissitudes of the environments. Since the 1970s, many authors have tried to explore the evolutionary history of the geological environments in the loess areas during the Quaternary in terms of clay mineral studies, and correspondingly, the studies of clay minerals in loess have gradually transferred to the quantitative stage from the previous qualitative stage. Zheng Honghan (1982) and Han Jiamao (1982) carried out quantitative investigations on the clay minerals in loess, using X-ray diffraction and petrochemical method respectively. The variation of the "crystallinity" parameter of illite was taken to discuss the environmental vicissitudes during the Quaternary. On the basis of the study of the clay minerals in loess along Heimugou section, a systematic summary on loess clay minerals is given in this section.

5.3.1 Clay Mineral Composition of Loess

Shown in Fig. 5.7 are X-ray diffraction patterns of clay particles in loess and paleosol samples of different ages along the Heimugou section. The patterns of untreated clay particle ($<2\,\mu m$) samples (Fig. 5.7A) are characterized by strong reflections at 14, 10, 7.1, 5.0, 4.7, 3.5 and 3.33 Å; some untreated samples show also peaks at 15 Å and $\geqslant 17$ Å. In contrast with paleosols, loess samples also show

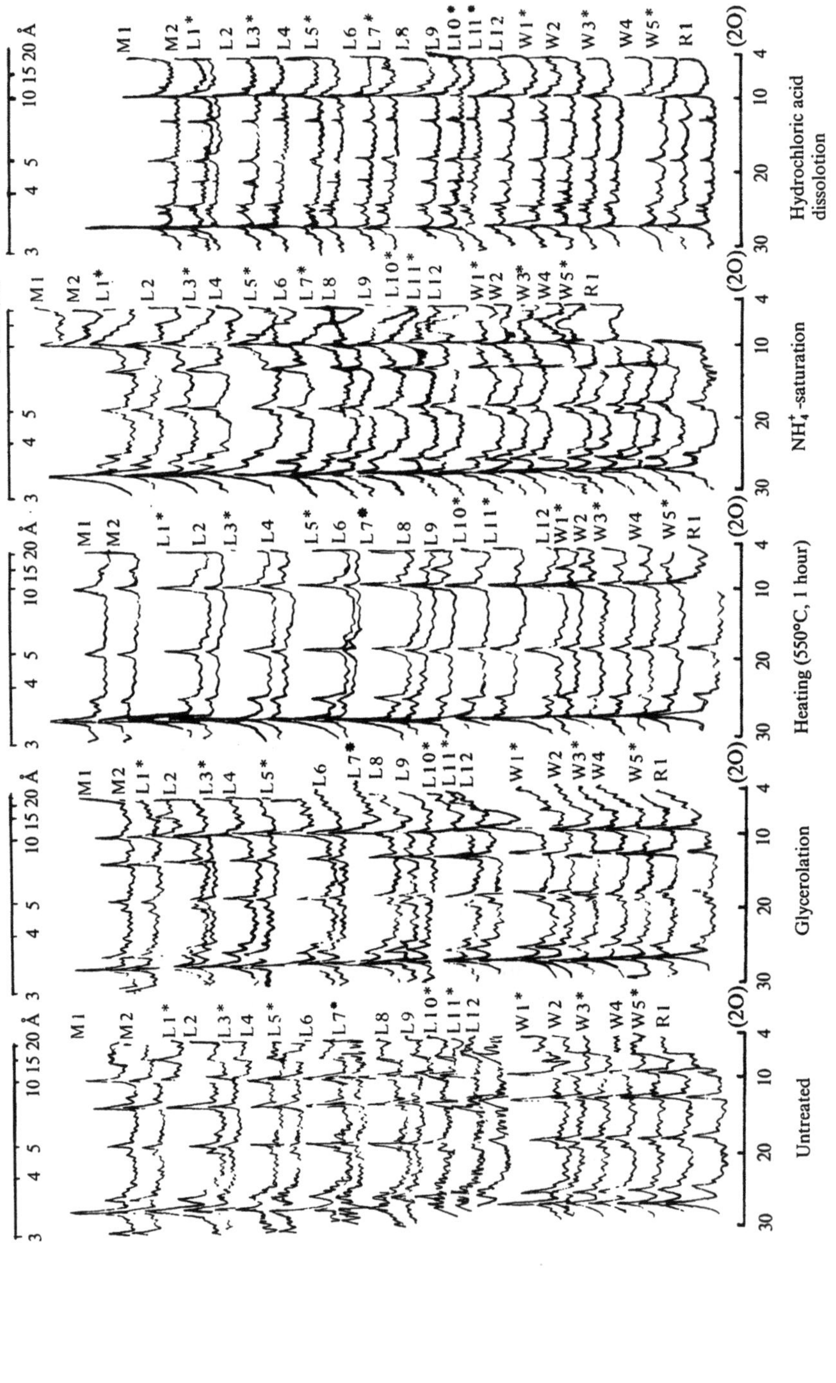

Fig. 5.7. X-ray diffraction patterns of clay particles in Heimugou loess, Luochuan, Shaanxi Province

Table 5.5. Contents of various minerals in clay fraction of Luochuan loess

Sample number	Kaolinite	Illite	Montmorillinite	Chlorite	Vermiculite	Calcite	Goethite	Others
Malan 1	16.8	48.7	3.7	5.7	4.7	16.1	2.9	1.4
Malan 2	14.8	45.6	4.3	5.4	5.0	20.9	2.7	1.3
Lishi 1[a]	18.2	57.4	7.9	2.6	7.9	0	4.6	1.4
Lishi 2	18.5	42.1	10.8	5.2	4.8	14.4	3.0	1.2
Lishi 3[a]	20.8	53.6	8.4	5.4	4.8	1.3	4.4	1.3
Lishi 4	13.0	55.7	5.3	4.5	5.3	11.6	3.1	1.5
Lishi 5[a]	17.4	63.2	1.1	6.4	5.6	1.0	4.2	1.1
Lishi 6	23.4	42.9	12.5	9.2	1.3	6.1	3.2	1.4
Lishi 7[a]	19.3	61.3	3.2	9.2	2.0	0	3.9	1.1
Lishi 8	16.4	51.2	7.3	7.1	3.8	10.3	2.7	1.2
Lishi 9	20.7	45.0	7.5	7.8	3.3	11.2	3.2	1.3
Lishi 10[a]	21.3	50.9	11.7	7.0	2.5	1.8	3.6	1.2
Lishi 11[a]	21.4	53.3	10.3	8.0	2.3	0	3.6	1.1
Lishi 12	17.7	55.2	13.0	8.1	1.9	0	2.9	1.2
Wucheng 1[a]	19.8	49.0	8.1	6.6	4.1	8.4	3.0	1.0
Wucheng 2	20.8	49.2	12.5	7.6	3.4	3.1	2.5	1.0
Wucheng 3[a]	20.3	50.0	11.2	5.3	4.8	4.3	3.0	1.1
Wucheng 4	14.7	52.7	7.3	4.3	6.8	10.7	2.2	1.3
Wucheng 5[a]	23.1	40.7	14.0	2.7	6.9	8.1	3.4	1.1
Red Clay	21.5	51.5	11.7	5.9	3.8	1.4	3.2	1.0

[a] Paleosol.

diffraction at 3.03 Å. From changes in the patterns of samples treated respectively by glycerine, heating (550 °C, 1 h), saturated ammonium chloride solution and concentrated hydrochloric acid (Fig. 5.7B–E), one can conclude that the major components of clay minerals in loess are illite, then kaolinite, montmorillonite, chlorite, vermiculite and small amounts of disordered and/or ordered mixed-layer minerals. The differences in clay minerals between paleosol and loess layers lie mainly in illite structure and in the relative contents of vermiculite and chlorite. The content of vermiculite increases in paleosols. These differences can also be verified by thermal analysis, chemical analysis and microscopic observation[8].

The variety of changes on minerals under ordinary temperature and pressure essentially result from the change of conditions in environments. The leads to the migration and concentration of chemical elements, which incurs the transformation of crystal structures in minerals. Based on this knowledge, and starting from the bulk chemical analyses of clay particle fractions, the authors carried out a quantitative calculation of the mineral composition in Heimugou loess, using a petrochemical method (Han Jiamao 1982). The results (Table 5.5) show that illite accounts for about 50%, kaolinite for about 20%, montmorillonite for about 10%, chlorite and vermiculite for about 5% of the clay fractions. In addition,

[8] Han Jiamao, 1982, Studies on clay minerals in loess and the Quaternary paleoenvironments. Dissertation thesis, Institute of Geology, Academia Sinica.

Table 5.6. The average contents of various minerals in clay fractions of loess and paleosols of various times in Luochuan

	Number of samples	Kaolinite	Illite	Montmorillonite	Chlorite	Vermiculite	Calcite	Goethite	Others
Malan Loess	2	15.8	47.1	4.0	5.6	4.9	18.5	2.8	1.3
Lishi Loess	6	18.3	48.7	9.4	7.0	3.4	8.9	3.0	1.3
Wucheng Loess	2	17.7	51.0	9.8	6.0	5.1	6.9	2.4	1.1
Paleosol in Lishi Loess	6	19.7	56.6	7.1	6.4	4.2	0.7	4.1	1.2
Paleosol in Wucheng Loess	3	21.0	46.6	11.1	4.9	5.3	6.9	3.1	1.1
Red Clay	1	21.5	51.5	11.7	5.9	3.8	1.4	3.2	1.0

there exists up to 20% of carbonate in loesses. These results are in good conformity with qualitative results obtained through X-ray diffraction analysis.

In order to make the comparison easily, the average contents of various clay minerals contained in Malan Loess, Lishi Loess, Wucheng Loess and in paleosols developed on Lishi Loess and Wucheng Loess are listed in Table 5.6. All the averages are derived from the quantitative calculations mentioned above. From Table 5.6, it is apparent that the contents of illite in loess of various times are comparatively steady, while the content of montmorillonite increases gradually from Malan Loess to Wucheng Loess. The maximum of chlorite is observed in Lishi Loess. The content of illite in the paleosol developed on Lishi Loess increases significantly, at the same time the content of montmorillonite seems to decrease. In addition, the paleosol contains a relatively high content of vermiculite. The variation of paleosols developed on Wucheng Loess is not as remarkable as that on Lishi Loess.

There exist a certain amount of non-clay minerals, main calcite and goethite in clay particles. The soil-forming processes caused the leaching of calcium and the relative enrichment of oxides and hydroxides of iron.

5.3.2 The Structural Variation of Illite in Loess

The name "illite" was first proposed by Grim et al. (1937) and since then it has been used as a general term for mica-like clay minerals which are widely distributed in argillaceous sediments. Recent investigations have confirmed that many species of illite may be interstratified (Bailey 1980). The term illite used here is defined as a mixed layer of mica and montmorillonite with less montmorillonite layers. It was MacEwan (1949) who proposed the mixed-layer structure built up by mica and montmorillonite layers. A well-established method used to investigate the mixed-layer characteristic of illite is the asymmetry of reflection (001) of X-ray diffraction patterns (Bradley 1954; Lucas 1963; Kubler 1961;

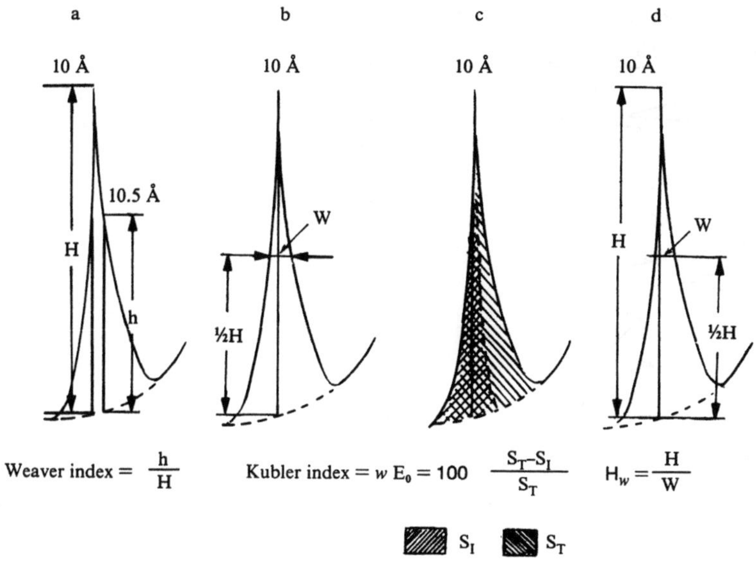

Fig. 5.8. Various "crystallinity" indices of illite

Kodama et al. 1969; Reynolds and Hower 1970). In view of the geological implications involved in the structural variations of illite, Weaver (1960), Kubler (1964), Weber (1972), Mellieres and Perez-Nieto (1973) developed the concept of "crystallinity" of illite (Fig. 5.8 a–c), to explain some geological observations, obtaining some remarkable results. Because the minerals with mixed-layer structures are, in fact, in metastable state, they are susceptible to changes in the surrounding environment, and thus probably reserve much geological information. Therefore, into the study on clay minerals in loess the authors also introduced the concept of "crystallinity" of illite. Owing to the difficulties often encountered in attempting to apply the above-mentioned various "crystallinity" indices to practical problems (e.g., the unwarranted precision, too small magnitudes of variation or overelaborate procedure in measurements), the authors adopted a new representation of the "crystallinity" of illite which is more accurate and also easier to measure. This is known as the height-to-width index of d_{001} of illite (simply called H_w index, see Fig. 5.8 d):

$$H_w = \frac{H}{W},$$

where H stands for the peak height (mm) of (001) reflection of illite at 10 Å, and W for the half-width (mm) of the same peak.

The measurement of H_w generally uses X-ray diffraction patterns of untreated samples with grain size smaller than 2 μm.

Taking illite as a mineral with mixed-layer structure into consideration, the index H_w, reflects the proportion of montmorillonitic layers contained in the mixed layer structure. The more the montmorillonitic layers are contained the more pronounced the asymmetry of the peak, so H_w becomes smaller. Converse-

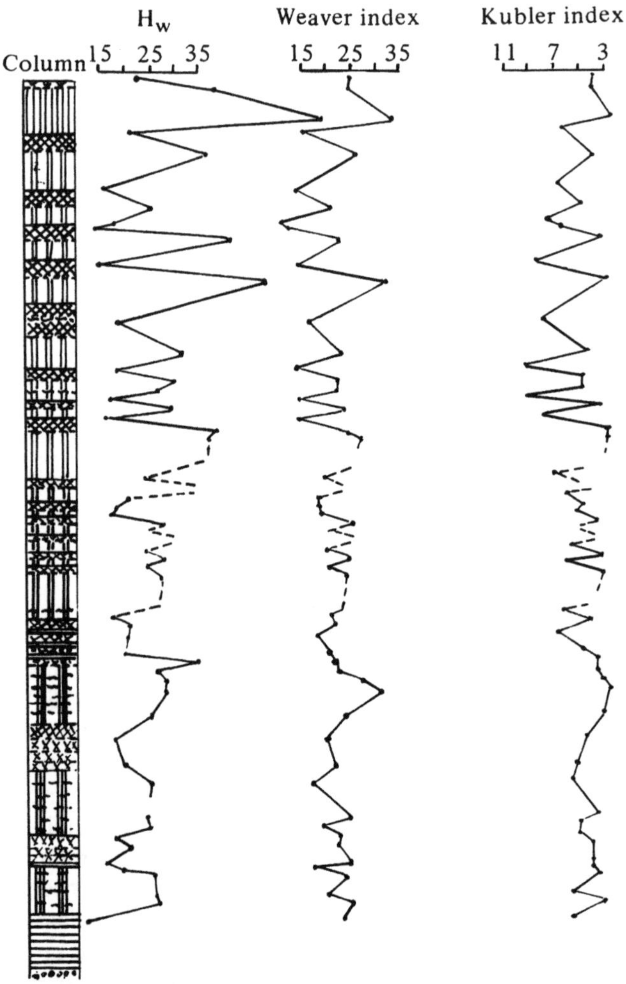

Fig. 5.9. Variation of H_w – "crystallinity" of illite along Heimugou section

ly, the less the montmorillonitic layers are contained, the more obscure its asymmetry, as H_w becomes correspondingly larger.

Figure 5.9 shows the variation of H_w in various samples from the Heimugou section. A periodic change can be seen in H_w along the section. In addition, both the frequency and the magnitude of variation of H_w were small during the Matuyama chron, and increased considerably in the Brunhes chron.

From Fig. 5.9 one can see that there is a remarkable difference between the illite in loess layers (with $H_w > 25$) and that in paleosol (with $H_w < 20$), indicating that the paleosol-forming process trended towards the increase of expansion layers in illite structures. Comparing loesses of various ages, H_w increases progressively from Wucheng Loess, through Lishi Loess, to Malan Loess. There is less difference in H_w between the paleosols developed on Lishi Loess and Wucheng Loess.

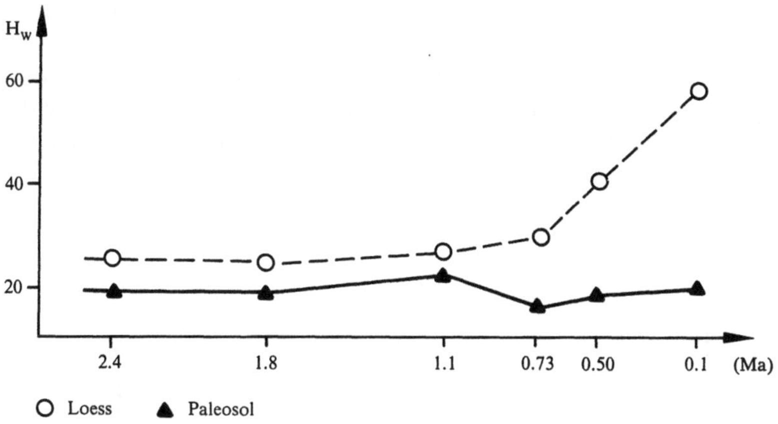

Fig. 5.10. Variation of "crystallinity" of illite with geological times

Plotting the H_w values ("crystallinity" indices) of illite in loess layers and paleosols near the major time boundaries on the time scale estimated in Section V, Chapter 3, the variation of H_w with times becomes obvious (Fig. 5.10). During the period of 2.4 – 1.1 m.y., the change in "crystallinity" of illite was not as great as during the period since 1.1 m.y., and especially since 0.73 m.y. This suggests that in the former half of the Quaternary Period the magnitude of variation in paleohumidity was still small, whereas over the past 1.1 m.y. it has become more and more pronounced.

5.3.3 A Discussion on the Geological Environments of Formation of Loess and Paleosol with a Special Emphasis on Clay Mineral Analyses

Since the affirmation that the "red bands" intercalated in "reddish clay" (named by Teilhard de Chardin and Young 1930) are buried paleosols and confirmation that there exist loess deposits older than Malan Loess, the paleoclimatic implication of the multiple overlap of loess and paleosol horizons has gained more and more attention from Quaternary geologists. Consequently, clay minerals have been taken as an index of the subdivision of climatic cycles (Liu Tungsheng et al. 1980a; Zheng Honghan 1982).

For the Luochuan area, amongst the many factors affecting weathering and pedogenic processes, the difference in composition and structure of the parent material is negligible. The basic geomorphologic pattern in the "Yuan" area changed insignificantly during the Quaternary, and thus the topographic influence is also subordinate. Because each and every loess-paleosol sequence has gone through a long enough period of time relative to the formation of soil sections, the factor of time is naturally unimportant. Obviously, the main reason for the difference between loess and paleosol, and particularly, for the difference in the composition and structure of clay minerals must be the paleoclimatic vicissitudes, and hence, the variation which was controlled by the paleoclimatic conditions.

The influence of climatic changes on clay minerals is difficult to ascertain. The evolution of mica-like clay minerals is closely related to the number of potassium cations in their unit cells. Burst (1958) summarized the research results concerning this problem and pointed our that for the aluminous mica group, with a decrease of potassium cation number, the mineral will change from muscovite

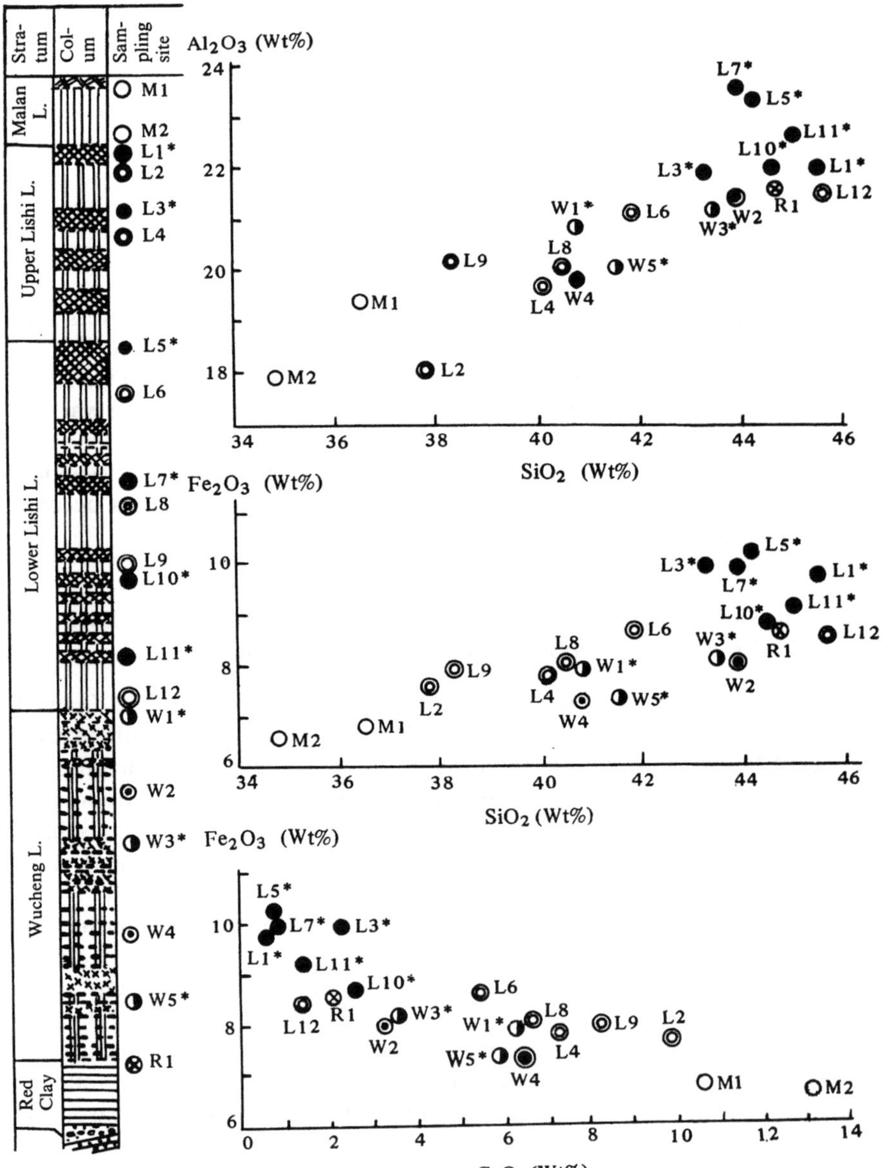

Fig. 5.11. The difference of environments of formation between loess and palesol based on chemical analyses of clay particles

($K^+ = 2$), to illite ($K^+ = 1-1.4$), to degraded illite, to mixed-layer clay ($K^+ = 0.2-1$) and finally to montmorillonite ($K^+ <0.2$). If we take the new concept of illite into account, does this decrease of potassium cation number imply a progressive increase of the expansion interlayers in the structure of illite? The leaching out of potassium cations is favorable for the formation of illites that involve more montmorillonite interlayers. The potassium content in soils are mainly related to the climatic conditions, especially the precipitation. The study on potassium contents in the soils developed on loess in the United States (Wells and Riecken 1969) shows that the total potassium content in clay of the leached horizons decreases with increasing precipitation. From the change of clay minerals in loess, especially the variation of "crystallinity" of illite, one can see that paleosols were formed in relatively humid environments, whereas loesses developed in relatively arid periods. Comparison of the forming environments of loess in different times shows that Wucheng Loess was formed in a slightly humid climatic environment, while Lishi Loess and Malan Loess were deposited in a much more arid climatic environment.

The chemical analyses of clay particles can also be a good indicator of the environments. In Fig. 5.11 plots of SiO_2 vs. Al_2O_3, SiO_2 vs. Fe_2O_3 and CaO vs. Fe_2O_3 are used respectively to infer the humidity conditions, the extent of chemical weathering, and the intensities of leaching processes of clay paricles. It is very clear that there is a remarkable difference between loesses and paleosols. There is a gradation from Malan Loess to Lishi Loess to Wucheng Loess. The difference between Lishi Loess and its paleosol is more remarkable than that between Wucheng Loess and its paleosol.

In summary, loess was formed in an arid environment with relatively weak chemical weathering and leaching, whereas paleosols were formed in a humid environment with relatively strong chemical weathering and leaching. During the early period of loess deposition the magnitude of climatic fluctuation was small, then became more and more considerable. The main reason for the increase of magnitude is that the climate became considerably arid. The climatic fluctuation involved not only the difference in heat amount, but also that in humidity; the amplitude of the former was small and the latter was great.

5.3.4 Source of the Clay Minerals in Loess

Recent investigations have shown that the dust from the Sahara can move a long distance and settle on the American continent (Pewe 1981), and the dust from the Asian deserts can also move more than 5000 km and float in the sky over Alaska (Rahn et al. 1981). An analysis of the synoptic process of dust storms for mid-April, 1980, which included an investigation of the dust material when fell on April 18, 1980 in Beijing (Liu Tungsheng et al. 1981, 1982) verified that the dust (which shows grain-size distribution and mineral composition closely similar to Malan Loess) mainly came from the delta area with Erenhot, Hami and Wuwei as its apices. In other words, they came from the deserts of Badain Jaran, Tengger, Ulan Buh and Hobq. This provides strong evidence that the loess material was of aeolian transport.

The source of clay minerals in loess is one of the problems to which the researchers of loess clay minerals have paid a good deal of attention. The clay minerals found in American loess are mainly illite and montmorillonite. Earlier studies suggested that they came from the glacial till, but montmorillonite is not the major clay mineral in moraines in North America. Therefore, it was believed that the montmorillonite might be derived from volcanic glass that previously existed in the loess (Swineford and Frye 1951; Beavers 1957). Ruhe and Olson (1980) carried out research on Farmdale Loess and Peoria Loess which were attributed respectively to early and late Wisconsinan. They maintained that the clay minerals in the former were related to Sangamon paleosol while the latter showed a variety of cases depending on the localities: the loess in the Ohio Basin was related to Sangamon paleosol, and that in the Wabash Basin to glacial till. Because the clay minerals in loess in West Europe (mainly fine and dispersive mica and chlorite-vermiculite) are closely similar to those in interglacial soil nearby, Swineford and Frye (1955) suggested that they came from the source region.

The complexity of clay mineral assemblages of loess in China, particularly the existence of a certain amount of kaolinite, puzzled some investigators for a time because neither the source region of loess in China (desert areas in Northwest China) nor the deposition region (the Loess Plateau) had a consistent humid and acidic environment favorable for the formation of kaolinite. Therefore, it is of great importance for both studies on clay mineral evolution and on loess itself to shed light on the source of clay minerals in loess in China.

The TEM observation has revealed that the grains of clay minerals in loess are mainly allotriomorphic and hypautomorphic, and dissolution and abrasion can often be observed on the grain surfaces. This implies a clastic origin of clay minerals in loess. Since the loess material came from deserts, loess and desert are cognate. According to Zhu Zhenda et al. (1980), the sand source of Chinese deserts are mostly alluvial, diluvial and lacustrine deposits associated with water flows. Among these deposits there were certainly some clay materials. It is likely that these clay materials were bound to the surfaces of other grains or in other forms and transported by wind from the source region to the deposition region.

Nevertheless, since late Tertiary the Great Desert Area in Northwest China has been generally arid (Scientific Expedition to Xinjiang, Academia Sinica, 1978), and it is very difficult to interpret the appearance of kaolinite in such a great amount in loess from the inorganic phase equilibria. Thus, the authors hold that it is necessary to re-emphasize the influence of biological activities on the formation and evolution of minerals. "For a long time, the phase equilibrium study on clay minerals including the synthetic work have taken into consideration only the influence of some ions or complexes of certain elements on the conditions of formation and phase transformation of clay minerals, while neglecting the influence of some organic materials on the phase relations among clay minerals. In nature, the superficial processes cannot be the ones involving only the interactions among inorganic materials, but inevitably include the influence of organic materials on the formation and phase transformation of clay minerals." (Translated from Gu Xiongfei et al. 1980). The study on the "litholac" in the Gobi Desert (Zhu Xianmo et al. 1982) is a vivid example of the influence of biological processes on the formation of clay minerals. They found that the "litholac" material includes not only bacteria and algae, but also some well-crystallized ferric oxides and clay minerals (e.g. kaolinite, montmorillonite and illite). The biological activities created the microenvironments necessary for the formation of clay minerals. This study undoubtedly contributes to our understanding of the source of clay minerals in loess. We conclude that the clay minerals in loess are mainly the primary debris and authigenic minerals derived from biological and other processes and then moved to the deposition regions.

5.3.5 Influence of Diagenesis on Clay Minerals in Loess

For aeolian sediments such as loess, the dust deposition can properly be considered a process of diagenesis.

Generally speaking, the evolution of clay minerals as depth, temperature and pressure increase is from a disordered to ordered structure, from incomplete to complete lattice and from more expansion layers to more non-expansion layers. The usual trend is from montmorillonite to mixed-layer minerals of montmorillonite-illite, then to illite (Perry and Hower 1970; Hower et al. 1976). Aoyagi and Kazama (1980) have calculated the temperatures and pressures at which these transformations can be realized. They are: at 100 °C and 900 kg/cm^2 and at 140 °C and 920 kg/cm^2 respectively. A calculation shows that even at the bottom of Heimugou section the required temperatures and pressures of transformation are still too high. The compositional analysis of clay minerals did not show this trend of diagenesis either. Near the lower part of the section, especially in Wucheng Loess, the content of montmorillonite increases. This is possibly associated with the long-term action of underground water enriched in magnesium cations. The major aquifer in the loess in Luochuan area is under the first sandy loess, while the hydrochemical type of the underground water belongs mainly to bicarbonate magnesium type[9].

Thus it can be seen that diagenesis in loess is still in the stage of compression. Furthermore, the ionic concentration, namely the concentration of magnesium ions, affects the composition of clay minerals in the lower part of the section, whereas the pressure and temperature are still beyond the range within which phase transformation will occur in clay minerals. Finally, the compositional and structural characteristics of clay minerals in loess provide accurate information about the geological environments of the deposition of loess. "Clayization" is often considered as a characteristic of loess diagenesis. It is likely that the compositions and structures of initial clay minerals have already undergone some subtle changes after the deposition of dust materials through the pedogenic processes. This includes changes in composition and structure reflected by the "crystallinity" of illite (H_w values), the increase of the amount of montmorillonite in the lower part of the section caused by the action of magnesium ions, the occurrence of small amount of authigenic kaolinite in well-developed

[9] Hydrogeological Temas No. 2 of Bureau of Geology and Mineral Resources of Shaanxi Province, 1981, Report on the hydrogeology for agricultural water supply in loess "Yuan" area in Luochuan, Shaanxi Province.

soil horizons, etc. Nevertheless, it is not the "clayization" that has given rise to the clay minerals seen in loess. It seems proper to consider the "clayization" as some change in chemical properties and crystal structures of the clay minerals coming from the source regions during the deposition and evolution of loess or during its diagenesis. It is doubtful that all clay minerals in loess were generated through the "clayization".

5.4 Geochemical Characteristics

5.4.1 The Main Chemical Components of Loess

The main chemical components of the Luochuan loess are: SiO_2 (>50%), Al_2O_3 (>10%), CaO (7.5%–10.5%), and subordinately Fe_2O_3 (3%–6%), FeO (0.4%–1.5%), MgO (1.5%–5.0%), K_2O (1.5%–2.5%) and Na_2O (1.2%–2.3%). There is a relationship between the chemical and mineralogical compositions of loess. As with the grain size and mineral composition of loess, the homogeneity of chemical composition is also manifested in the comparison between the main chemical components of Luochuan loess and those of loesses from other areas along the middle reaches of the Huanghe River. Nevertheless, some slight variations are still noticed (Wen Qizhong et al. 1964). In addition, some differences are also recognized in element distribution for different grain-sized fractions. The ratios of SiO_2/Al_2O_3, FeO/Fe_2O_3, CaO/MgO and CaO + K_2O+Na_2O/Al_2O_3 tend to increase in the loess section from the bottom upwards, but the ratio of K_2O/Na_2O shows an opposite tendency (Fig. 5.12). The variation curves of these ratios in the section fit the curve of $CaCO_3$.

In the weathering process of loess, the Fe_2O_3, Al_2O_3, K_2O, MgO and SiO_2 became relatively concentrated, and CaO and Na_2O were considerably leached. The leaching value of CaO is positively correlative with the accumulation value of Fe_2O_3 (Wen Qizhong et al. 1981 a), but fluctuations are observed in the individual layers of loess and paleosol. According to the leaching and accumulation values for each layer of loess, the Luochuan loess can be divided into weakly weathered loess, moderately weathered loess, relatively strongly weathered loess and intensely weathered loess (Wen Qizhong et al. 1981). Furthermore, the difference in the degrees of leaching and accumulation of oxides in paleosol during the weathering process may reflect differences in the range of paleoclimatic fluctuation during the warm and wet soil-forming stages.

5.4.2 Carbonates in Loess

In the loesses of different periods at Luochuan, $CaCO_3$ varies from 3.6% to 20.9%, averaging 11.6%. In the section from the bottom upwards, the content of $CaCO_3$ tends to increase in fluctuation rhythmically (Fig. 5.12). Nonhomogeneous distribution of $CaCO_3$ is noticed in different grain-sized fractions. $CaCO_3$ is contained mainly in 0.05–0.005 mm silt grains (Wen Qizhong et al. 1964). As to

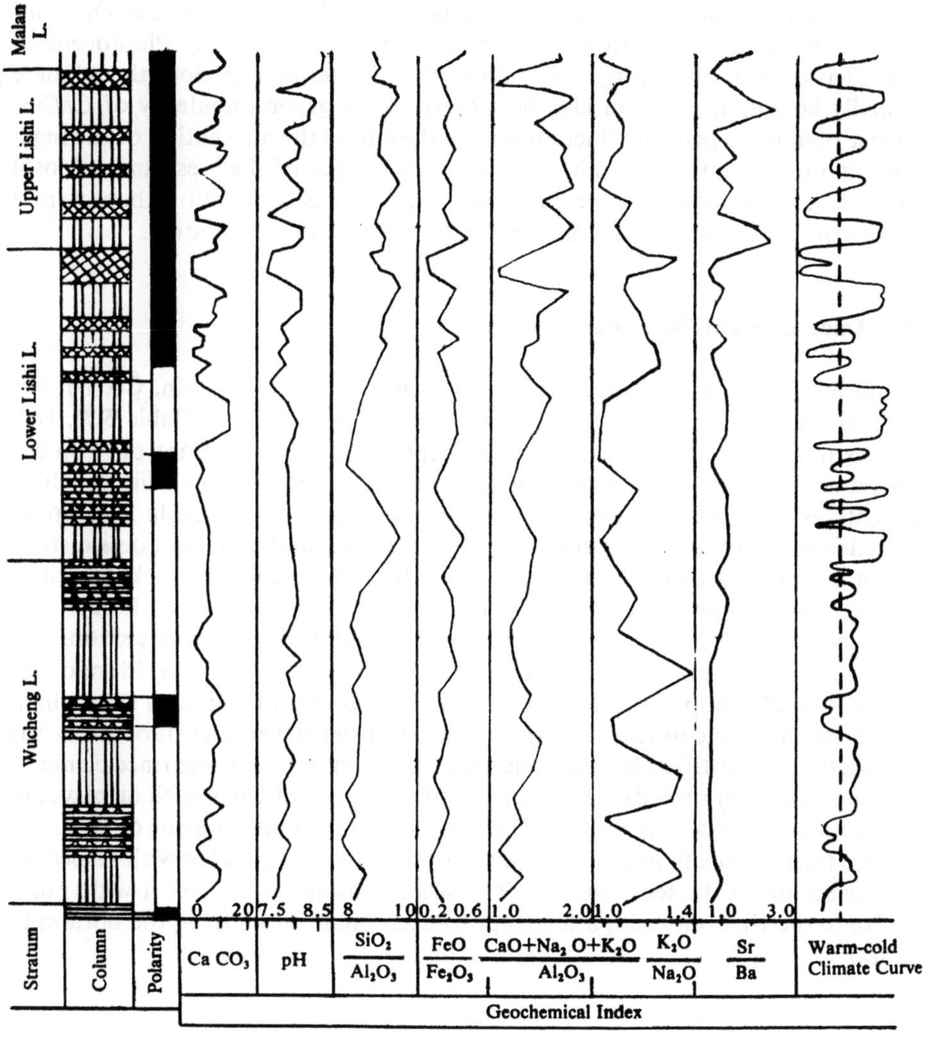

Fig. 5.12. The curves of geochemical indices in Luochuan loess section

the existing form of $CaCO_3$ in loess, two types can be distinguished according to its genesis: the primary and the secondary. In terms of the characteristics of crystallization of carbonate minerals in loess, phanerocrystalline and microcrystalline $CaCO_3$ are recognized. The primary carbonate occurs in the form of detrital carbonate minerals, while the secondary carbonate has a variety of forms, such as secondary coarse-grained phenocrystal, fine-grained aggregate, dessiminated microcrystalline carbonate, pellicle, dermatic crust, pseudomuceli-um and nodule. Some differences are also noticed in the coexisting form and distribution of loesses of different periods. Intercalated in the Luochuan loess section are 45 layers of carbonate concretion. There are also differences in distribution, occurrence and trace composition for concretions in loesses of different periods (Wen Qizhong et al. 1964). There is a general tendency of $CaCO_3$ to increase in the loess section from the bottom upwards (from Early Pleistocene to Late Pleistocene), indicating that since the Quaternary period the climate gradually became dry. It can also be seen from the general tendency of $CaCO_3$ increase that in the periodic fluctuations on the curves the alternation of the maximum and minimum values is in coincidence with that of the loess and paleosol sequence. This suggests that in the process of loess accumulation there were a number of small climatic fluctuations in humidity and temperature.

5.4.3 Trace Elements in Loess

The authors studied the following ten trace elements (Zn, Cu, Mn, Co, Ni, Pb, P, Ti, Sr and Ba) in loess samples from the Luochuan section (Table 5.7). Differences in the content of these trace elements in loess depend not only on the mineral composition of loess, but on the inherent geochemical behavior of these elements as well. Although different elements in loess are variable in content, some elements have certain correlations. For example, in the Malan Loess distributed in the areas along the middle reaches of the Huanghe River, marked positive correlations are shown between Zn, Cu and Mn (Diao Guiyi et al. 1982).

In regard to their main distribution characteristics in the loess section, two groups of trace element can be recognized: the first group (Co, Ni, Pb and Cu) and the second group (Zn, Mn and Ti). The first group of elements show slight differences in their content, though they are somewhat variable throughout the loess section, as justified by relatively close mean values for these trace elements in loess of different periods. The frequency distribution of Co and Ni displays approximately a normal distribution, showing an almost homogeneous distribution of these trace elements in the loess section. Pb is relatively stable with respect to its distribution in the loess section. The second group varies over a wide range relative to the first one, with a tendency to decreasing values from the bottom to the top.

Table 5.7. The analysis results of several trace elements of the Luochuan loess (ppm)

Element	Range of content	Average content	Standard deviation	Coefficient of variation	Average value of loess	Clark value of soil
Zn	71 – 147	91	12.5	0.137	77	50
Cu	13 – 34	25	3.3	0.133	25	20
Mn	486 – 878	691	82.7	0.120	626	850
Co	14 – 28	21	2.6	0.128	21	10
Ni	27 – 50	40	4.4	0.110	39	40
Pb	15 – 40	30	3.6	0.120	30	10
P	131 – 742	388	113.9	0.294	429	800
Ti	3417 – 4676	4058	308.3	0.076	3587	4600
Sr	152 – 288	195	34.4	0.177		
Ba	500 – 647	579	39.1	0.068		

The average contents of various trace elements in paleosol are all higher than those in loess, and this indicates varying degrees of enrichment of the trace elements in paleosol. More obvious enrichment of Zn, Mn and Ti is observed in paleosol, with the average content being 10% higher than that of loess layers.

The F/Cl ratio in the Luochuan loess varies from 0.15 to 0.40 with a tendency toward decreasing values from the bottom upwards (with the exception of the fluctuation of F/Cl in paleosol). The F/Cl ratio in paleosol ranges from 0.20 to 0.75, averaging 0.42. In accordance with the variation of F/Cl ratios the pedogenesis of the paleosol can be divided into four types: F/Cl$<$0.3 (S_2, S_{10}, S_{12} and S_{13} in the section), typical of weak pedogenesis; 0.3$-$0.4 (S_6, S_7, S_9, S_{11}), of moderate pedogenesis; 0.4$-$0.5 (S_3, S_4, S_8 and S_{14}), of relatively strong pedogenesis; and $>$0.5 (S_1, S_5) of intense pedogenesis. The variation of F/Cl in the layers of paleosol in the section from the bottom upwards reflects the difference in pedogenetic intensity during warm and wet soil-forming stages. The variation of F/Cl ratio in the Luochuan loess section is consistent with that of $CaCO_3$ and pH, and the distribution of SiO_2/Al_2O_3 and FeO/Fe_2O_3. By comparing the average values of F/Cl ratio for loess of various periods, we can see that the F/Cl ratio shows remarkable variations above and below the boundary between the Lishi Loess and Wucheng Loess. The F/Cl values above the boundary are greater than the average, whereas the F/Cl values below the boundary are less than the average.

5.4.4 Geochemical Features of Paleosol

As evident in Table 5.8, the average values of the major elements Si, Al, Fe and K in paleosol are higher than those in loess, but values of Ca, Fe^{2+} and Na in the paleosol are lower than those in loess.

The distribution of the elements in the paleosol section is characterized by the increase in Al, Fe and K from the bottom to the top, with the maximum peak values appearing in the clayey layers and the minimum peak values in the calcium layer of the parent-loessic beds. In contrast, the content of calcium carbonate is lowest in the clayey layers, but it is remarkably enriched in the calcic and the parent-loessic beds. The trace elements Zn, Cu, Mn, Ti and Co are higher than those in the parent material of loess. These elements are most enriched in the clayey layers.

Paleosols of different occurrences and properties may reflect the differences in pedogenetic intensity during weathering. With the enhancement of pedogenesis of paleosol, Sr/Ba and the ratios of various oxides (except for K_2O/Na_2O) tend to decrease, but the F/Cl ratio and the content of Fe_2O_3 show a tendency to increase. All of these geochemical features reflect the differences in bioclimatic environment for paleosols of various types during pedogenesis. In the weathering process of loess into paleosol, the intense leaching-out of carbonates led to obvious accumulation of Fe_2O_3 and Al_2O_3. With the enhancement of paleosol pedogenesis, the leaching-out of CaO became more and more intense, leading to higher and higher enrichment of Fe_2O_3.

Table 5.8. Chemical compositions of different types of paleosol in the Luochuan section

	Element content (%)								Element content (ppm)										Layer order of paleosol
	Si	Al	Fe^{3+}	Fe^{2+}	Ca	Mg	K	Na	P	Ti	Zn	Cu	Mn	Co	Ni	Pb	Sr	Ba	
Black loam	26.60	6.10	2.32	0.90	6.27	1.19	1.68	1.24	0.031	0.35	73	51	530	21	33	40	–	–	S_0
Calcareous drab soil	28.55	7.11	3.29	0.73	2.51	1.28	2.16	1.14	0.033	0.42	95	24	703	21	43	26	370	1300	S_3, S_{10}, S_{12}
Drab soil	29.44	7.25	3.32	0.64	1.65	1.26	2.28	1.20	0.046	0.44	104	25	747	20	41	31	330	1900	S_2, S_6, S_8, S_9, S_{11}, S_{13}, S_{14}
Luvic drab soil	30.39	7.20	3.41	0.62	1.57	1.17	1.90	1.10	0.034	0.43	93	26	734	22	41	30	330	1900	S_1, S_4, S_7
Drab-brown soil	30.06	7.64	3.76	0.76	0.71	1.21	2.01	0.93	0.040	0.45	105	27	825	21	44	32	360	3400	
Buried weathering bed	27.85	6.91	3.46	0.47	3.29	1.33	2.23	1.00	0.036	0.40	87	29	744	21	46	30	390	1200	
Average value of paleosol	28.77	7.04	3.24	0.69	2.67	1.24	2.04	1.10	0.037	0.42	93	30	714	21	42	32	330	1900	
Average value of loess	27.60	6.55	2.84	0.73	4.29	1.29	1.93	1.15	0.040	0.39	88	25	651	20	38	29	552	1235	
Clark value of earth crust[a]	29.00	8.05		4.65	2.96	1.87	2.50	2.50	0.093	0.45	83	47	1000	18	58	16	340	650	
Clark value of soil[a]	33.00	7.13		3.80	1.37	0.63	1.36	0.63	0.080	0.46	50	20	850	10	40	10	300	500	

[a] From A.P. Vinogradov (1962).

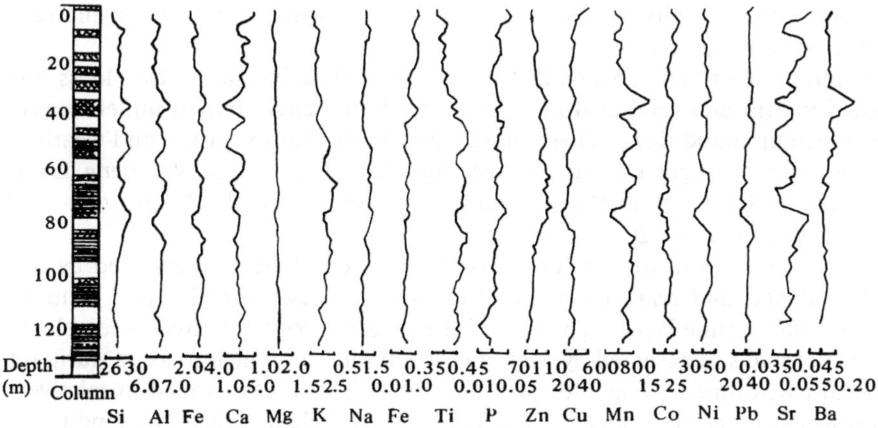

Fig. 5.13. Fluctuation curves of chemical elements in Luochuan section (Zn, Cu, Mn, Co, Ni, Pb in ppm; the rest in percentage)

5.4.5 The Relationship Between Element Geochemistry and Paleoclimatology

In order to explore the evolutionary trend of chemical elements in the loess section, calculations were made by using the five-term moving average analysis. As shown in Fig. 5.13 elements such as Ca, Sr, Fe^{2+}, etc. in the loess section show an evolutionary tendency increasing from the bottom to the top, and $CaCO_3$ also tends to increase in the manner of rhythmical fluctuation; Fe^{3+}, Al, K, Mn and Ti tend to decrease in the same direction. Such an evolutionary trend of the chemical elements in the loess is related not only to the distribution of mineral components and the variation in grain size of loess, but also to the paleoclimatic environment and geochemical medium conditions. This general tendency of the evolution of chemical elements is considered to be related to climatic change from wetter to drier after loess accumulation.

In this general trend of evolution, Al, Fe^{3+}, K, Mn and Ti are remarkably increased, whereas Ca and Sr are highly decreased in paleosol layers S_1, S_5, S_8, S_{10} and S_{14}. Clearly, the variation curves for the chemical elements in these paleosol layers are of obvious fluctuation, in good coincidence with the cycle of sedimentation or the stratigraphic boundary.

5.4.6 Medium Conditions of Loess Accumulation

The pH values of the Luochuan loess section vary from 7.5 to 8.6, indicating a weakly alkaline medium condition. As illustrated in the distribution curves, the variation of pH in the loess section is in pace with that of $CaCO_3$.

The Eh values of the Luochuan loess fall into the range 420–470 mV, with an average of about 440 mV, marking an oxidating environment.

In addition, loess has a high content of $CaCO_3$, but a low content of organic matter. Amino acids present in loess are predominated by acidic ones. The K_i

values in the loess section range from 6.48 to 9.65, with a tendency to increase from the bottom upwards.

The above-described geochemical indices are a clear indication that loess was deposited in arid and semi-arid areas in an oxidating environment under weakly alkaline medium conditions. These indices also show that oxidation and weathering were much stronger during the accumulation stage of the Wucheng Loess before 1.20 – 1.40 m.y. than those during accumulation of the Lishi Loess and Malan Loess of 0.01 – 1.20 m.y.

In the loess section, the argillic horizons of paleosol are characterized by low $CaCO_3$ and pH, and relatively high Al_2O_3. Of the clay minerals, illite contains more montmorillonite layers. The top of the paleosol contains fossil assemblages of terrestrial snails represented by *Metodontia hausaiensis*, which reflect a warmwet ecological environment. This indicates that the paleosol was formed in relatively wet forest-prairie terrain under weakly acidic-weakly alkaline medium conditions.

5.5 Rare-Earth Elements

In this section the focus of the study is placed on the REE contents, REE distribution patterns and relevant REE ratios in loess and its different particle-sized fractions from the Luochuan section, Shaanxi Province, and a comparison has also been made of the REE distribution characteristics of Luochuan loess with those of other sediments of different genetic types so as to gain a deeper understanding of the formation conditions and material source of loess.

Ten loess samples were analyzed, taken from the studied section, including two samples from the first and second sandy loess. In addition, three samples of paleosol each from the first, the fifth and the eighth palesol, and three samples collected from the buried weathered-layer in the Wucheng Loess were also analyzed. For comparison, one sample of lacustrine sediment from the Qingchi, Luochuan, and one sample of modern moraine from the Tailan River, Uygur Autonomous Region of Xinjiang, were also analyzed.

Instrumental neutron activation analysis (INAA) was adopted in the analysis of the samples. The analytical precision of this method is about 10%.

5.5.1 Content and Distribution of Rare Earth Elements in Loess

Listed in Table 5.9 are the contents of individual rare earth elements and their total amounts ($\sum$ REE) in various samples $\sum$ REE values for loess of various periods vary within a range of 160 – 190 ppm. The small range of variation of REE indicates the homogeneity of REE distribution throughout the loess section. The average REE content of loess in close to that of sedimentary rocks. Results of loess analysis from the Luochuan section indicate that there is a general tendency for REE to decrease from old to young samples. This feature is related to the grain size and mineral composition of loess throughout the loess section. In paleosol REE varies within a range of 190 – 210 ppm, much higher than that in loess.

Table 5.9. REE analysis results of Luochuan loess and other sediments (ppm)

Sample[a]	horizon	La	Ce	Pr	Nd	Sm	Eu	Gd	Tb	Dy	Ho	Er	Tm	Yb	Lu	ΣREE	LREE/HREE	Eu/Sm	Sm/Nd	Eu/Eu	Ce/Ce
$80M_3$	L_1	30.9	61.4		41.4	5.88	1.20		0.742					2.83	0.462	168	7.84	0.204	0.142	0.752	0.838
$80M_{13}$	L_1	10.1	56.9		39.5	5.67	1.21		0.794					2.74	0.465	161	7.46	0.213	0.144	0.759	0.804
$80L_7$	L_2	31.3	59.3		43.0	5.77	1.11		0.740					2.65	0.470	167	8.03	0.192	0.134	0.704	0.795
$80L_{28}$	L_5	32.8	63.5		29.9	6.17	1.16		0.803					2.89	0.475	171	8.06	0.188	0.155	0.686	0.849
$80L_{38}$	L_6	31.8	60.0		39.6	6.10	1.15		0.861					2.49	0.493	166	7.74	0.189	0.154	0.665	0.819
$80W_7$	L_{11}	34.0	63.0		45.8	6.43	1.17		0.917					2.77	0.482	180	7.87	0.192	0.140	0.641	0.780
$80W_{35}$	W_{L-1}	33.5	71.9		40.6	6.92	1.40		0.973					2.65	0.487	184	8.14	0.202	0.170	0.715	0.942
$80W_{49}$	W_{L-2}	31.9	67.6		40.0	6.63	1.30		0.703					2.59	0.472	177	7.45	0.196	0.166	0.767	0.918
$80L_{35}$	L_9	31.5	62.4		42.4	6.03	1.15		0.805					2.74	0.491	171	8.14	0.191	0.142	0.687	0.834
$80W_{24}$	L_{15}	31.5	58.1		40.4	5.88	1.08		0.713					2.63	0.487	163	8.19	0.184	0.146	0.685	0.792
Average		31.9	62.4		41.3	6.15	1.19		0.805					2.70	0.478	171	7.89	0.194	0.149	0.706	0.837
80Q-1	Lacus.	28.4	58.0		32.5	4.78	1.20		0.882					1.91	0.399	151	7.88	0.208	0.178	0.712	0.916
TR-1	Moraine	50.8	108		61.0	10.4	1.54		1.65					4.11	0.833	278	7.66	0.148	0.170	0.250	0.937
OC-1	shell I	5.9	15.0	2.6	13.0	4.1	1.3	5.5		5.2		3.6		3.3		61		0.317	0.315	0.959	0.841
ES-1	Marine	10.6	45.8	5.42	20.3	4.27	0.82	4.23	0.58	3.22	0.66	1.30	0.15	1.57	0.12	109	8.22	0.192	0.210	0.663	0.957
	F. dust	25.7	32.0	6.83	24.7	4.27	1.10	3.43		3.75	1.4	2.55		2.10		108		0.258	0.173	0.961	0.534
$80L_2$	S_1	35.6	73.3		43.7	6.68	1.28		0.914					2.96	0.526	192	8.03	0.192	0.153	0.685	0.900
$80L_{31}$	S_5	37.1	71.2		51.6	7.21	1.35		0.907					3.12	0.567	201	8.19	0.187	0.140	0.689	0.797
$80L_{49}$	S_8	36.8	72.2		47.9	6.97	1.33		0.973					3.06	0.537	197	7.93	0.191	0.146	0.678	0.838
$80W_{28}$	W_{S-1}	32.8	66.2		37.0	6.62	1.33		0.921					2.61	0.473	173	7.42	0.201	0.179	0.716	0.910
$80W_{40}$	W_{S-2}	33.0	65.3		35.6	6.71	1.34		0.849					2.57	0.474	170	7.57	0.199	0.188	0.734	0.907
$80W_{57}$	W_{S-3}	30.3	62.8		35.3	6.53	1.27		0.928					2.28	0.466	164	7.36	0.194	0.185	0.688	0.924

[a] 80Q-1 collected from Qingchi, Luochuan; TR-1 from Tailan River Velley; OC-1 from Oceanic crust; ES-1 from continental shelf of East China Sea; others from Luochuan section.

Table 5.10. REE partition in different grain size fractions of loess

Grain size		REE content (ppm)								
Original No.	(mm)	La	Ce	Nd	Sm	Eu	Tb	Yb	Lu	ΣREE
$80M_3$	whole rock	30.9	61.4	41.4	5.88	1.20	0.742	2.83	0.462	168
$80M_{3-1}$	>0.063	7.13	11.9	7.04	1.25	0.323	0.148	0.620	0.108	33
$80M_{3-2}$	<0.02	31.2	63.6	42.2	6.11	1.26	0.827	2.78	0.436	173
$80M_{3-3}$	<0.001	31.0	51.2	31.1	6.27	1.25	0.827	2.73	0.463	148
Gu-26	clay	41.5	69.7	54.7	9.68	1.97	1.22	3.00	0.579	215

In order to explore the existing forms of rare earth elements in loess, we analyzed the REE contents of different grain-sized fractions of the Malan Loess. Analysis results (Table 5.10) indicate that the minimum REE occurs in >0.063 mm fractions, indicating the depletion of REE in this grain-sized fraction. The existence of a tiny minority of minerals containing rare earth elements is suggested for the same grain-sized fraction. Moreover, in the >0.001 mm fraction the REE content is similar to that of the whole rock of loess. This seems to suggest that only a small amount of REE has been absorbed on clay minerals in loess. Thus in loess REE are concentrated mainly in detrital minerals present in the silt-sized fraction. Further studies are needed.

5.5.2 REE Distribution Patterns and Characteristics of Luochuan Loess

Figure 5.14 indicates that the Malan, Lishi and Wucheng Loesses are chiefly rich in the Ce-family elements. Their REE distribution patterns are all similar, with negative slopes. These similar REE distribution patterns indicate the same material source for the Malan, Lishi and Wucheng Loesses, though they accumulated in different periods.

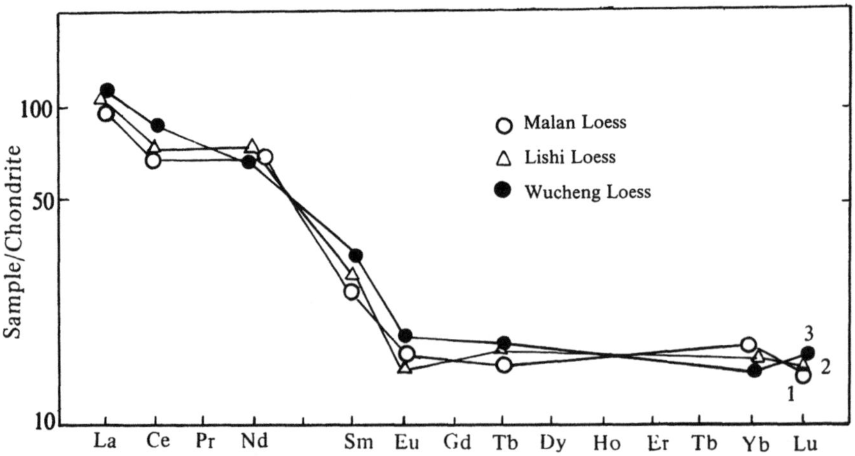

Fig. 5.14. REE distribution patterns in loesses of different periods

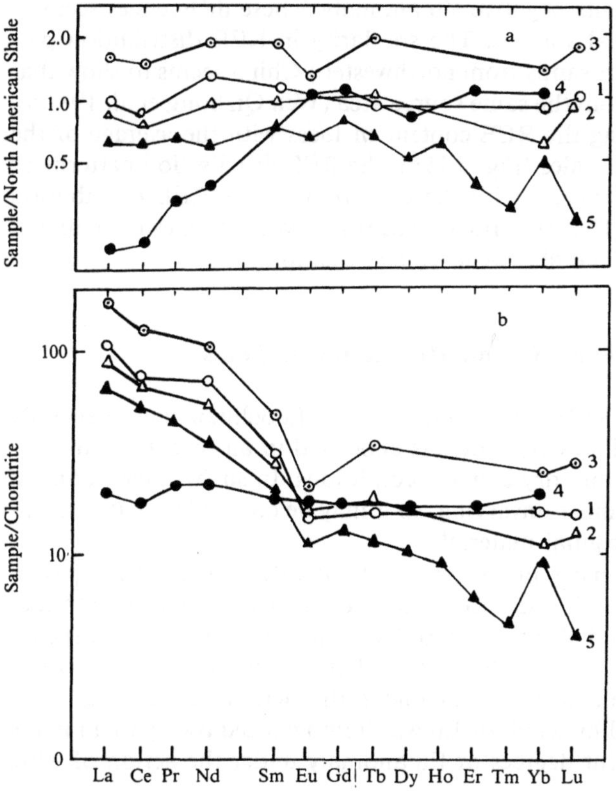

Fig. 5.15. REE distribution patterns in loess and other sediments. *1* Loess; *2* lacustrine sediment; *3* modern moraine; *4* oceanic crust; *5* East China Sea continental shelf

In the Luochuan Loess section, the REE distribution patterns of paleosols are also similar to one another, as well as to those of the whole rock of loess (Wen Qizhong et al. 1984). This shows that both paleosol and loess have the same material source. Paleosol still retains the REE distribution characteristics of the original material – loess, indicating no intense REE fractionation during pedogenesis. From the REE distribution characteristics of loess, it may be concluded that the formation of paleosol in loess section represents a short intermission of oxidation (or weathering) during the process of loess accumulation rather than another sedimentary event.

For comparative study, sediment samples of other genetic types (e.g. lacustrine sediment, moraine, etc.) were also analyzed for rare earth elements. Figure 5.15b shows that all the REE distribution patterns of various sediments are characterized by negative slopes with the exception of the REE distribution curve for the oceanic crust, but the differences lie in the extent of REE fractionation in various types of sediment. The modern moraine exhibits remarkable negative Eu anomaly and the deposit of the East China Sea continental shelf also shows negative Eu anomaly (Wang Xianjue et al. 1982). The extent of REE (the Y-family) fractionation is greater than that in loess, showing remarkable positive Yb anomaly; but the lacustrine sediment exhibits negative Yb anomaly. These differences seem to reflect their different material sources. The similarity in REE distribution patterns between loess and desert sands from northwestern China seems to show that both of them were derived from the same source area (Wen Qizhong et al. 1981 b).

In addition, by comparing the REE content of loess with the average of the REE in 40 North American Shales (Fig. 5.15 a), the REE distribution patterns of loess are relatively smooth curves, with the exception of the REE distribution curve for the oceanic crust. The REE fractionations in various types of sediment are also similar to those of the above-mentioned example.

5.5.3 Geochemical Implications of Some REE Ratios in Loess

Table 5.9 shows that the LREE/HREE ratio in the Luochuan loess generally ranges from 7.5 to 8.2. We know that loess is deposited in an oxidizing environment under weakly alkaline and dry climatic conditions. In such an environment the LREE/HREE ratio is almost unvarying, so it may be concluded that loess inherits the features of the original material.

There are marked differences in Eu/Sm ratio for different genetic types of sediment. The ratio of Eu/Sm in loess from the Luochuan section is close to that of sedimentary rocks (0.2) and also similar to that of fallen atmospheric suspended particles (Wen Qizhong and Yu Suhua 1979). This seems to show that loessic materials entered the sedimentary system under the action of running water before they were transported by wind, or Eu was homogenized owing to a mixing of the materials with different degrees of Eu anomaly under the action of wind as the carrier.

In REE distribution patterns of loess the shape of Eu is close to a smooth pattern, indicating that loess has no evident negative Eu anomaly as its parent material. This seems to reflect an arid and semi-arid bioclimatic environment at the time of loess accumulation, even remarkable negative Eu anomaly did not take place in the paleosol. These observations show that although the temperature was high in the period of paleosol pedogenesis, the humidity at that time was not as high as that in South China, where laterite ws formed with distinct negative Eu anomaly (Tang Songliu et al. 1980).

The ratios of Ce/Ce in the Luochuan loess section are all less than unity, indicating that loess inherited the features of the original material. This then seems to indicate that loessic materials in the source area were also formed in an arid and semi-arid bioclimatic environment under weakly alkaline conditions.

5.5.4 Correlations Between REE and Elements of Fe, Al and Ca in Loess

From calculated correlation coefficients we know that La, Ce, Sm, Eu, etc. are positively correlated with TiO_2, Al_2O_3, Fe_2O_3, and K_2O, but negatively correlated with FeO and CaO.

The results of regression analysis also show the same correlations. For example, the regression coefficient of Fe_2O_3 is 19.30, and its correlation coefficient (γ)

is 0.91, indicating a relatively good linear correlation between REE and Fe_2O_3. A regression coefficient of −4.14 for CaO indicates a negative correlation between REE and CaO (Wen Qizhong et al. 1984).

5.6. Amino Acids

Amino acids are always present in the lithosphere, the hydrosphere and the atmosphere, and even in extraterrestrial substances such as meteorites. The distribution, origin and evolution of amino acids in a geologic body and their geochemical implications have become major research subjects in the fields of biogeochemistry, organic geochemistry, paleontology, geochronology and the Quaternary research.

The aim of this section is to explore the biogeochemical significance of amino acids in the study of stratigraphic correlation, depositional environment, paleoclimate, through investigating the distribution of amino acids in loess.

5.6.1 Sampling and Analytical Preparation

The samples were collected from the Heimugou section at Luochuan. In order to examine the variation of amino acid composition, 32 samples were collected from the upper and the lower sections, including Pleistocene loess of different stages, such as Malan Loess, upper Lishi Loess, lower Lishi Loess, Wucheng Loess and a sample from Pliocene Red Clay. The samples were desiccated and pulverized after their outer layers had been peeled off. Ten grams of pulverized samples were placed in the solution of 5 N HF−0.1 H HCl and followed by hydrolysis with 6 N NCl solution at 110 °C for 24 hours under refluxing (Blunt and Warnke 1980).

Since a large amount of inorganic salts in the hydrolyzed solution can interfere with the identification of amino acids, they must be removed from the hydrolyte by filtration and ion exchange column with Dowex 50W-X8 cation resin. The 835−50 model amino acid analyzer equipped with a 2.6×150 mm column packed with 2619 resin was employed to determine amino acids. Running conditions of the automatic amino acid analyzer include buffer pressure 80−110 kg/cm^2, ninhydrin pressure 20−35 kg/cm^2, temperature of column 53 °C, and time period 73 minutes.

5.6.2 Results and Discussion

Results of this study show that more than 20 kinds of amino acid such as aspartic acid, threonine, serine, glutamic, glycine, alanine, cysteine, valine, methionine, isoleucine, leucine, tyrosine, phenylalanine, lysine, histidine, arginine, etc. were isolated from samples of the Luochuan section. The samples from different beds contain different kinds and amounts of amino acid. The results of this study also

Table 5.11. Amounts of amino acids in loess of different ages

Name of stratum	Depth (m)	Total amount of amino acids (μg/g)	Amount of acidic amino acids Asp, Glu (μg/g)	Amount of alkaline amino acids Lys, His, Arg (μg/g)	Ratio between acidic amino acids and alkaline amino acids
Malan	0.8 – 8.7	152.84	56.00	12.09	4.7
Lishi (upper)	9.1 – 34.5	63.03	11.00	5.60	2.2
Lishi (lower)	36.5 – 77	51.82	44.07	2.99	1.3
Wucheng	84.5 – 124	43.6	3.82	2.22	1.7

Table 5.12. Amount of amino acids in Clear Lake sediment (Blunt 1981)

Depth (m)	14.57	15.54	27.43	79.55
Total amounts (*n*Mol/wet sample 1 g)	8674	4114	2093	329

Table 5.13. Organic carbon content in Luochuan loess

Sampling No.	L_7	L_{17}	L_{22}	L_{24}	L_{38}	L_{44}	L_{47}	W_7	W_{15}	W_{16}	W_{32}	W_{40}	W_{57}
Organic carbon	0.08	0.05	0.07	0.07	0.03	0.08	0.07	0.09	0.08	0.09	0.08	0.07	0.05

show that as a general rule the total amount of amino acids decreases with increasing burial depth relative to samples. In other words, the older the sample is, the smaller the total amount of amino acids the sample will contain (Table 5.11). Table 5.11 shows that the average total amounts of amino acids for the Malan Loess are 152.84 μg for the depth less than 8.7 m, 63.03 μg for the upper Lishi Loess 9.5 – 34.5 m deep, 51.82 μg for the lower Lishi Loess 36.5 – 77 m deep, and 43.6 μg for the Wucheng Loess beneath 84.5 m. The variation rule of amino acids is coincident with the former results studied by other scientists. For example, an amino acid study of sediments of Clear Lake (Blunt 1981) has shown that the total amount of amino acids decreases with the increase of burial depth relative to samples (Table 5.12). In addition to the analysis of amino acids, the organic carbon content was also analyzed in a number of loess samples from the upper to the lower of the Luochuan section in order to confirm the variation of the rule of amino acids. Table 5.13 shows that with increasing depth the percentage of organic carbon, relative to the samples, has a slight variation which does not follow any rule. This indicates that there is not much difference in amount of organic matter among the samples of the whole Luochuan section. Therefore, the fact that the total amounts of amino acids decrease with the increase of burial depth of samples is mainly due to the diagenesis of organic matter.

We note that the amounts of aspartic acid, glutamic, threonine and serine in loess appear to decrease rapidly with the increase in their burial depths and are related to the age of loess. The degree of loss for each amino acid is different.

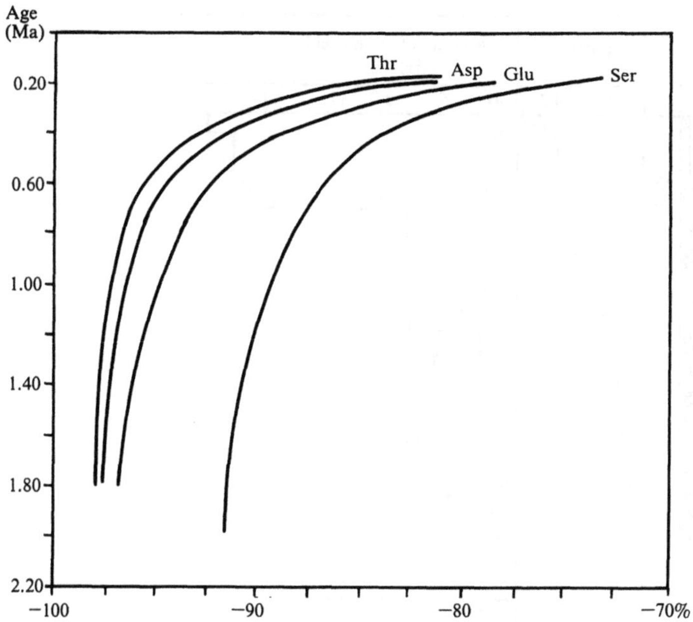

Fig. 5.16. Relation between percentage loess of aspartic acid, glutamic, threonine and serine and their ages

However, a common feature for these four amino acids is that the loss of each amino and began to decrease after the loess had accumulated over one million years. The relationship between the amino acid content contained in loess samples of different age and the percentage loss of amino acids related to content in loess 10000 old is shown (Fig. 5.16). All amino acids lost 90%−97.5% after the loess age over one million years, and that the loss rate of amino acids with various ages is different for different amino acids. Threonine decreases most rapidly with increasing age and is followed by aspartic acid, glutamic and serine. The plot drawn from the percentage loss of the four amino acids versus loess age is significant, for it may be applied to the dating of loess in China. As Quaternary loess is widespread in northwestern China, and its primary sedimentary environment, climate and development were quite different, loess in different localities might contain diverse amounts of amino acids. Thus, it is impossible to date loess by the amount of amino acid loss. However, as the percentage loss of amino acid with increasing age is independent of the initial amino acid abundance in loess, the plot might be used to date loess after the percentage loss of amino acids in the loess is determined. Although the possibility of applying this method needs further confirmation, it is indeed an interesting problem worth exploration.

Although the amount of amino acids in loess appears to decrease gradually from Malan Loess to Wucheng Loess, a slight variation of the amount of amino acid in the section is noted. The variation is found to be related to the occurrence of paleosol in the section. The results indicate that the amount of amino acids in paleosol is always greater than that in the nearby loess. The distribution of

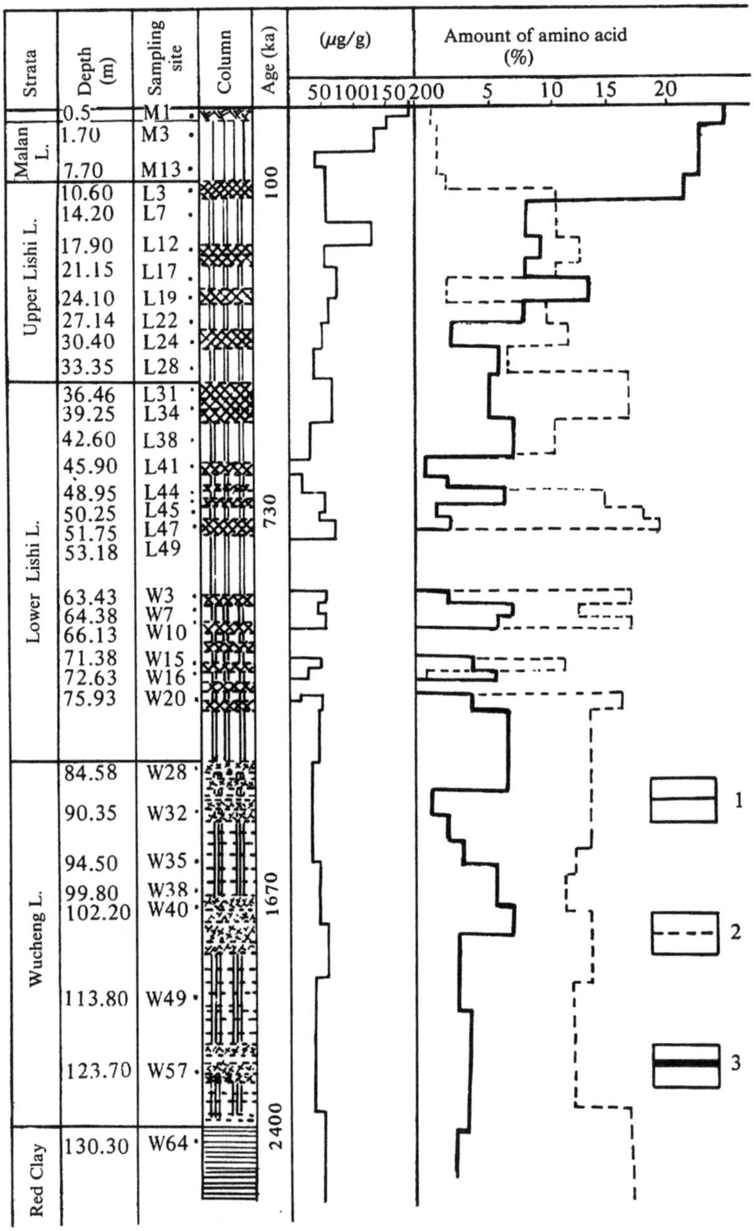

Fig. 5.17. The total amounts of amino acids and the percentage of phenylalanine and aspartic acid in the Luochuan section

amino acids in the section and their variation are shown in Fig. 5.17. The differing amounts of amino acids in loess and paleosol may be due to the different development of organisms during loess accumulation and paleosol formation. This difference in development of loess and paleosol organisms may be explained by the diverse climatic environments in which loess and paleosol formed. Previous studies indicate that the regions of deposition and supply for loess were all in oxidation environments with arid and semiarid climates during loess sedimentation[10], and the paleontological fossils in loess confirm that the ecological environment of the loess deposition region in the period of loess formation was steppe (Liu Tungsheng et al. 1966). But the paleosol was formed from loess after the loess had been weathered by chemical and physical processes. The warm, humid environment with relatively intense weathering in which the paleosol was formed was favorable for the development and preservation of organisms. Therefore, the amino acid abundance difference between loess and paleosol shows that organism development in loess was different from that of paleosol during its formation. The environment and climate which exerted an influence on organism development in loess was different from that in paleosol during its formation.

It is well known that amino acids, according to the number of the amino group in the molecule relative to the number of the carboxyl group, can be classified as acidic, alkaline and neutral. The alkaline environment of a sediment is favorable for the preservation of acidic amino acids. The average amounts of acidic amino acids and alkaline amino acids in the Wucheng Loess, Lishi Loess and Malan Loess are shown in Table 5.11. The fact that acidic amino acids are present in relatively greater amounts than alkaline amino acids in the whole Luochuan section indicates that the depositional environment of Luochuan section has a bias to an alkaline environment. This agrees with the geochemical medium condition conclusions of loess deposition drawn by Liu Tungsheng et al. (1981). The variation of acidic amino acids throughout the Luochuan section, from Malan Loess to Wucheng Loess, shows that the decreasing rate of acidic amino acids is far higher than that of alkaline amino acids. Although the environment of the whole Luochuan section has an alkaline bias, the ratio of acidic amino acids to alkaline amino acids is 4.7, 2.2, 1.3 and 1.7 respectively from the younger loess layers to the older ones. This suggests that environments of various layers are not the same, and the alkaline nature of depositional environment becomes weak gradually from Malan Loess to Wucheng Loess.

Neutral amino acid amounts in Lishi Loess and Wucheng Loess, such as tyrosine and phenylalanine, are rather high. An exception is found in Malan Loess. The variations in phenylalanine and aspartic acid amounts in the Luochuan section are plotted in Fig. 5.17. By comparing phenylalanine acid with aspartic acid in the older loess, it is obvious that the older loess contains higher amounts of phenylalanine, while the amounts of aspartic acid tend to decrease (Fig. 5.17). As protein of different organisms may consist of various kinds of amino acids, the high amounts of tyrosine and phenylalanine in the Wucheng Loess and Lishi Loess may reflect the development of organisms whose protein

[10] According to the loess in the middle reaches of the Huanghe River with the exception of loess of other subtypes.

composition is of more phenylalanine and tyrosine. In addition, slightly higher amounts of phenylalanine are associated with the paleosol in the section. This relationship implies that paleosol was formed under a warm, humid climate which was favorable for the development and growth of the specific organism whose protein might contain more phenylalanine.

This study represents the first attempt to understand the occurrence of amino acids in the loess of China. We find that the amount of amino acids in the loess of the Luochuan section decreases with the increase of the burial depth of loess, that the amount of amino acids in paleosol is higher than that in nearby loess, and that the rate of acidic amino acids decrease from the upper to the lower part in the Luochuan section is greater than that of alkaline amino acid decrease. The loess of the Luochuan section, with the sole exception of Malan Loess, has more tyrosine and phenylalanine. All the characteristics of amino acid distribution in the Luochuan section loess indicate that diagenesis of amino acids becomes intensified from younger loess to older loess, and that the Quaternary climate not only experienced fluctuations from dry and cold to humid and warm, but also changed gradually to a drier climate from the Early Pleistocene to the Late Pleistocene. The results of the present study are in accord with the paleoclimatic change reflected by the increase of $CaCO_3$ from the upper loess to the lower loess of the Luochuan section and the ratio of oxides in the Luochuan section as well as the ratio of F/Cl (Wen Qizhong et al. 1981b; Liu Yulan et al. 1981).

5.7 Paleosol

Paleosol (or fossil soil) as a product of the natural environment in the geological or historical periods indicates the effects of various factors on the earth's surface. These include soil-forming factors such as parent material, climate, biocommunity, topography, including hydrographical condition, and time of soil formation. Paleosol can be considered as a good geological record of paleoclimate and paleoenvironment.

The history of paleosol investigation in the loess of China reflects the gradual understanding of paleosol and its significance for geological environment. J. Thorp (1935) first noted the existence of red-brown paleosol in loess of northwestern China. Song Daquan (1950) studied the paleosol in the Xiashu loess, Nanjing. Zhu Xianmo (1958), Ma Yungchi (1958), Shi Yuanchun (1958) and Liu Tungsheng (1959) confirmed that red earth with "red bands" is old loess containing buried soil. Zhu Xianmo (1965) and Liu Tungsheng et al. (1964) classified paleosol into two basic types of black loam and cinnamon soil from a viewpoint of pedogenesis. Zhu Xianmo holds that loess is formed under the climatic condition of dry-cold, but the cinnamon paleosol represents warm-slightly humid climate. Zhu Haizhi (1963), An Zhisheng and Wei Lanying (1978, 1980) and Tang Keli (1981) investigated the micromorphology of paleosol in the loess of China. Lu Yanchou and An Zhisheng (1979) subclassified the paleosol of Luochuan cinnamon soil series into five different paleosol categories. An Zhisheng and Wei Lanying (1980) suggested that paleosol S_5 of the brown cinnamon soil category with strong pedogenesis in the Loess Plateau could be used as an independent stratigraphic unit.

Sections from the middle reaches of the Huanghe River, where loess and paleosol alternately occur, constitute a loess-paleosol sequence from the Pleistocene to the Holocene. Additionally, westward to the northern slope of Tianshan Mountains, Xinjiang, eastward to Miaodao Islands between the Huanghai Sea and the Bohai Sea, and northward to Huangshan of Harbin, as well as southward down to middle-lower reaches of the Changjiang River, where loess is developed, associated paleosols are present.

The occurrence of paleosol is conditioned by the paleotopographical surface. On the loess "Yuan", paleosols occur horizontally or a little out of the perpendicular in the loess section (Fig. 3.2). In the loess "Liang" and "Mao" areas and small intermontane basins, paleosols occur usually inclined or wavy in section as the underlying topography varied. The occurrence of unconformable contact between paleosols in valley districts is mainly characterized by "overlapping". Paleosol phenomena of turning, branching, superimposing or converging often appears at contact locale on different terraces in the river valley.

In the loess-paleosol sequence, two or more than two paleosols are often superimposed upon another. Such an overlapping of paleosols can be called a paleosol complex. Paleosol S_5 of Luochuan is over 5 m in thickness and consists of three layers of paleosols sometimes sandwiched by two thin loess layers, indicating the development of a paleosol complex (Fig. 5.18). Sometimes the paleosols are transitional to the loess overlying in grain size and micromorpho-

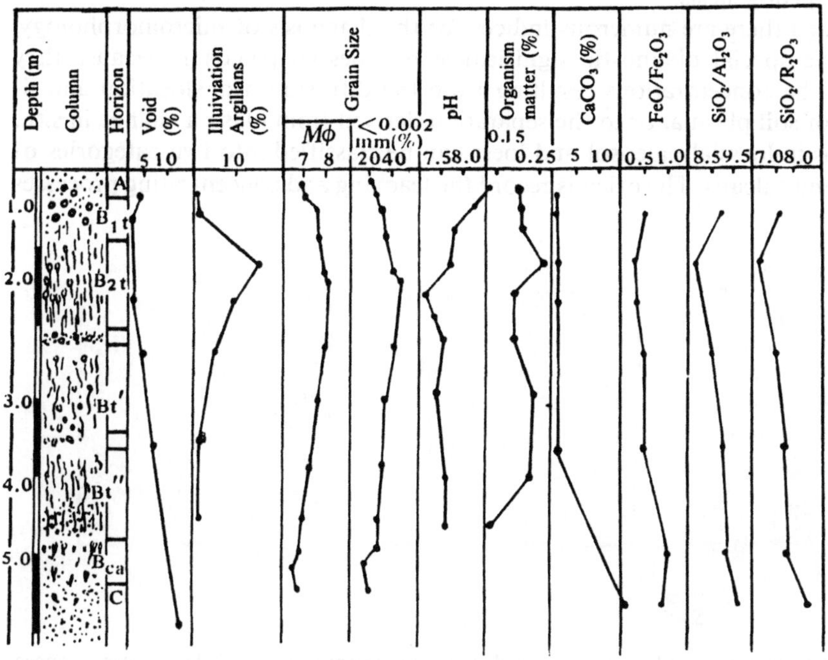

Fig. 5.18. Comprehensive section of Luochuan paleosol S_5

logical characteristics. Typical loess, more or less, is marked by traces of weak pedogenesis. Paleosol is the result of strong pedogenesis. Therefore, loess layers or paleosols could be explained as the result of the prevalence of the processes of accumulation of dust (loess) on the one hand and soil formation (degree of pedogenesis) on the other hand.

5.7.1 Micromorphology of Paleosol and Loess

R. Brewer (1964) described the physical constitution of soil material reflected by the spatial arrangement of soil particles and associated void, as soil fabric. Loess generally exhibits a granular fabric, in which skeleton grains are connected with voids and grains are directly in contact with one other or a few plasma are in joints, which corresponds to a cementation type of contact-embedding (Fig. 5.19).

Strongly weathered loess and paleosol are generally characterized by porphyritic fabric. In the fabric, coarse grains such as skeleton grains, etc. are connected with plasma, leaving small voids between grains. Distribution of skeleton grains (in the form of porphyritic crystal) in plasma are present as well, forming a fabric similar to the embedded type of cementation. If clayey plasma is increased, or if argillans occur in large quantity, and skeleton grains seem to be embedded by materials in the form of clayey colloids, such a fabric is called a colloporphyritic fabric. The number of fine grain materials in the porphyritic or colloporphyritic fabric compared with granular fabric indicate the strengthening of weathered pedogenesis.

Although there are numerous indices for the diagnosis of micromorphology, cutans have special diagnostic significance for loessial paleosol, because they belong to the concentrations resulting from the diffusion, translocation and illuviation of soil plasma, or to the separation derived from plasma change in situ. The cutans in loessial paleosol and loess can be classified into two categories of argillans and calcans. The calcans record the leaching and concentrating processes

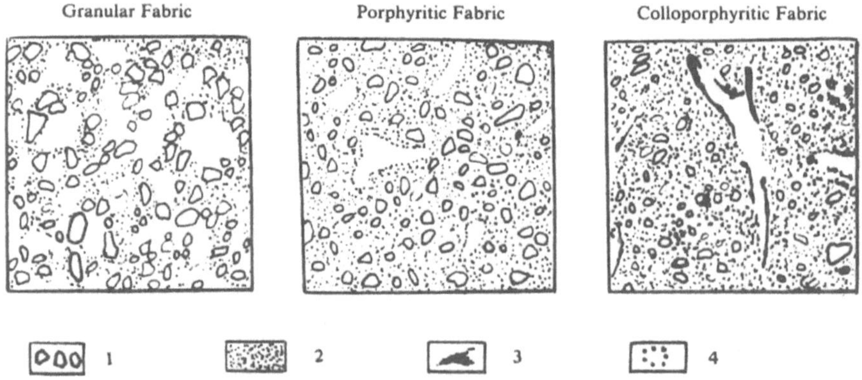

Fig. 5.19. The fabric types of loess and paleosol. *1* Skeleton grain; *2* plasma; *3* clayey plasma concentration; *4* voids or vughs

of carbonate during soil formation, whereas argillans record an argillation in situ or translocation and illuviation of clay grains.

5.7.2 The Classification of Paleosols

Rate of weathering as expressed by micromorphology, chemical composition, clastic minerals and clay minerals as well as development characteristics of the soil profiles in the Luochuan section are an important index for the classification of paleosol. The classification of loessial paleosol categories depends on the understanding of acid-alkaline reaction and decalcification or illuviation of carbonate in different horizons of paleosol, argillation of paleosol, residual argillation in situ or mechanical movement and concentration of clay particles as recorded by the type and quantity of argillans. On the basis of comparison of pedogenetic characteristics between Luochuan paleosols and recent soils such as black loam, cinnamon soil and brown soil, Luochuan paleosols (S_0-S_{14}) since 1.1 m.y. be classified into five different categories (Fig. 5.20) as follows: brown cinnamon paleosol (S_5), leached cinnamon paleosol (S_1, S_4, S_7), cinnamon paleosol (S_2, S_6, S_8, S_9, S_{11}, S_{13} and S_{14}), carbonate cinnamon paleosol (S_3, S_{10} and S_{12}) and black loam (S_0). In brown cinnamon paleosol S_5, the upper and the lower argillic horizons (Bt and Bt″) are characterized by dark reddish brown color, stable nutlike and prismatic structures, a high content of clay fraction and dark brown coatings distributed on the walls of voids and vughs as well as on the surfaces of vertical prismatic body. The middle argillic horizon (Bt′) has a reddish-

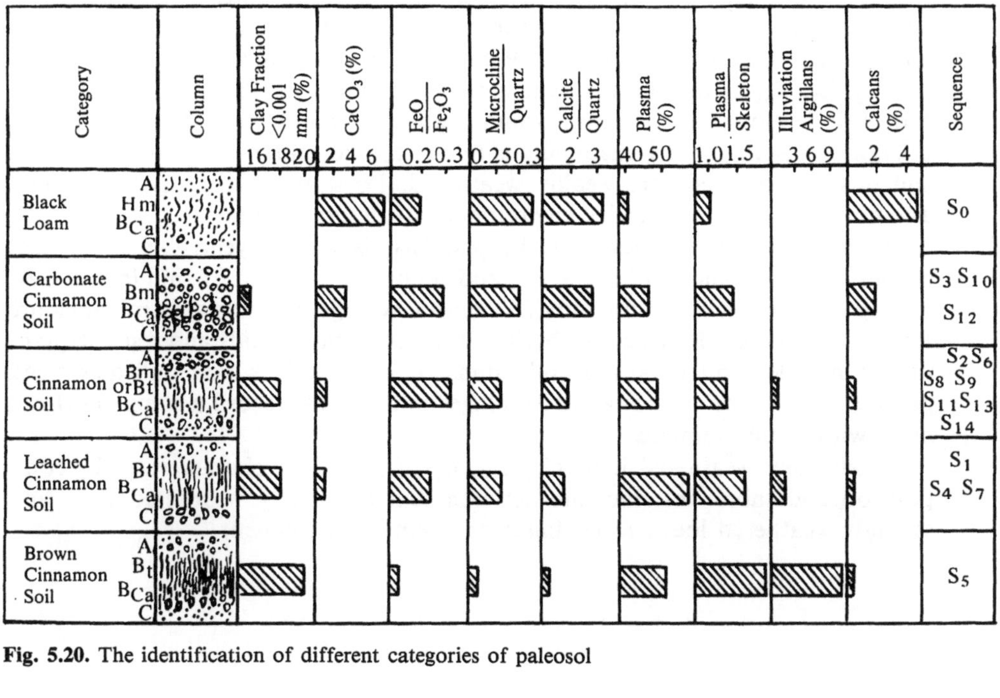

Fig. 5.20. The identification of different categories of paleosol

drab color and prismatic structure, but the content of clay fraction is lower. In Bt horizon, contents of clay fraction, Fe_2O_3 and Al_2O_3 are higher, carbonate has been leached out, and illuviated ferriargillans in forms like sinter or fluid colloform occupy 12.8% of the thin section area, which indicates apparent translocation and illuviation of clay grains and channelled colloporphyritic fabric. Samples from Bt horizon of S_5 with highest magnetic susceptibility in Luochuan paleosols reflect the movement of iron. But the horizon is weakly alkaline and the movement of iron sesquioxides within the soil is not as apparent as in the recent brown soil in Shandong Province. Sometimes, illuviated argillans are decreased within Bt′ and Bt″ horizons where diffusion and stress cutans are increased; thus paleosol S_5 is not only different from cinnamon soils which contain carbonate and have undergone an argillic process, but also from brown soils in which obvious mechanical movement and concentration of clay material and sesquioxide have taken place under acid conditions. Paleosol S_5 should be considered as a transitional type between brown soil and cinnamon, and is thus called brown cinnamon soil. Brown cinnamon paleosol S_5 records the climatic optimum about 0.5 m.y. on the Loess Plateau in the past 1.1 m.y.

5.7.3 Weathered Types of Loess

The composition and fabric of individual loess layers record different intensities of physical and chemical weathering and pedogenic formation from which aeolian dusts with silt-sized grains endured during accumulation. Thus the history of differences in the depositional environment of loess are preserved. Features such as grain size and chemical composition, $CaCO_3$ content and micromorphology of different loess layers that mirror weathering intensity, Luochuan loess formation in the last 1.1 m.y. can be generally graded. Weakly weathered loess (L_1, L_2, L_5, L_6, L_9 and L_{12}) is light grey-yellow in color, silty, porous and loose with well-developed vertical joints and a granular fabric. Carbonate content is 12% on an average. Carbonate cutans reflect the Secondary carbonation in situ and unstable minerals are weakly weathered. Loess containing *Cathaica* snail assemblages record a cold-dry environment of desert-steppe.

Strongly weathered loess (L_7, L_8 and L_{10}) is reddish-yellow in color, less porous, fine silty and hard in nature, and occurs in the form of thin layers. Carbonate content accounts for only 5% on an average and unstable minerals are strongly weathered. It has a porphyritic fabric containing voids and a small quantity of nodules and cutans or subcutans of carbonate, which indicates weak pedogenesis. *Metodontia* snail assemblages in this loess record a steppe environment with a mild climate.

Moderately weathered loess (L_3, L_4, L_{11}, L_{13}, L_{14}) is characterized by composition and fabric intermediate between that of weakly weathered loess and strongly weathered loess, reflecting a dry-steppe environment.

5.7.4 Paleoenvironmental Significance of Paleosol and Loess

The characteristics of modern geographical distribution of recent cinnamon soil series can be used for a discussion of the paleoenvironmental significance of Luochuan cinnamon soil series. Recent black loam is the zonal soil developed in the semi-arid steppe in warm-temperate zones in northern China. The vegetation in the recent zone of cinnamon soil is dominated by semixerophilus forest and steppe vegetation. In the zone of recent brown soil, called brown forest soil, the vegetation is characterized by deciduous broadleaf and xerophilous forests. The annual mean value of climatic parameters of the zones of recent black loam, cinnamon soil and brown soil of Shandong Province (Institute of Pedology, Academia Sinica, 1978) naturally reflects the paleoclimatic conditions recorded by buried black loam, cinnamon paleosol and brown cinnamon paleosol in Luochuan section respectively. The paleoclimate of carbonate cinnamon paleosol can be expressed by mean climatic conditions in the transitional zone from recent black loam to recent cinnamon soil. The mean climatic condition in the transitional zone from recent cinnamon soil to brown soil may be a representative of leached cinnamon paleosol. Based on the above assumption, it is possible to reconstruct bioclimatic zones and approximate values of climatic parameters in the past, which are recorded in five different paleosols of Luochuan (Table 5.14).

Table 5.14. Climatic record of Luochuan paleosol

Recent soil	Luochuan paleosol	Horizon	Bioclimatic zone	Past climate		Value difference[c]	
				A.M.T. (°C)[a]	A.M.P. (mm)[b]	A.M.T. (°C)	A.M.P. (mm)
Black loam	Black loam	S_0	Steppe	8–10	300–500	approximate	approximate
	Carbonate cinnamon soil	S_3 S_{10} S_{12}	Steppe to xerophilous forest	9.5–12	400–600	+2.5	+100
Cinnamon soil	Cinnamon soil	S_2 S_6 S_8 S_9 S_{11} S_{13} S_{14}	Xerophilous forest and steppe	11–14	500–700	+3.5	+200
	Leached cinnamon soil	S_1 S_4 S_7	Xerophilous forest	11.5–14	600–700	+4	+280
Brown forest soil	Brown cinnamon soil	S_5	Xerophilous forest to deciduous broad-leaf forest	12–14	700–800	+4	+350

[a] Annual mean temperature.
[b] Annual mean precipitation.
[c] Annual mean temperature (9 °C) and annual mean precipitation (400 mm) of recent black loam zone represent the mean value of recent climatic elements of Luochuan area, while that recorded by paleosol is mentioned in the text. The value difference means the difference between past and recent values.

As Luochuan is situated in the zone of recent black loam, recent mean climatic conditions of Luochuan are expressed by the annual average temperature and annual mean precipitation of the zone. Thus the comparison of the assumed mean climatic condition deduced from diagnostic features recorded in paleosols of various periods in the past 1.1 m.y. with those of the recent is of interests (Table 5.14).

Grey calcareous soil on the west side of recent black loam zone is the transitional soil zone from steppe to desert. Annual mean precipitation in this zone is about 200 (150) – 300 (350) mm (Nanjing Institute of Pedology, Academia Sinica, 1978). In a brown calcareous soil zone surrounding the semi-desert, the annual mean temperature is about 2 – 7 °C and the annual mean precipitation is about 100 (150) – 280 mm. It is assumed that the dry-humid degree in the transitional zone between brown calcareous soil and grey calcareous soil represents the mean dry-humid condition during the deposition of the loess or the weakly weathered loess which designated as a product of the desert-steppe and steppe environment. The precipitation of 200 mm can be taken as a minimal value during the deposition of the loess and the weakly weathered loess.

Late Pleistocene Malan Loess is typical weakly weathered loess. The climate deteriorated in the Dali glacial period of Late Pleistocene in China and serious deterioration of climate occurred 15 000 years ago. Yang Huairen and Xu Xin (1980) estimated that the annual mean temperature of the late glacial period in eastern China was 7 – 13 °C lower than that of the present, which is in accordance with the distributional record of *Picea* and *Abies* forests during late glacial period. Sun Dianqing et al. (1982) and Cui Zhijiu (1982) suggested that annual mean temperature of the Late Pleistocene was 9 °C lower than that of the present, based on pollen data and periglacial phenomena. Annual mean temperature of Weinan region in the south of the Loess Plateau at 23 000 – 20 000 yr B.P. was calculated to be 8 °C lower than that of the present by the Institute of Botany, Academia Sinica (1966), based on the geographical distribution of *Picea* and *Abies* forests, but the climate became more deteriorative in the late Dali glacial period. It seems 8 – 10 °C lower than present temperature as a changeable value of the late glacial interval in the Loess Plateau, and the annual mean temperature during the deposition of weakly weathered loess or Malan Loess was 0 °C or below 0 °C. The values of climatic elements recorded by moderately weathered loess and strongly weathered loess can be estimated interpolately from the range of temperatures and precipitations between that of weakly weathered loess and black loam.

In the Luochuan loess-paleosol sequence alternate occurrence of three types of loess in different weathering degress and five categories of paleosol with different pedogenetic intensities have recorded the history of long-term change of natural environment in the Loess Plateau in the last 1.1 m.y. The variational sequence of Luochuan climatic elements in the last 1.1 m.y. can be derived from the climatic values of different types of loess and paleosol (Fig. 6.6), providing a possible amplitude of long-term climatic change in the Loess Plateau. For example, the fluctuation of annual mean temperature reaches 12 – 14 °C or more and annual mean precipitation reaches 500 – 600 mm. A diagram of paleoclimatic reconstruction from data of paleosols and of snail, mole-rat and pollen fossils (see

Chap. 4) show that the paleoclimate of warm-humid intervals, i.e., forming of paleosols in the Loess Plateau, might resemble the recent mean climatic conditions to the north of a line of Qingling Mountains-Huaihe River, but not like that as the south of the Changjiang River.

CHAPTER 6

Loess and the Environment

6.1 Dust Deposition and the Modern Dust Storm

The entire process of occurrence-development-extinction of dust storms, namely the weather process of dust storms, has provided a basis for the geological investigation of wind processes of loess (uplift, transportation and deposition). The weather process of dust storms serves as the physical basis for the aeolian theory of loess origin.

China is one country where dust storm events were documented in earliest times. As early as 1150 B.C., dust fall was recorded in the book "Zhu Shu Ji Nian" (Bamboo Slip Annals). Later, large amount of "dust rain" records appeared in various historical literature. The so-called "dust rain" occurred in North China, and showed a close relation to loess deposition in China (Wang Jiayin 1965; Liu Tungsheng et al. 1964). However, all of these records are mainly descriptions of phenomena, and lack practical observations and analyses. F. v. Richthofen (1882) and V. A. Obruchev (1958) both held an aeolian origin of loess and paid special attention to observations of dust storm events. They searched for the relationship between loess origin and atmospheric motion and tried to find out evidence to support the aeolian origin of loess from the dust storm events.

Since the 1970s, Chinese geologists, in cooperation with meteorologists, have made some petrological investigations of dust which fell in Beijing and analyzed the weather process of dust storms by using satellite cloud maps. They have preliminarily revealed the relationship between the dust storm and atmospheric motion, bringing the study of the wind transportation mechanisms from a qualitative stage to an experimental observation stage (Liu Tungsheng et al. 1981, 1982, 1984).

China covers a vast territory. North China is the area where dust is raised, transported and deposited. This area includes the source region of sand and dust in the western China deserts and the deposition region of sand and dust in the upper and middle reaches of the Huanghe River and the Changjiang River. Therefore, North China is one of the best areas for the study of wind geological action.

6.1.1 General Characteristics of Dust Storms and Their Bearing on Loess

Sand (dust) storms, raised sand, suspensive dust and dust haze refer to various extents of wind-drift sand phenomena, which reflect the amount of sand and dust in the air or the size of dust particles. A sand (dust) storm usually refers to horizontal visibility less than 1 km due to the particle material in the air; raised sand refers to horizontal visibility between 1–10 km, while suspensive dust and dust haze refer to horizontal visibility of less than 10 km, but with a relatively fine grain size of particle material in the air. The suspensive dust is mainly composed of fine sand and silt dust with grain-size usually between 0.1 and 0.001 mm, but dust haze mainly refers to floating dust in the air with grain size usually less than 0.001 mm.

In order to study the aeolian mechanism of the loess origin, the authors have adopted the term dust storm. A dust storm is not a purely synoptic phenomenon but a geological event too, that is responsible for the creation of modern loess. It is composed of a series of wind-drift sand phenomena including the whole process from the sand-uplift stage to the transportation and deposition stage of the raised sand, suspensive dust and dust haze. Study of the uplift, transportation and deposition of sand and dust in various stages in space and time, and study of its climatic background and meteorological conditions are helpful to quantitative exploration of the aeolian mechanism of loess origin.

6.1.1.1 Distribution of Dust Storms in Time and Space

1. Spatial Distribution. From a comparison of the diagram showing the yearly day-numbers of suspensive dust and sand storms with the known source and deposition regions of loess (The Distribution Map of Loess in China), it is obvious that: (a) the area with the yearly sand storm day-numbers more than 5 (with the exception of Qinghai-Xizang Plateau) covers basically the source regions of modern dust (loess material); (b) the area with yearly sand storm day-numbers between 1–5 days lies roughly in the south of the above area and is the major deposition region; (c) the line along which the yearly suspensive dust day-number equals 5 coincides with the southern boundary of the deposition region and lies in the middle and lower reaches of the Changjiang River. It is close to the zero-day line of sand storms. The area between 5 days of sand storms and 5 days of suspensive dust is the "dust rain" concentration area, and is also the major distribution area of loess in China (Fig. 6.1).

Thus it can be seen that the area with more than 5 days of sand storms per year is the major source region of dust (loess material), whereas the "dust rain" concentration area is the deposition region of dust (namely, the deposition region of loess material).

2. Temporal Distribution. On the basis of meteorological data from China, the sand storm usually occurs in spring and winter. The counties along the boundary between loess and desert in Shaanxi Province are given as an example. From east to west, the counties Fugu, Shenmu, Yulin, Hengshan, Jingbian and Dingbian give, respectively, the yearly frequencies of sand storms as 7.4, 9.9, 13.7, 14.9, 22.3 and 36.1 (times), and highest frequency is observed in April – sand storm frequency this month accounts for about 1/3 to 1/2 of the total frequency in one year. The "dust rain" in historical times was also concentrated in spring and

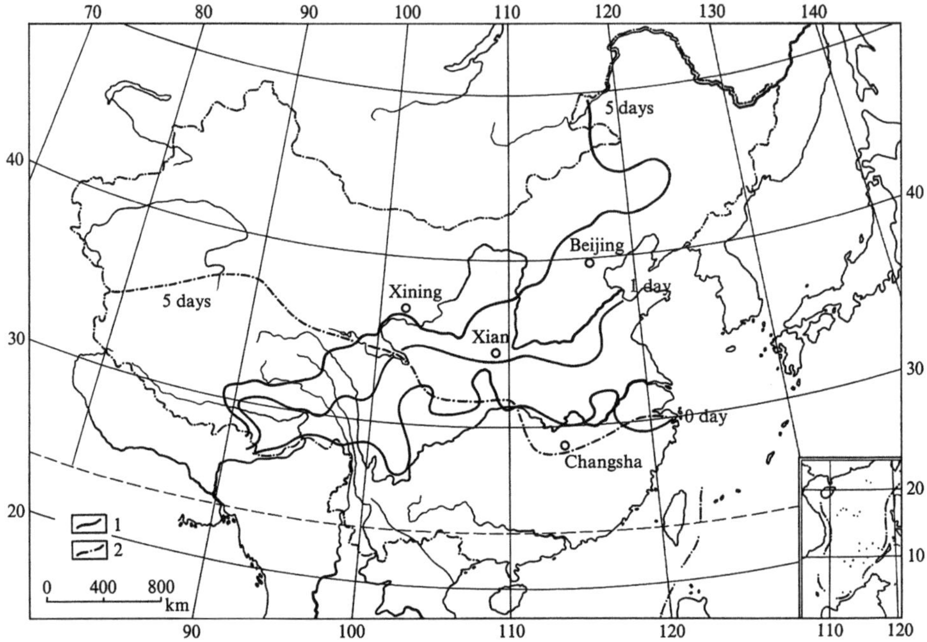

Fig. 6.1. Yearly suspensive dust (2) and sand storm (1) day numbers. (Central Meteorological Bureau 1978; Zhang De'er 1984)

winter, and the times of "dust rain" in April accounted for approximately 26% of the annual value (Zhang De'er 1982).

6.1.1.2 Dust Storm and Its Weather Background

1. Temperature Background of Dust Storms. The modern dust storm often appears in the low-temperature winter half year, but the highest frequency of sand storms is in April when the monthly average temperature increases rapidly from about 3 °C in March to above 10 °C (for the above-mentioned counties in Shaanxi Province). However, the frozen layer, up to 7–125 cm thick, still characterizes the earth surface at this time. For example, in Hami (Xingjiang), and Dunhuang (Gansu), lying in the source region of dust storms, though the average temperature is as high as 0 °C in April, the absolute lowest temperature is still below 10 °C. Therefore, the highest frequency of occurrence of sand storms is still within the low-temperature period. In addition, on the basis of long-term temperature background, Zhang De'er's research on the temperature background of "dust rain" in the past 500 years shows that the frequency of "dust rain" increased in the low-temperature time intervals.

2. Humidity Background of Dust Storms. Dust (sand) storms usually occur in relatively arid areas and relatively dry winter and spring. Their highest frequency of occurrence is in April. According to the Synoptic Atlas of The People's of China (Central Meteorological Bureau 1978), in the area with relative humidity less than 40% it reaches a maximum value in April. Fugu and other counties in Shaanxi Province have their minimum relative humidity (40% – 50%) in March to June. On the basis of the long-term humidity background data, the high frequencies of "dust rain" often corresponded to dry periods in the last 500 years (Zhang De'er 1984). Thus, periods with high frequencies of sand storms are coincident with dry and cold periods, while the most favorable time intervals for the sand storm occurrence correspond to the most dry seasons and seasons when the frozen earth melts. Therefore, loess should be produced in dry and cold climatic conditions.

6.1.2 Analyses of Several Dust Storm Events

Dust storms occurring respectively on March 17 – 20, 1958; April 17 – 21, 1980; April 27, 1983 and April 22 – 25, 1976 can be taken as examples to show the aeolian processes of loess origin and the influence of atmospheric movement on these processes.

1. The Dust Storm Event in 1980 (Liu Tungsheng et al. 1981, 1982). According to the report of the Central Meteorological Bureau and the satellite cloud maps, at 12:00 p.m. on April 17, 1980, the embryo of a cyclone was already formed between 100°– 120°E and 40°– 60°N. Under atmospheric pressures lower than 300 mb in the area covering Hami, Erenhot and Wuwei, the R_i value was below 0.2, indicating the occurrrence of turbulent mixing. The westerly jet stream moved downward with the accompanying downward transport of the momentum, and resulted in wind speeds up to 12 – 20 m/s. Thus sand and dust were raised, dropping visibility down to less than 8 km, even to 0.3 – 0.6 km in some regions. Sand and dust were rolled up by the ascending currents to high altitudes of 300 – 500 mb and the speed of the westerly jet stream was as high as 40 – 50 m/s. Therefore, it took only 10 hours to transport the sand and dust from the source regions to the sky above Beijing lying 1500 km away. On April 18, the cyclone was fixed and the sand and dust were transported by the westerly jet stream to the high altitudes (7000 – 7500 m) over North China, and then fell. On April 19, the cyclone moved eastwards, the frontal cloud belt moved eastwards and southwards, and the sand and dust arrived in Changchun in the east and at the Huaihe River and the middle and lower reaches of the Changjiang River in the south. The distribution altitudes of the sand and dust were up to 6000 m in Hebei and Shanxi Provinces, and up to 3000 – 3500 m in the middle and lower reaches of the Changjiang River. On April 20, the cyclone shifted more eastwards and faded gradually, meanwhile the sand dust drifted southeastwards and was distributed in the area covering Changsha, Hunan Province and Ningbo, Zhejiang Province in the middle and lower reaches of the Changjiang River, then arrived in Korea. According to the report of meteorological observatories, the suspended dust reached the southern part of Japan on April 21 (Fig. 6.2).

According to the synoptic maps, the surface cold front shifted southeastwards, and the sand and dust fell down to the earth surface just behind the front.

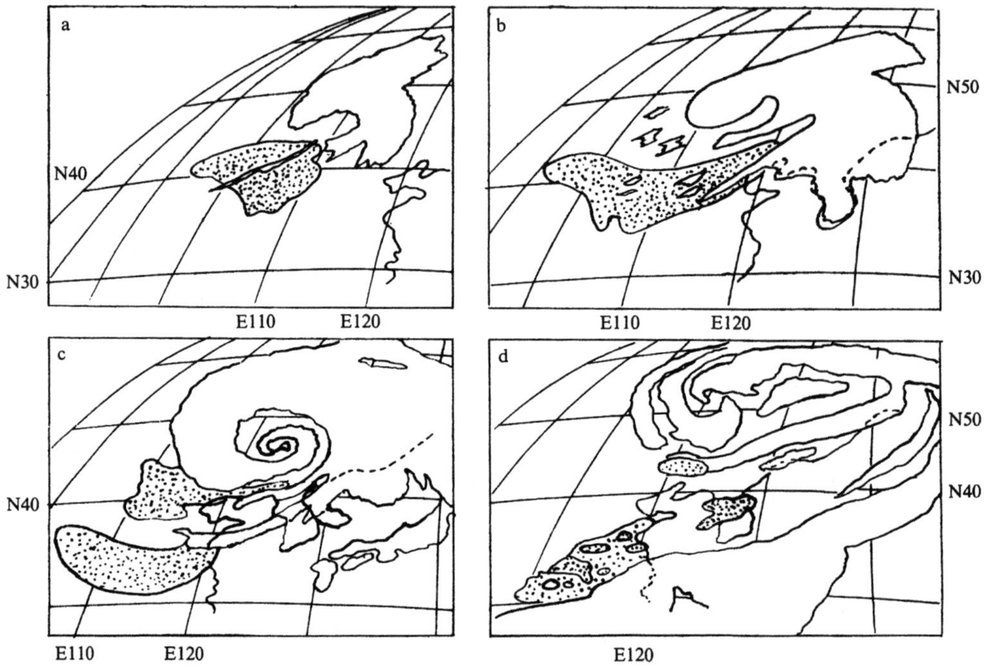

Fig. 6.2a–d. A sketch of GMS satellite cloud map. **a** Generation of the cyclone and the sand-dust area on April 17; **b** the first period of the fixed cyclone and the sand-dust area on April 18; **c** the second period of the fixed cyclone and the sand-dust area on April 19; **d** the fade-out of the cyclone and the sand-dust areo on April 20

2. The Dust Storm Event on April 22–25, 1976 (Rahn et al. 1981). The moving path of the cold front and the dust fall in China was approximately the same as the dust storm event on April 17–21, 1980. On April 22–23, at approximately 500 mb the cyclone swept over the Gobi Desert and Loess Plateau. In the initial 24 hours the speed of ascending current was 5 cm/s and a great deal of sand and dust were carried up to high altitudes and then transported to the middle and lower reaches of the Changjiang River. On April 24, sand and dust arrived in Japan. From April 23 to May 1, a small amount of dust was transported near the U.S. coast by intense upper wind and then moved to Barrow, Alaska, traveling over 12000 km. Some sand and dust were directly carried by intense upper wind over Barrow, where the dust was distributed at 1–2 km above sea level.

3. The Dust Fall Event in Beijing on March 17–20, 1958 (Liu Tungsheng et al. 1964; Zhang De'er 1984). Dust storms occurred on March 17 in both southern Xinjiang and Dunhuang, Gansu Province. On March 18, dust storms appeared mainly in the west of Gansu Province, where a strong ascending current with a speed of 13.2×10^{-3} mb/s in its center carried the sand and dust up to the high altitudes of 3000–5000 m. The west wind carrying the sand and dust shifted eastwards with a speed of 11.8 m/s. On March 19–20, suspended dust appeared in Beijing and dust fall occurred.

All the above examples show that ascending currents occurred in the source regions of sand and dust and carried the sand and dust up to high altitudes of 3000–10000 m, followed by the westerly transport to eastern areas.

4. The Dust Storm on April 27, 1983. At 8:00 a.m., on April 26, a cold front characterized the area covering Urumqi, Xinjiang. At 8:00 a.m. on April 27, a gale blowing with the wind speed of 18 m/s (near the earth surface) was recorded, and sand and dust were rolled up into the air as the dust storm progressed. The gale arrived in Zhongwei at 2:30 p.m. of the same day, then in Yanchi and Dingbian with wind speeds as high as 24–25 m/s. A vertical "sand-and-dust wall" was created at the gale outset, and as a cloud of sand and dust flew up visibility was recorded as less than 20 m. According to estimates of the top of the "sand-and-dust wall", the top altitude of the dense cloud of sand and dust was up to 3000 m in the air. At 8:00 p.m. on April 27, the cold front swept over the Qinling Mountains. Because the dense "sand-and-dust wall" shifted with the gale southwards, the sand and dust were mainly kept in the lower part of the troposphere and transported by lower winds. Because of the weakened westerlies at that time, no dust fell in Beijing during this dust storm event.

The mineralogical studies on the dust fall in 1958[11] and in 1980 (Liu Tungsheng et al. 1981) showed that dust contains many minerals, and that quartz + feldspar + calcite account for more than 90% – very similar to the mineral composition of loess. Concerning grain size, both the dust fall in Beijing and Malan Loess in Luochuan has silt particles as their major fractions, indicating that both of them are cognate dust deposits and the dust fall in Beijing is a real loess deposition.

The above dust storm events indicate that the moving paths of sand and dust during the dust storms of 1958, 1976 and 1980 were in the middle and upper parts of the troposphere and the sand and dust were transported a long distance by the westerlies from the source regions to North China Plain, Pacific Ocean and even to Alaska, in the United States. In the 1983 dust storm event the cloud of sand and dust was very dense near the earth's surface, resulting in a surging "sand-and-dust wall", thus belonging to the type of sand and dust which are transported by lower winds in the troposphere, and hence the transportation distance was also relatively short. When the sand and dust storm occurred in the source region, sand and dust moved with the surface wind and subsided with the weakening wind. Therefore, the dust storms can be classified into two categories: (1) High altitude type: sand and dust raised by dust storms can reach the middle or upper part of the troposphere and are mainly transported by westerlies with the transportation distance up to more than 10000 km; (2) Low altitude type: the raised sand and dust are below 3000 m, and are mainly transported by the surface wind. They subside in-situ with the weakening wind.

---

[11] Zhang Shuyuan (1964) A study on the wind-drift dust in Beijing (unpubl.).

6.1.3 Dust Storms and Atmospheric Motion in Various Scales

Regarding the relationship between the theory of aeolian transportation and deposition of loess and atmospheric motion on various scales, V. A. Obruchev (1958) suggested that "cold loess" is related to the anticyclones centering in the continental glacial sheets, while the "hot loess" is related to the heated local cyclones in plain areas. V. A. Obruchev (1958) suggested also that the loess in China is distributed in the area where the oceanic monsoon meets the continental monsoon, and the emphasized the function of the atmospheric vertical turbulent currents. Liu Tungsheng et al. (1979) noted the importance of westerlies for the transportation of loess materials. The following is a description of the effect of atmospheric motion in various scales on dust storms.

1. Cold High Pressure Activity. The activities of cold high pressures in middle and high latitudes areas are very frequent. On the average, one cold current event occures every 4 days and is an important factor giving rise to strong NW winds. The intensity of high pressure activity weakens in spring but increases in frequency. At that time, the frozen earth in northern areas gradually melts and thus sand and dust are easily raised by wind. The frequency of dust storms reaches its maxima in April.

2. Westerlies Activity. The westerlies lie over the interior arid zone in Central Asia and over the Loess Plateau. They change remarkably with seasons. In the winter half year, they are divided into two branches by the Qinghai-Xizang Plateau and the westerly jet stream can reach as far as 25 °–35 °N. Above the Pacific Ocean the westerlies then change location frequently: at 30 °–40 °N in January, 30 °–35 °N in April and 45 °N in July. In spring, the trough and ridge activity of the westerlies is very active. The cold currents shifting southward from Siberia create favorable conditions for the development of intensification of cyclones and anticyclones, resulting in the increase of surface wind speed and the occurrence of dust storms.

The westerlies have a great ability to transport sand dust of high level type dust storms.

3. Activities of Summer Monsoon. China is a country where the monsoon prevails throughout. According to Gao Youxi et al. (Geographical Department, Zhongshan University et al. 1979), the dry monsoon weakens significantly in March and the wet monsoon appears in South China in April and affects Central China in mid-April. The wet monsoon brings in vapor over the oceans and the vapor integrates with dustparticles in the air, forming water droplets or ice crystallites. The dust subsides together with the water droplets and ice crystallites. Thus rain droplets carrying loess are often observed in North China in April and May. The vapor cleans out the dust particles in the air and prevents the dust from moving southward. Therefore, the wet monsoon is one of the factors determining the southern boundary of dust fall at 30 °N, and consequently, the southern boundary of loess distribution reaches about 30 °N.

From the foregoing discussion on dust storms, the following preliminary conclusions may be drawn:

a) In view of the great significance of westerlies in raising and transporting sand and dust, wind action should include not only the action of the currents in the low level of the troposphere but also the action of the currents in the middle and high levels of the troposphere.

b) The joint action by low level currents and the westerlies may give rise to various aeolian facies. Some examples include: (1) The erosion and deposition by low-level currents have created deserts and stone deserts in which the fine materials have been transported elsewhere; (2) the mixed action of low-level currents and westerlies has given rise to the loess zone, which can be further divided into several subzones, including the sandy loess subzone, loess subzone and clayey loess subzone from NW to SE (Zhu Haizhi 1959; Su Lianyi et al. 1959); (3) the action of westerlies has resulted in the deposition of fine silt particles in the North Pacific Ocean, especially in its abyssal regions, and in Arctic Alaska, where extremely small amounts of aeolian clays may have been deposited but are difficult to detect. As a result, various aeolian facies bands have been formed from Northwest China eastwards to the Pacific Ocean.

c) The uplifting of the Qinghai-Xizang Plateau divides the westerlies in China into two branches. The southern branch, namely the subtropical westerly jet stream usually appears at 25°–35°N. This change in atmospheric circulation exerts a great influence on the occurrence and development of high atmospheric pressures in Siberia. The strong north wind shifts straight down southwards, leading to an arid climate in the South and creating favorable conditions for the deposition of Xiashu loess in the middle and lower reaches of the Changjiang River during the mid-Pleistocene. Because the loess in China has migrated southwards to about 30°N, the continental loess and the aeolian sediments at 30°–40°N in the Pacific Ocean have been integrated into a single geological entity.

6.1.4 Dust Storms and the Source Region of Loess

According to the boundary between yearly zero and 5-days of sand storms, the line with average relative humidity in April less than 40%, and the distribution of deserts, loess and mountainous regions (see Meteorological Atlas of the People's Republic of China), we can determine the source region of loess in China. This source region, with a total area of approximately 350–400 km^2, is bounded by Mt. Kente and Mt. Tangnu in the north, by the Da Hingan Mountains in the east, by the Great Wall and Kunlun Mountains in the south and by the Pamirs Plateau in the west. South of this region is the deposition region of loess. Its southern border coincides with the line of yearly average of 5 days of suspensive dust (Fig. 6.3).

The source region of loess covers the areas with semi-arid, arid and extremely arid climates. The yearly aridity index is between 2 and 16, and even as high as 29–60, and the yearly precipitation is less than 200 mm for most parts of the region, and can reach 400 mm for some parts. Minimum precipitation is

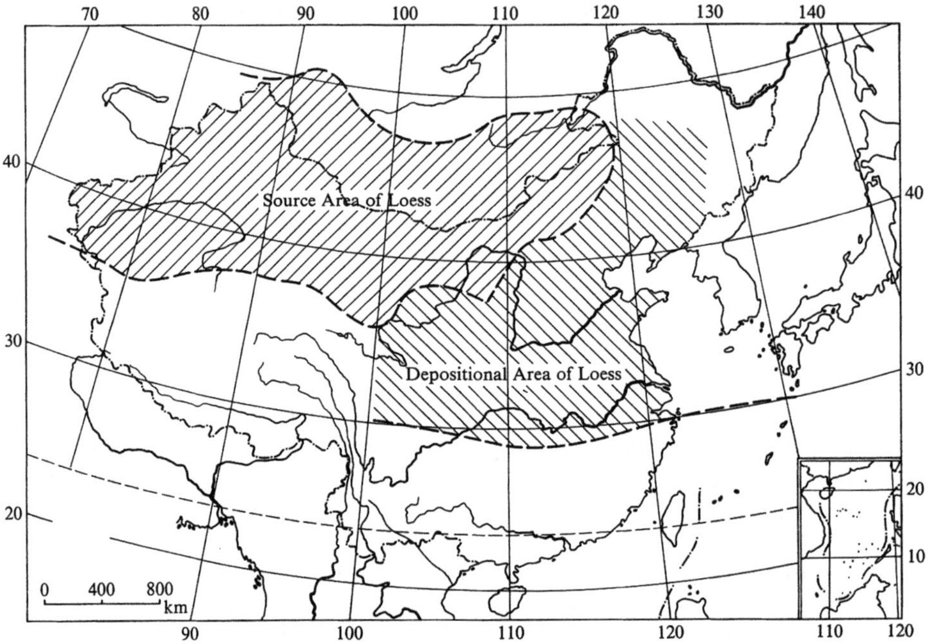

Fig. 6.3. Source region and deposition region of loess

10–30 mm in this region. Gale days are usually over 10 and can be as high as 50–100 days per year. In the southern part of the Taklamakan Desert, Xinjiang, the wind-drift sand days account for one third of the total days in a year and may even be 145 days. In 1959, in Minqin, a county located at the margin of the Tengger Desert, wind-drift sand days were 148, accounting for 41% of a year, and during March, April, May and June, every month the wind-drift sand days accounted for more than half a month (Zhu Zhenda et al. 1980). During the period of wind-drift sand, wind transports sand and dust repeatedly along the earth surface to the deposition region. Even in southern Xinjiang where the northeast wind prevails, sand and dust can also be transported by the westerlies to Beijing (see "Dust Fall Event in Beijing on March 17–20, 1958").

6.2 Amount and Rate of Loess Deposition

The paleomagnetic polarity column of the Luochuan loess-paleosol section has identified several stratigraphical ages, but up to the present no available methods for accurately dating the individual loess and paleosol layers existed. Based on boundary ages of paleomagnetic chrons or subchrons, however, it seems possible to evaluate the ages of individual loess and paleosol layers from their depositional amounts.

6.2.1 Depositional Amount of Loess

1. The bulk density (g/cm^3) of each sample is calculated from

$$D = W/V \ ,$$

where W is the sample weight, V is the sample volume.

2. The average bulk density ($\bar{D}$) of individual loess or paleosol is calculated from

$$\bar{D} = \frac{\frac{1}{2}D_1S_1 + \frac{1}{2}D_1(S_1+S_2) + \frac{1}{2}D_2(S_2+S_3) + \ldots + \frac{1}{2}D_n(S_n+S_{n+1}) + \frac{1}{2}D_nS_{n+1}}{\sum_{i=1}^{n+1} S_i} \ ,$$

where S_1 is the distance from the sample located uppermost in the stratigraphical portion to the top surface of the portion; S_n is the distance between the two adjacent samples ($n-1$th and nth samples); S_{n+1} vs the distance from the sample located lowermost in the stratigraphical portion to the bottom surface of the portion.

If the sample amount collected in each layer is sufficient and/or the distances at which the samples collected are approximately the same, then an arithmetic mean will adequately apply and the above formula is reduced to

$$\bar{D} = \sum_{i=1}^{n} D_i/n \ ,$$

where n is the number of samples.

3. The depositional amount of individual layer AQ (g/cm^2) is given by

$$AQ_i = \bar{D}_i S_i \ ,$$

where S_i is the layer thickness.

By summing up the depositional amount in unit area of each layer, an integrated depositional amount in unit area of individual stratigraphical portion is obtained as

$$AQ = \sum_{i=1}^{n'} AQ_i \ ,$$

where n' is the number of layers.

In the calculation we chose the samples in two column sections to determine the natural bulk densities of loess and paleosol. One group of the samples is that for paleomagnetic measurements, consisting of 72 samples collected from the outcrop sections at Heimugou, Luochuan in 1975. The other, also for paleomagnetic measurements, consists of 263 samples collected from Qun 22 boring holes near Yaotun, Luochuan, and additional 6 loess samples and 10 calcareous nodule samples as well as 2 black loam samples collected from the outcrop sections at Heimugou in 1983.

These paleomagnetic samples all experienced a long-term natural air-dry process, and thus preserved well in a natural state. Their bulk densities are slightly greater than those of loess obtained by using rather complicated methods. Cal-

Table 6.1. Thickness, bulk density, depositional amount and model age of individual layers of Qun 22 Core, Luochuan

	Thickness (cm)	Bulk density (g/cm^3)	Depositional amount (g/cm^2)	Accumulated time ($\times 10^3$ yr)		
				Model 1	Model 2	Model 3
S_0	200	1.40	280	10	10	10
L_1 S_1	800 240	1.50 1.85	1200 444	140	146	142
L_2 S_2	420 320	1.74 1.86	731 595	245	255	249
L_3 S_3	300 160	1.76 1.84	528 294	309	323	315
L_4 S_4	340 230	1.76 1.88	598 432	391	408	398
L_5 S_5	400 510	1.86 2.01	744 1025	531	554	540
L_6 S_6	600 150	1.75 1.84	1050 276	635	664	647
L_7 S_7	260 120	1.76 2.02	458 242	691	722	703
L_8 S_8	160 210	1.98 1.92	317 403	745	767	755
L_9 S_9	1140 100	1.84 1.90	2098 190	896	903	902
L_{10} S_{10}	200 150	1.88 1.89	376 284	940	942	944
L_{11} S_{11}	130 80	1.88 2.00	244 160	966	966	970
L_{12} S_{12}	100 180	1.86 2.02	186 364	1027	1027	1033
L_{13} S_{13}	140 120	1.89 1.94	265 233	1084	1084	1090
L_{14} $W_{S\text{-}1}$	680 1100	1.89 1.97	1285 2167	1482	1482	1485
$W_{L\text{-}1}$ $W_{S\text{-}2}$	1180 900	1.92 1.87	2266 1683	1937	1937	1936
$W_{L\text{-}2}$ $W_{S\text{-}3}$	640 580	1.99 2.02	1274 1172	2218	2218	2216
$W_{L\text{-}3}$	740	1.98	1465	2387	2387	2384

Table 6.2. Thickness, bulk density, depositional amount and model age of individual layers of Heimugou natural section, Luochuan

	Thickness (cm)	Bulk density (g/cm^3)	Depositional amount (g/cm^2)	Accumulated time ($\times 10^3$ yr)	
				Model 4	Model 5
S_0	40	1.40	56	10	10
L_1	720	1.48	1066	142	145
S_1	270	1.65	446		
L_2	600	1.49	894	239	244
S_2	130	1.69	220		
L_3	400	1.53	612	329	335
S_3	250	1.65	413		
L_4	350	1.56	546	418	425
S_4	260	1.81	471		
L_5	400	1.58	632	550	561
S_5	490	1.81	887		
L_6	510	1.63	831	647	660
S_6	160	1.76	282		
L_7	270	1.72	464	709	722
S_7	131	1.80	236		
L_8	138	1.78	246	766	776
S_8	254	1.79	455		
L_9	860	1.61	1385	894	898
S_9	140	1.75	245		
L_{10}	210	1.77	372	943	945
S_{10}	150	1.67	251		
L_{11}	75	1.72	129	970	970
S_{11}	130	1.67	217		
L_{12}	150	1.79	269	1033	1033
S_{12}	120	1.78	214		
L_{13}	150	1.69	254	1096	1096
S_{13}	130	1.78	231		
L_{14}	480	1.70	816	1482	1482
W_{S-1}	1218	1.77	2156		
W_{L-1}	980	1.81	1774	1870	1870
W_{S-2}	645	1.89	1219		
W_{L-2}	995	1.85	1841	2261	2261
W_{S-3}	630	1.85	1166		
W_{L-3}	720	1.86	1339	2435	2435

Table 6.3. Average bulk density, depositional amount and rate of sections of Qun 22 core and Heimuogou natural outcrop, Luochuan

Strata		Average bulk density (g/cm^3)		Depositional amount (g/cm^2)		Depositional rate (g/cm$^2 \times 10^3$ yr)	
		Qun 22	Natural	Qun 22	Natural	Qun 22	Natural
Black loam		1.40	1.40	168	56	16.8	5.6
Malan loess		1.50	1.48	1200	1060	12.63	11.46
Lishi loess	Loess	1.82	1.62	8880	7450	12.32	10.78
	Paleosol	1.92	1.75	4942	4568	12.17	11.20
Wucheng loess	Loess	1.96	1.84	5005	4954	8.67	7.70
	Paleosol	1.95	1.82	5022	4541	8.69	7.70
Red clay		1.99	1.73				
Loess		1.83	1.68	15085	13470	10.83	9.44
Paleosol		1.93	1.78	9964	9109	19.13	9.12
Whole section		1.87	1.72	25329	22635	12.61	9.30

Table 6.4. Average bulk density, depositional amount and rate of individual magnetostratigraphical age stage of Qun 22 core, Luochuan

Time sequence ($\times 10^3$ yr)	Average bulk density (g/cm^3)		Depositional amount (g/cm^2)		Depositional rate (g/cm$^2 \times 10^3$ yr)	
	Model 2	Model 3	Model 2	Model 3	Model 2	Model 3
0–730	1.77	1.77	8884	9121	12.17	12.49
730–970	1.87	1.87	4029	3736	16.79	15.57
970–1870	1.92	1.92	7813	7869	8.68	8.74
1870–2400	1.98	1.98	4491	4491	8.68	8.74

Table 6.5. Possible depositional amount of loess in the loessial area of China and in middle reaches of the Huanghe River

	Loessial area of China					Middle reaches of the Huanghe River						
Average bulk density (t/m^3)	1.72											
Loess area (m^2)	4.4068×10^{11} [a]					2.756×10^{11} [b]						
Average thickness (m)	25	50	75	100	125	25	50	75	100	125	150	175
Depositional amount ($\times 10^{13}$(t))	1.89	3.79	5.68	7.58	9.47	1.19	2.37	3.56	4.74	4.93	7.11	8.30
Formula	AQ (Depositional amount) = $\bar{D}$ (Average bulk density) $\times$ $\bar{T}$ (Average thickness) $\times$ A (Loess area)											

[a] Liu Tungsheng et al. (1965).
[b] Liu Tungsheng et al. (1964).

culations were made for samples from both boring hole and outcrops by using the method mentioned above. The results obtained are given in Tables 6.1, 6.2, 6.3 and 6.4. Weighted corrections were applied to bulk density values of paleosol layers by taking into account the contributions of calcareous nodule layers. The average bulk density values given in Table 6.4 were all corrected in this manner.

Luochuan is located in the middle of the Loess Plateau. Samples from this area may have bulk density values greater than those from the northwestern region, but smaller than those from the southeastern area. Since there are no more data available from other sections, the bulk density values of natural section of the Luochuan loess may serve as a basis for the calculation of the total depositional amount. As the thicknesses of the loess in different regions vary widely, it is difficult to determine the real average thickness of loess layers in the loessial area of China or in the middle reaches of the Huanghe River. But it is possible to calculate the total depositional amount of the loess beds in these areas by using a series of different average thickness values. The results obtained are shown in Table 6.5.

As shown in Table 6.5, since the Pleistocene the possible depositional amount of loess is $1.89-9.47\times10^{13}$ tons in the loessial area of China and $1.19-8.30\times10^{13}$ tons in the middle reaches of the Huanghe River.

In the loessial area, there is a critical problem of soil erosion. In addition to natural erosions, human activities also speed up the process. According to the data made by the Yellow River Water Conservancy Board, 1600 million tons of silt are carried into the Huanghe River from the Loess Plateau each year (Jing Ke and Chen Yongzong 1983). Distributing this amount to the whole loessial area, we calculated the erosion rate to be $363.08\ g/cm^2\times10^3$ yr, corresponding to 39.04 times the average depositional rate ($9.30\ g/cm^2\times10^3$ yr) since the Pleistocene. Supposing all the silts are from the loessial area, the soil erosion would be expected to significantly influence the environment of this area. At the current erosion rate, all the loess deposits will be eroded in $11.8-59.2\times10^3$ yr. If the dust accumulation in this period is taken into consideration and the depositional rate is assumed to equal the average rate since the Pleistocene, the loess deposits may only last $0.3-1.52\times10^3$ yr longer. Of course, in view of the climatic changes during the Pleistocene, it is quite unlikely that the present postglacial or "interglacial period" will last as long as 60×10^3 yr. A few thousand years hence, it will probably by replaced by a glacial dry-cold climatic period when the dust accumulation will be greatly enhanced and erosion weakened. It is obvious that soil erosion will exert a great influence on the ecological balance in the Loess Plateau of China, and thus, special attention must be paid to the water conservancy of this area.

6.2.2 Depositional Rate and Stratigraphical Ages of Loess

The method with which the depositional rate of loess is evaluated has an important bearing on the reliability of the ages of loess. At present, a common approach to evaluating the average depositional rate ($cm/10^3$ yr) is to divide the depth, when the stratigraphical portion in question lies, by the time interval between the paleomagnetic boundaries of this portion. This approach disregards the dif-

ference in depositional rate between accumulation period and pedogenesis period, because the thickness of paleosol, though generally thinner than that of loess, represents a relatively long interval. By assuming a similar average dust accumulation rate for each loess-paleosol assemblage, errors due to different depositional rate can be eliminated in the calculation based merely on single loess and paleosol layers. This also favors partitioning the time intervals in each paleomagnetic age-stage pertinent to individual loess-paleosol assemblage in accordance with the relevant depositional amount ratios of one to each other. Differences between individual loess and paleosol layers in porosity and compactness have been taken into account in calculating ages from depositional amount. Thus this dating method should be more reliable than that based on depth.
The method consists of the following measures:

1. Calculate depositional amount using the method discussed above.
2. Determine the starting and ending ages of individual layers or assemblages. In each age-stage governed by the related polarity chron or subchron in the stratigraphical section, the time interval spanned by each layer (assemblage) is calculated from the formula

$$T_i = (A\,Q_i/A\,Q)\cdot T\ ,$$

   where T_i is the time interval spanned by ith layer (assemblage); T is the time interval spanned by the age-stage in question. By summing up the time intervals spanned by each layer (assemblage) in the order from new to old, the starting and ending ages of each layer are obtained.
3. Determine depositional rate ($g/cm^2 \times 10^3$ yr). The magneto-stratigraphic boundaries presently available for the Luochuan section, though roughly consistent, still show differences, and thus several different age models were chosen for the calculation. Three age models, for example, were utilized in calculating the ages of the Qun 22 core section.

Model 1: According to the paleomagnetic ages of the Qun 22 core (Heller and Liu Tungsheng 1982), the relevant B/M boundary is at 53.05 m; the top and bottom boundaries cf. the Jaramillo subchron are at 67.3 m and 72.5 m, respectively; the top and bottom boundaries of Olduvia subchron are at 107.4 m and 113.1 m, respectively.

Model 2: According to the paleomagnetic ages estimated from Beihanzhai loess section (Liu Tungsheng and An Zhisheng 1984), the relevant B/M boundary corresponds to the position of 51 m in the Heimugou section. The boundaries and positions of Jaramillo and Olduvai subchrons are the same as those of Model 1.

Model 3: According to the paleomagnetic age of Heimugou natural section (Heller and Liu Tungsheng 1984), the relevant B/M boundary is at 52.3 m. The top and bottom of Jaramillo subchron boundaries are corresponding to depths 69 m and 72.2 m in the Qun 22 core, respectively. The boundaries and positions of the Olduvai subchron are the same as those of Model 1.

The starting and ending ages of individual loess-paleosol assemblages obtained from Models 1, 2, 3 are given in Table 6.1.

The following two models were used in calculating the depositional ages of natural sections.

Table 6.6. Age ($\times 10^3$ yr B.P.) of starting of loess-paleosol assemblages

	Estimatic age I[a]		Estimatic age II[b]		Kukla (1977)		Shackleton and Opdyke (1973)	
	Method I[c]	Method II[d]	Method I[c]	Method II[d]	Termination	Age	$\delta^{18}O$ stage	Age
S_0	10.0	10.0	10.0	10.0	I	11.5	1	13
L_1	109.1		104.9				2–4	75
S_1	145.8	162.8	144.6	152.8	II	128	5	128
L_2	206.2		224.2				6	195
S_2	255.3	271.5	243.8	258.2	III	240	7	251
L_3	298.9		298.3				8	297
S_3	323.2	339.1	335.0	351.9	IV	330	9	347
L_4	372.6		383.6				10	367
S_4	408.3	422.9	425.5	439.9	V	400	11	440
L_5	469.3		482.2		VI	480	12	472
S_5							13	502
							14	542
	554.5	556.6	560.8	568.3	VII	570	15	592
L_6	641.2		634.8				16	627
S_6	664.0	666.8	659.9	665.0	VIII	630	17	647
L_7	701.8		701.3				18	688
S_7	721.8	722.6	722.3	722.8	IX	710	19	706
L_8	743.0		741.9				20	729
S_8	767	765.7	775.9	773.1	X	810	21	782

[a] From Qun 22 core.
[b] From Luochuan section.
[c] Calculated from accumulated volumn.
[d] cf. Lu Yanchou (1981).

Model 4: The relevant B/M boundary is located at the top of paleosol S_8. The boundary of Jaramillo subchron is located from 0.45 m below the top surface of loess L_{10} down to the bottom of paleosol S_{11}. According to the lower and bottom boundaries of Olduvai given by Heller and Liu Tungsheng (1984), the Olduvai subchron is believed to include from top to bottom of Wucheng Loess W_{S-2}.

Model 5: The B/M boundary is located 0.5 m below the top surface of loess L_8. The other polarity boundaries are similar to Model 4.

The starting and ending ages of individual loess-paleosol assemblages obtained from Models 4 and 5 given in Table 6.2.

On the basis of the paleomagnetic ages estimated from the Beihanzhai section, the starting and ending ages of individual loess-paleosol assemblages, counted from the B/M boundary, were obtained for both the natural (1975) and Qun 22 core section (Table 6.6).

For all of these models, the ages of individual strata below the Olduvai subchron were calculated by assuming a depositional rate similar to that pertinent to the interval from the start of the Jaramillo subchron (0.97 m.y.) to the start of the Olduvai subchron (1.87 m.y.).

6.2.3 Discussion

1. Depositional rate of Luochuan section (Table 6.3). The total depositional rate of Luochuan loess-paleosol sequence is estimated at 10.61 $g/cm^2 \times 10^3$ yr for Qun 22 and 9.30 $g/cm^2 \times 10^3$ yr for the Luochuan section.

2. Starting age of loess deposition and comparison between the age based on depositional-amount and the age based on depth. Loess in known to have started to deposit 2 m.y. ago in China (Heller and Liu Tungsheng 1984). The results obtained in this section were 2.387 m.y. for Models 1 and 2; 2.38 m.y. for Model 3 (Table 6.1); and 2.435 m.y. for Models 4 and 5 (obtained from the Heimugou section, Table 6.2).

The starting age can also be determined by plotting the data of the Qun 22 core. Assuming that the depositional rates of loess-paleosol assemblages are similar to those pertinent to the interval from the start of the Jaramillo subchron (0.97 m.y.) to the start of Olduvai (1.87 m.y.), the starting age is then obtained from Fig. 6.4 and Fig. 6.5 to be 2.41 m.y. and 2.39 m.y., respectively. Assuming that the bottom boundary of the Olduvai subchron coincides with the lower boundary of W_{S-2} paleosol assemblage in the core, and determining the time intervals spanned by the W_{L-2}, W_{S-3} and W_{L-3} stratigraphical portions allows one to estimate the starting age of Chinese loess at about 2.3 m.y.

Calculations and plots made for different sections and models indicate that the starting age of loess deposition in China is 2.3 – 2.43 m.y. ago with the majority around 2.4 m.y.

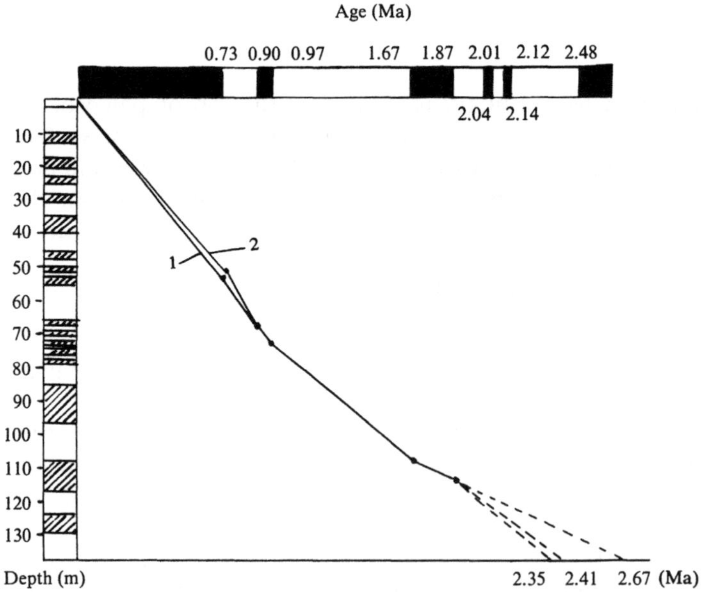

Fig. 6.4. Correlation between the depth of Luochuan loess-paleosol sequence and time. *1* B/M boundary is located at the top of S_8; *2* B/M boundary is located at the mid-upper portion of L_8

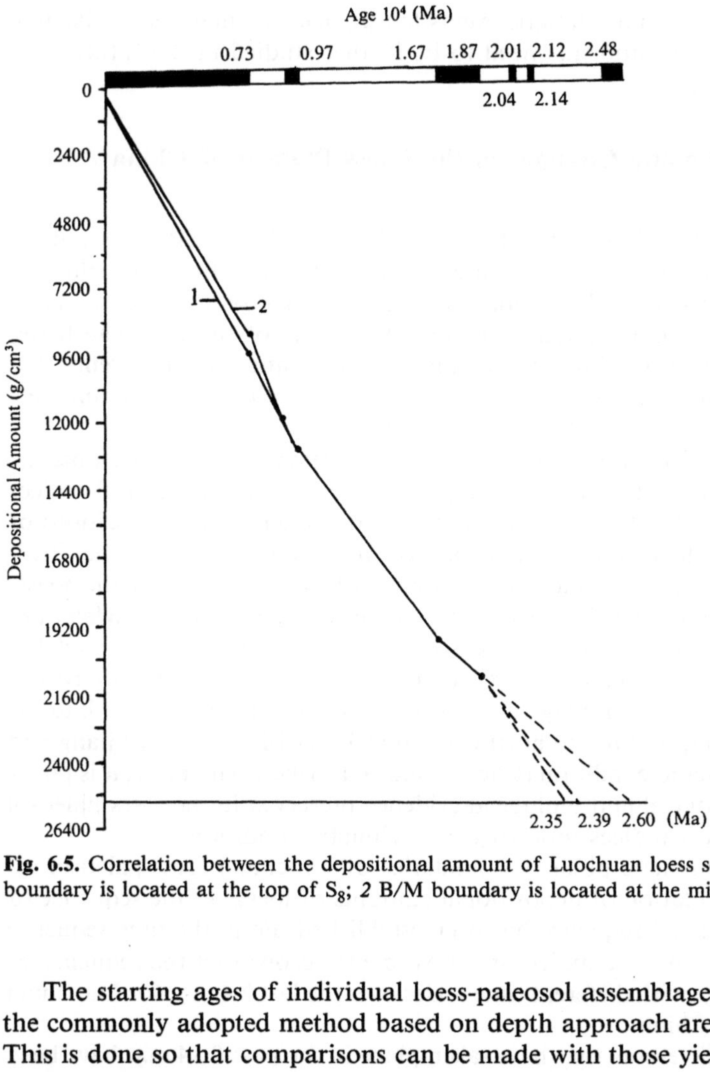

Fig. 6.5. Correlation between the depositional amount of Luochuan loess section and time. *1* B/M boundary is located at the top of S_8; *2* B/M boundary is located at the mid-upper portion of L_8

The starting ages of individual loess-paleosol assemblages obtained through the commonly adopted method based on depth approach are given in Table 6.6. This is done so that comparisons can be made with those yielded by the depositional amount.

3. Comparison of the starting ages of loess-paleosol assemblages above the B/M boundary obtained by different models with the average of glacial cycle termination age (Kukla 1977) and the oxygen isotope-stage ages (Shackleton and Opdyke 1973).

Data shown in Table 6.6 indicate that the starting ages of loess-paleosol assemblages in the Qun 22 core and the Luochuan section are fairly close to the average termination age given by Kukla.

Conversely, the starting and ending ages of individual loess-paleosol assemblages obtained by the depositional-amount approach for both the Qun 22 core and the Luochuan section (1975) are in better agreement with Kukla's average termination ages and $\delta^{18}O$ stage ages from core V28–238 than those obtained by the depth approach. There still exist deviations which are most likely due to a number of factors, such as climatic events taking place prematurely or lagging in different areas. Such problems need to be further studied in the future.

6.3 Long-Term Climatic Changes in the Loess Plateau of China

Loess is widely distributed in China at 30°–40°N. In the loess-paleosol sequence a number of paleoclimatic records are preserved which may reflect paleoclimatic changes in northern China. The Luochuan loess-paleosol sequence of Shaanxi Province, 130 m thick and located in the middle reaches of the Huanghe River, has preserved information about the major geological and climatic events that have happened in the past 2.4 m.y. It is recognized as one of the most complete geological-climatic records in the Northern Hemisphere.

The grain size distribution, $CaCO_3$ content, heavy mineral assemblages, micromorphological characteristics, susceptibility and other geochemical indices (Fe_2O_3/FeO and SiO/Al_2O_3 ratios, and amino acid content and composition) of loess-paleosol provide geological evidence of paleoclimates, while the fossil assemblages of land snail, pollen and mammalian fauna buried in the loess-paleosol provide biological evidence of paleoclimates. For example, the *Metodontia* assemblage of fossil snails in paleosols, which is still alive and stretches southward down the Changjiang River valley, reflects a warm-humid environment, and the *Cathaica* assemblage in loess, which distributed in the south to the Huanghe River valley, and in the northwest up to Inner Mongolia, Xinjiang and Tibet, reflects a dry-cold condition (Chen Deniu et al. 1982). Therefore, it is possible to turn the geological and biological evidence preserved in the loess-paleosol sequence into indirect indices reflecting paleoclimatic conditions.

Based on the magnetic polarity boundaries of Luochuan loess-paleosol sequence and the calculation of depositional amounts of loess, a time sequence of Luochuan loess-paleosol sequence has been established. From the time sequence and various indirect climatic indices it is possible to reconstruct the sequence of paleoclimatic change with different time scales over the last 2.4 m.y. for Luochuan and even northern China.

The paleoclimatic events that have taken place in the past 0.14 m.y. have been discussed in Section V, Chapter 3. In the following we will focus attention on those of the last 1.1 m.y. A preliminary comparison will also be made between these events and the paleoclimatic records of deep-sea sediments.

6.3.1 Variational Sequence of Climatic Elements (Annual Mean Temperature and Annual Mean Precipitation) of Luochuan for the Past 1.1 Million Years

It is a difficult task to quantitatively reconstruct the paleoclimatic conditions of the Loess Plateau, and there is no effective method to redetermine the paleoclimatic element values. We have evaluated the paleoclimatic element values (annual mean temperature and annual mean precipitation), which might be reflected by individual paleosol layers in the portion from S_0 to S_{14}. This is based

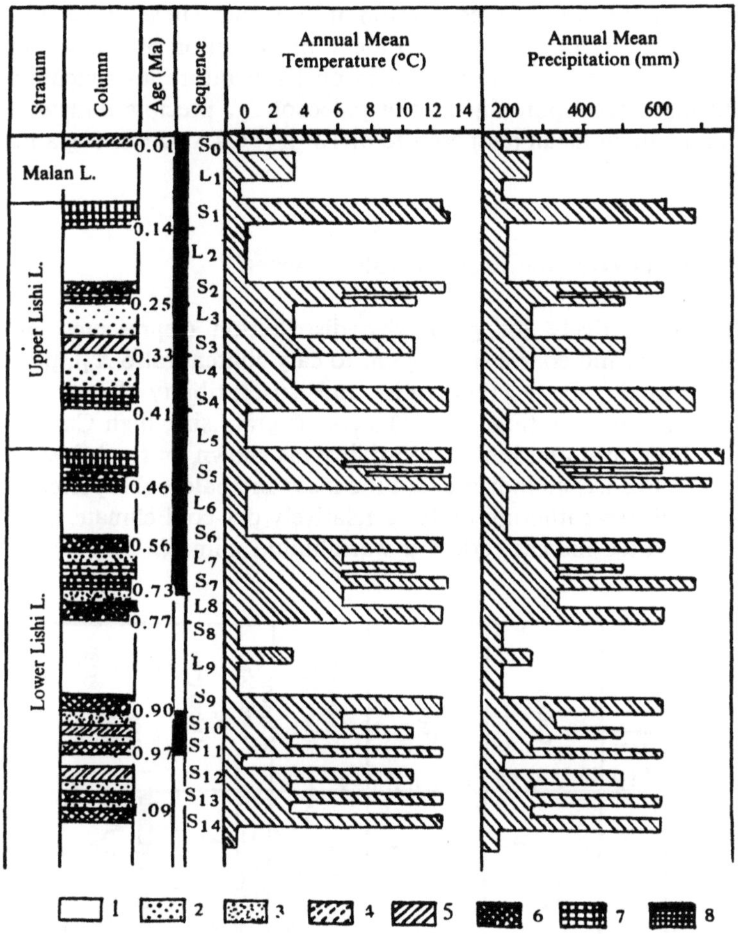

Fig. 6.6. Variation sequence of climatic elements of Luochuan since the last 1.1 m.y. *1* Weakly weathered loess; *2* moderately weathered loess; *3* strongly weathered loess; *4* black loam; *5* carbonate cinnamon soil; *6* cinnamon soil; *7* leached cinnamon soil; *8* brown cinnamon soil

only on the modern geographical distribution characteristics of five different types of paleosols developed in the past 1.1 m.y.

The annual mean precipitation reflected by weakly weathered loess (ca. 200 mm) is similar to the modern transitional zone between black loam and desert soil. It has been estimated that the annual mean temperature during the deposition of weakly weathered loess was 0 °C. This is based on an estimated temperature decrease of 8–10 °C at least in northern China from 0.15–0.18 m.y. (see Sect. VII, Chap. 5). From the temperature and precipitation ranges inferred from weakly weathered loess to black loam, the climatic elements recorded by the moderately and strongly weathered loesses were evaluated through interpolation. A variational sequence of the climatic elements of Luochuan during the last 1.1 m.y. was obtained in accordance with the time sequence of individual loess and paleosol layers as well as their climatic element values (Fig. 6.6).

The annual mean temperature and precipitation of Luochuan since 1.1 m.y. exhibit regular fluctuations. The range in annual mean precipitation may amount to about 600 mm with the range of the annual mean temperature approaching 14 °C. At about 0.5 m.y., corresponding to the paleosol S_5, precipitation was about 800 mm with an annual mean temperature of 12–14 °C (An Zhisheng and Wei Lanying 1980).

6.3.2 Paleoclimatic Curves Over the Last 2.4 Million Years

Curves were plotted respectively for grain size distribution (represented by medium grain diameter M_ϕ and coarse silt content to caly content ratio K_d), percentage content of $CaCO_3$, Fe_2O_3/FeO ratio and susceptibility K of the Luochuan loess-paleosol sequence (Fig. 6.7). The coarse grain size, high $CaCO_3$ content, small Fe_2O_3/FeO ratio and low susceptibility (as shown on the left side of the horizontal axis in Fig 6.7) all indicate that the loess and paleosol experienced rather weak biochemical weathering under a relatively dry-cold climate. The opposite conditions of these indices reflect a warmhumid climate. The curves

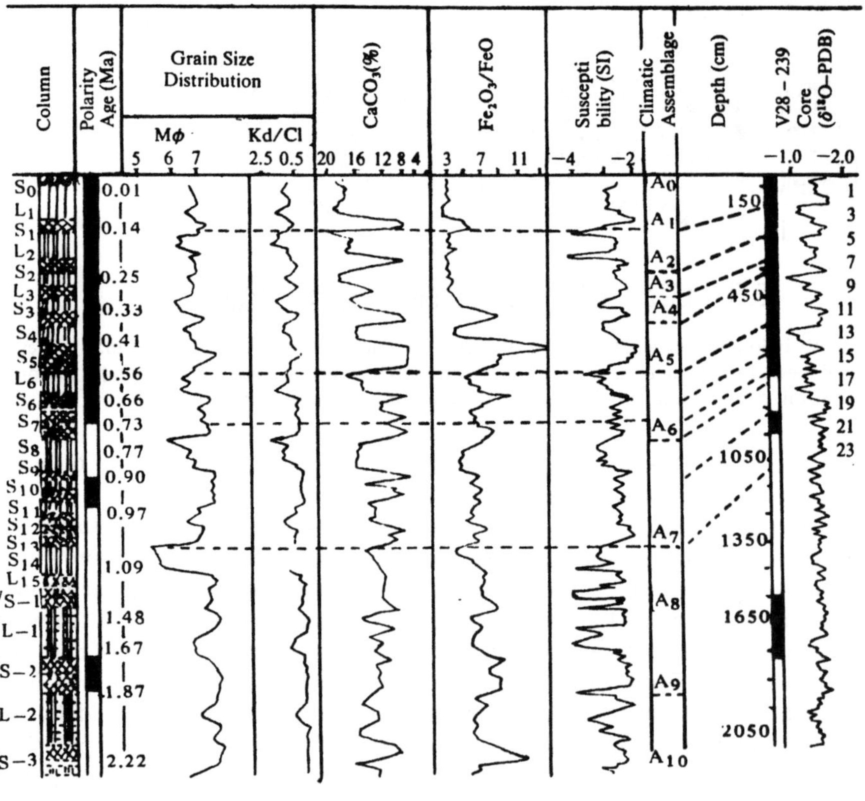

Fig. 6.7. Curves showing variations of indirect climatic indices of Luochuan loess-paleosol sequence since the last 2.4 m.y.

demonstrating the cross-sectional variation of individual indirect climatic indices above show approximately a synchronous variational trend. These may possibly serve as proxies to reflect the climatic changes of Luochuan over the last 2.4 m.y.

An analysis of the curves has yielded the following conclusions: the Matuyama Chron (0.73 – 2.4 m.y.) was characterized by relatively frequent fluctuations in climate with rather warm-humid climatic conditions, while the Brunhes Chron (in the past 0.73 m.y.) featured relatively dry-cold average conditions and clear periodic climatic fluctuations with some larger amplitudes. Loess is an aeolian deposit, and constitutes the parent material of paleosol. Both of them are indicators of climatic and environmetal changes. As a consequence, the paleoclimatic change cycles should be built up in terms of paleoclimate. From the analysis of individual indirect climatic indices of loess and its adjacent paleosol layers (Fig. 6.7), it was concluded that the climatic change from cold (reflected by loess) to warm (reflected by paleosol) was more drastic than from warm to cold. Therefore it seems more reasonable to attribute paleosol and the overlying loess to a single paleoclimate assemblage which represents a climatic change cycle starting from a warm period and ending at cold period. The portion from S_4 to S_0 can thus be divided into (from top to bottom): (1) semi-assemblage A_0 (S_0) and (2) assemblages A_1 (S_1, L_1), A_2 (S_2, L_2), A_3 (S_3, L_3) and A_4 (S_4, L_4). From S_5 to the section, paleosol layers (or weathering layers) occur in a form of densely distributed assemblages, while the loess layers between the assemblages are fairly thick. The paleosol assemblage and its overlying loess layer can therefore be assigned to a single climatic change assemblage with a relatively large time scale. In this way, the following climatic change assemblages were defined (form top to bottom): A_5 (S_5, L_5), A_6 ($S_8 - S_6$, L_6), A_7 ($S_{14} - S_9$, L_9), A_8 (W_{S-1}, L_{15}), A_9 (W_{S-2}, W_{L-1}), A_{10} (W_{L-2}) and A_{11} (W_{L-3}) (semi-assemblage). These paleoclimatic assemblages can serve as a basis for defining paleoclimatic/stratigraphical groups in the loess strata division.

In the above 11 paleosol-loess assemblages, paleosol layers (or assemblages) are intermediated by 11 weakly-weathered or moderately-weathered loess layers with thicknesses greater than 2 m, indicating that in Luochuan area there were at least 11 dry-cold wind-borne dust accumulation periods (or dry-cold climatic periods) with a relatively long duration. Among them, at about 1.15 m.y. represented (by lower sandy loess L_{15}), 08 m.y. (by upper sandy loess L_9) and $95 - 10 \times 10^3$ yr (by Malan Loess L_1), the climate appeared to be much drier and colder. Correspondingly, there were at least 11 warm-humid climatic periods in the Luochuan area since 2.4 m.y. Taking paleosol layers (or asemblages) for example, around 1.8 m.y. (by W_{S-2}), 0.5 m.y. (by S_5) and 0.1 m.y. (by S_1), the climate was optimum or relatively optimum.

Investigation of fossils and analysis of indirect climatic indices show that since 2.4 m.y. the long-term climatic changes represented by the Luochuan loess-paleosol sequence have borne a multi-cycled feature with repeated climatic fluctuations from warm-humid to dry-cold. On the whole, the climate is becoming increasingly dry.

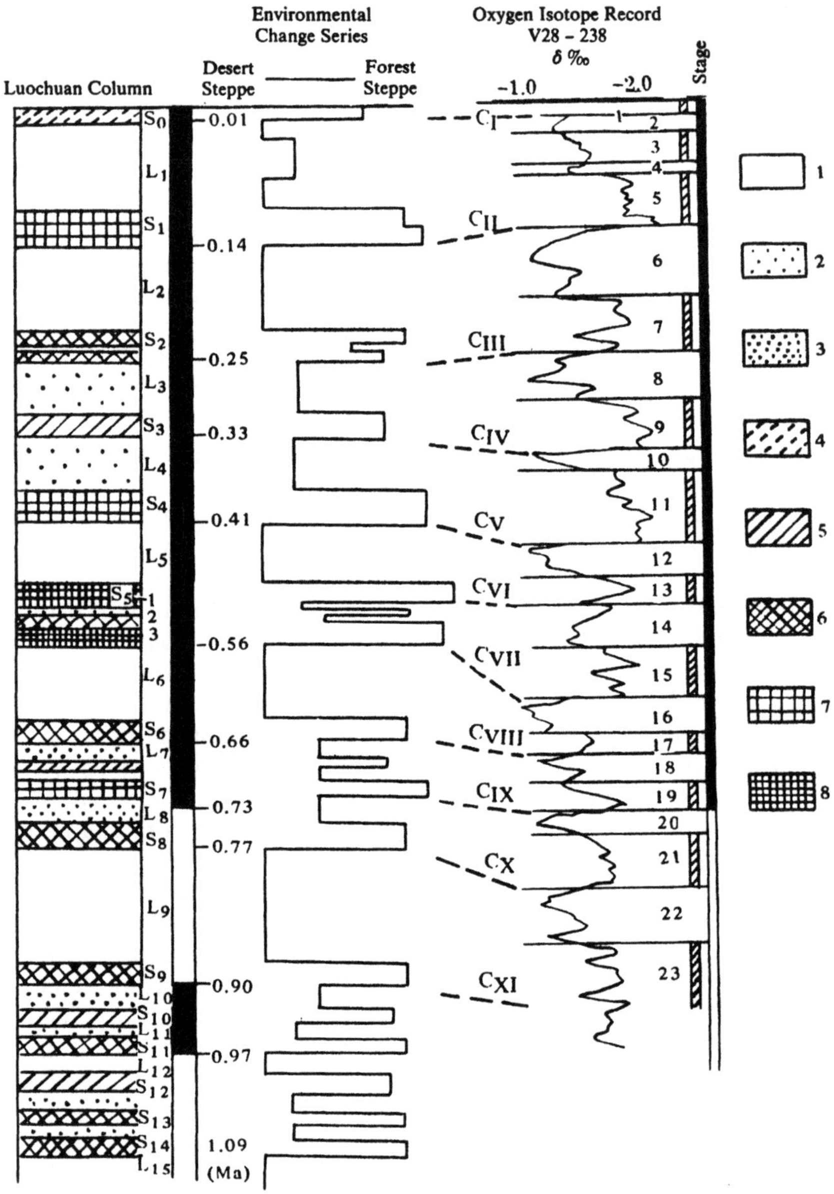

Fig. 6.8. Comparison between Luochuan natural environmental change sequence and V28–238 oxygen isotope records since the last 900000 yr. *1* Weakly weathered loess; *2* moderately weathered loess; *3* strongly weathered loess; *4* black loam; *5* carbonate cinnamon soil; *6* cinnamon soil; *7* leached cinnamon soil; *8* brown cinnamon soil

6.3.3 Comparison Between Pleistocene Land/Sea Climatic Records

In the Luochuan loess section from S_{14} to S_0, 5 different types of paleosol layers are developed, which alternate with three types of loess layers of different weathering extents. Based on the bioclimatic and environmetal characteristics (see Sect. VII, Chap. 5), we have developed a natural environmental change sequence (Fig. 6.8). In Fig 6.8 eleven climatic change cycles from warm-humid to dry-cold were recorded; namely, C_I, C_{II}, ... C_{XI} from S_9 to S_0. Curves of the change sequence indicate the average amplitude of the environmental changes of different periods. On this basis, and with reference to the indirect climatic indices change curves, comparisons were made between the Luochuan loess-paleosol sequence and the $\delta^{18}O$ climatic records preserved in deep-sea cores V28−239 (Shackleton and Opdyke 1973, 1976) in the last 0.9 m.y. (Fig. 6.8). Here S_0 is comparable to oxygen isotope stage 1.

The climatic curve for the loess-paleosol sequence of Luochuan during the past 2.4 m.y. can also be correlated (Figs. 6.7 and 6.8) with the results of Pacific Ocean deep-sea cores studied by Shackleton and Opdyke (1973, 1976) and Kukla (1970). Here loess L_1 correlates with stage 2, 3 and 4 of the oxygen isotope curve, and paleosol S_1 correlates with stage 5. Other earlier loess layers correlate with even-numbered stages and paleosols with odd-numbered stages of the oxygen isotopic curve. The polygenetic soil S_5 is tentatively correlated with stages 13, 14 and 15. This correlation from land to sea shows that the Quaternary climatic fluctuations revealed by the Luochuan sequence are synchronous with those in the deep sea and hence of worldwide significance. This is one very important conclusion reached from the study of Chinese loess.

6.4 Spectral Analysis of Luochuan Loess-Paleosol Sequence

The Luochuan loess-paleosol sequence, known to have developed in the past 2.4 m.y., exhibits multi-cycled climatic changes. These are evidenced by alternations of different types of loess and paleosol, as well as regular fluctuation of other geological and biological indirect climatic indices as the depth increases (Fig. 6.7). Particularly for the loess-paleosol sequence of the last 1−0.77 m.y., a wealth of data with indirect climatic indices make it possible to conduct multi-spectral analyses.

6.4.1 Time Models of the Loess-Paleosol Sequence

Ages of individual loess and paleosol must be first determined prior to the establishment of the time sequence of the loess-paleosol sequence. According to the age interval defined by Jaramillo Subchron and B/M boundary ages and based on the depositional amount, the age at which each paleosol layer started to develop. i.e., the boundary point age between individual loess-paleosol assemblages, can be determined (see Table 3.9). In addition, two different time models

Table 6.7. Time sequence of loess-paleosol sequence of Luochuan in the past 1 m.y.

Sequence	Model I ($\times 10^3$ yr B.P.)	Modell II ($\times 10^3$ yr B.P.)	Sequence	Model I ($\times 10^3$ B.P.)	Modell II ($\times 10^3$ yr B.P.)
S_0	10	10	S_6	660	660
L_1	95	95	L_7	700	705
S_1	140	40	S_7	720	720
L_2	200	200	L_8	742	740
S_2	250	250	S_8	770	770
L_3	300	300	L_9	870	822
S_3	330	330	S_9	900	900
L_4	375	375	L_{10}	920	920
S_4	410	410	S_{10}	940	940
L_5	460	446	L_{11}	960	957
S_5	555	555	S_{11}	970	970
L_6	642	630	L_{12}		

were designed for evaluating the top-surface age of each paleosol (or the bottom-surface age of the relevant loess).

Model 1: Within the time interval spanned by each paleosol-loess assemblage, i.e., the different between starting ages of two adjacent paleosols, the top-surface age of each paleosol (or bottom-surface of loess) is determined by assigning a duration to the relevant layer in accordance with its depositional amount (g/cm^2) (Table 6.7).

Model 2: Since the depositional rate of loess is greater than that of pedogenesis, it seems unreasonable to assign durations to loess or paleosol layers based on a mere depositional amount. If the comparison made by the author between the Luochuan loess-paleosol sequence and deep-sea sediments is reliable (see Fig. 6.8), the ratio of cold/warm climatic stages in each climatic cycle recorded by deep-sea sediments should, more or less, reflect the corresponding ratio of cold (loess) periods to warm (paleosol) periods for each climatic cycle (paleosol-loess assemblage) of Luochuan. The average ratio of cold/warm stages from oxygen isotope content in deep-sea cores V28–238 and V28–239 can then be taken as the ratio of time lengths spanned by the loess and paleosol in the relevant climatic cycle of Luochuan. The top-surface age of each paleosol (or bottom-surface age of the associated loess) can thus be determined (Table 6.7). As before, the age values of loess-paleosol assemblages estimated from magnetic boundaries and depositional amounts were adopted as the starting ages of the paleosols.

After the starting and ending ages of each paleosol (or loess) have been determined, the indirect climatic indices of the layer, which vary with the depth, can be converted through interpolation into the data with time scales, thus providing a foundation for multispectral analysis. Meanwhile, if time models I and II are taken as the scale, then through a smoothing treatment the variational sequence of individual indirect climatic indices recorded in the Luochuan loess-paleosol of the last 1 m.y. can be plotted. The indirect climatic indices contained in the analysis were K_d (data were taken from Chap. 5, Sect. I), Fe_2O_3/FeO (see Chap. 5, Sect. IV), $CaCO_3$ content in percent (according to Lu Yanchou 1981) and

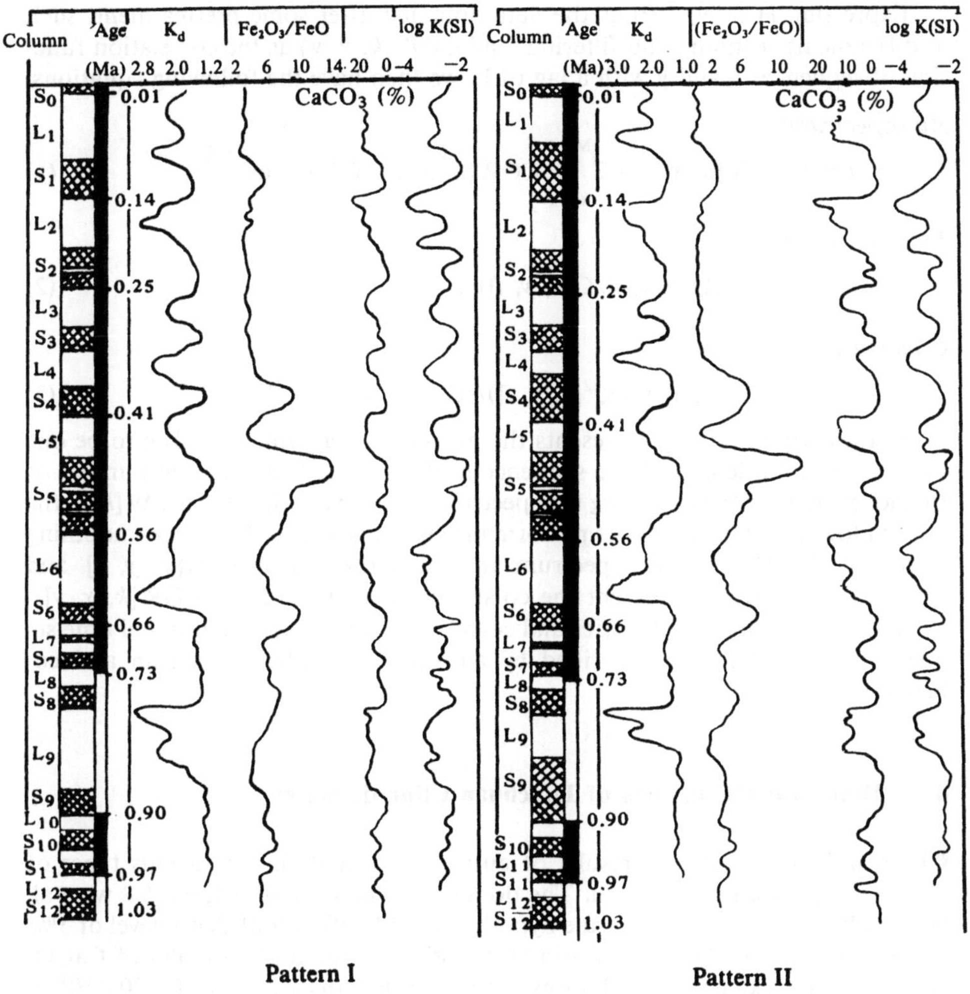

Fig. 6.9. Variation sequence of indirect climatic indices of Luochuan series in the past 1 m.y. K_d (Silt to clay ratio)

log K (Heller and Liu Tungsheng 1984) (Fig. 6.9). For each curve, the left side of the horizontal axis shows dry-cold climatic conditions, and the right side warm-humid conditions.

6.4.2 Approach to Multispectral Analysis

Multispectral analysis is recognized as a valid method to study the oscillation characteristics and correlations of different time sequences. The method adopted here is basically similar to that of Jenkins and Watts (1968) as well as of Lu Longhuan et al. (1984).

Expressing H (1:N, 1:P) as the data obtained after some pretreatment, such as differencing, running and filtering, and COR (k, x, y) as the correlation function of sequences x and y with a lag of k, we then have the following equations.

Autospectrum:

$$Gx,\ i = 2.\ \mathrm{COR}\ [o,\ x,\ x] + 2 \sum_{k=1}^{M-1} \mathrm{COR}\ [k,\ x,\ x] \cdot \mathrm{W}[k] \cdot \ \cos \frac{\pi k i}{\mathrm{L}}\ . \tag{1}$$

Phase-spectrum:

$$\Phi [s,\ y,\ i] = \mathrm{tg}^{-1}(\mathrm{Q}[x,\ y,\ i]/\mathrm{C}[x,\ y,\ i])\ . \tag{2}$$

Coherence:

$$\mathrm{R}[x,\ y,\ i] = ((\mathrm{C}^2[x,\ y,\ i] + \mathrm{Q}^2[x,\ y,\ i])/(\mathrm{G}[x,\ i] \cdot \mathrm{G}[y,\ i]))^{1/2}\ . \tag{3}$$

In these equations, M represents the maximum lag, which is taken to be one third of the total length of the sequence in this paper; L is the total number of frequency points for calculating the spectrum (L = 3 M in this paper); W [k] is the lag window (in time domain its performance is equivalent to Hanning smoothing treatment for the sample spectrum in frequency domain); C [x, y, i] and Q [x, y, i] respectively represent the co-spectrum obtained from COR [k, x, y].

The oscillation characteristics and correlations of a multidimensional time sequence can be perfectly described by autospectrum, phase-spectrum and coherence.

6.4.3 Dominant Oscillations of Luochuan Climatic Series

Given in Table 6.8 are the results of autospectral analysis for various types of climatic sequence of Luochuan. The dominant scales given in Table 6.8 were all verified by a "non-dominant scale hypothesis test" with a confidence level of 5%.

As evidenced in Table 6.8, during the last 1 m.y. the four variables of $CaCO_3$ etc. all exhibit extensive oscillations with cycle lengths between $10-90 \times 10^3$ yr. The dominant oscillation features obtained from Models I and II, as well as different sequence durations are approximately the same. The most pronounced oscillations are those of quasi−0.08, quasi−0.04, and quasi−0.02 m.y.

Fig. 6.10 illustrates the autospectral characteristics of a 1 m.y. sequence, which was generated through a five point running mean so that oscillations with cycle lengths less than 25×10^3 yr could be filtered out. It is quite clear that there exist two dominant scales (quasi−0.08 and quasi−0.04 m.y.).

In discussing the phase correlation of the dominant oscillation, we used a range of $70-90 \times 10^3$ yr to represent the quasi−0.08 m.y. oscillation and $35-45 \times 10^3$ yr to represent the quasi 0.04 m.y. oscillation.

As far as the quasi−0.08 and quasi−0.04 m.y. cycle lengths are concerned, for different variables the coherences are mostly above 50%, showing that they are closely correlated. The average phase differences of the dominant oscillations related to the two models are generally less than 10×10^3 yr. For $CaCO_3$ content in particular, the phase differences between the two models are even smaller. All

Table 6.8. Dominant oscillations of individual variables ($\times 10^3$ yr)

Model	Sequence duration	$CaCO_3$ (%)	$-\log K$ (SI)	K_d	Fe_2O_3/FeO
Model I	1000	6.7, 3.6, 2.8, 2.1*, 1.9*, 1.7, 1.5, 1.4, 1.2	7.2, 5.5, 4.0, 3.2, 2.8, 2.4*, 2.0*, 1.9*, 1.5, 1.4	6.7*, 4.9*, 3.7, 2.3, 2.0*	8.4*, 4.6, 2.3*
	770	3.6, 3.7, 2.1*, 1.9, 1.6, 1.3, 1.2	7.1, 4.0*, 3.1, 2.4*, 1.8*, 1.5, 1.4	7.8*, 4.9, 2.3	8.2*, 4.6
Model II	1000	4.0*, 2.0*, 2.2, 1.8*, 1.7, 1.5, 1.2, 1.0	7.2, 4.5, 3.6*, 2.7*, 2.2, 1.8, 1.7*, 1.5, 1.3	8.4*, 5.3, 4.2, 3.0, 2.5, 2.2*, 1.7, 1.4, 1.2, 1.1, 1.0	9.2*, 4.7, 2.5, 2.3, 1.9, 1.5, 1.4, 1.1*
	770	4.0*, 2.7*, 2.3, 1.8*, 1.6, 1.5, 1.2, 1.0	7.1, 4.5, 3.5*, 2.6, 2.2, 1.9, 1.5, 1.3	7.8, 5.4, 4.1*, 3.1, 2.1, 1.4, 1.2	8.7*, 2.4, 1.8, 1.5, 1.4, 1.1

* Items marked with asterisks were also verified through a "red noise test" with a confidence of level 5%.

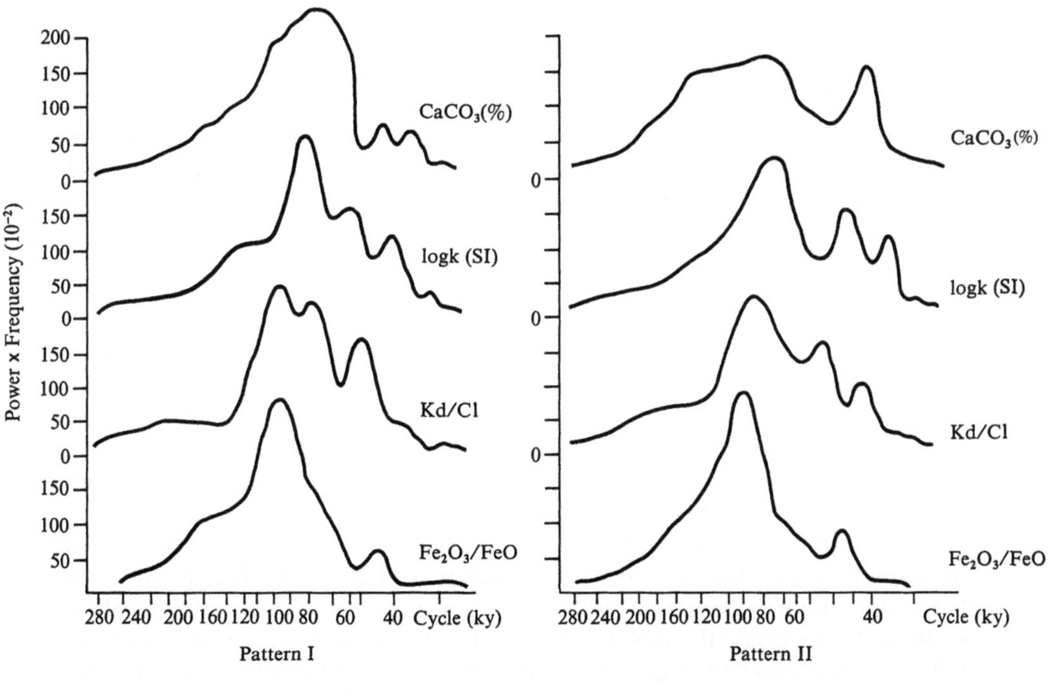

Fig. 6.10. Autospectral characteristics of the 1 m.y. sequence of the Luochuan section

Table 6.9. Average phase correlation of dominant oscillation of different variables

Φ (Degree)	COH (%)	$CaCO_3$ (%)		$-\log K$ (SI)		K_d		Fe_2O_3/FeO	
	Model I	0	100	16	79	7	81	185	75
Quasi – 0,08 m.y.	Model II	0	100	–1	87	–29	76	183	86
	Average	0	100	8	83	–11	78	184	81
	Model I	0	100	93	52	66	50	154	54
Quasi – 0.04 m.y.	Model II	0	100	46	77	22	79	158	59
	Average	0	100	70	65	44	65	156	57

these findings imply that the spectral characteristics of the time sequences established from time models I and II are essentially proximate to one another. Given in Table 6.9 are phase difference and coherence values with respect to the other three variables for quasi – 0.08 and quasi – 0.04 m.y. oscillations of $CaCO_3$ content. Within incycle range of quasi – 0.08 and quasi – 0.04 m.y., the coherences of individual variables are fairly large. The phase difference in quasi – 0.08 and quasi – 0.04 m.y. oscillations of $CaCO_3$ and Fe_2O_3/FeO are approximately 180°, indicating reversed oscillation phases. The average phase differences between $CaCO_3$ and $-\log K$, and between $CaCO_3$ and grain size quasi – 0.08 m.y. oscillations do not exceed 20°, indicating that their quasi – 0.08 m.y. oscillations occurred almost simultaneously. For quasi – 0.04 m.y. oscillations, the $CaCO_3$ phase led the grain size phase by 5×10^3 yr and led the $-\log K$ phase by 8×10^3 yr. In other words, the peak of the $CaCO_3$ quasi – 0.04 m.y. oscillation occurred closer to the present as compared with the peak ages of grain size and the negative logarithm of susceptibility oscillations.

6.4.4 Discussion

1. Domindnt Scales of Oscillation. According to the time model either I or II, for various indirect climatic indices there generally exist two dominant scales of quasi – 0.08 (0.079 – 0.09) and quasi – 0.04 (0.035 – 0.45) m.y. For sequences without running treatment, there were still quasi – 0.02 m.y. oscillations.

Only dominant scales are discussed here; individual autospectral peaks associated with different climatic variables are not covered. This is because the measured data still deviate to some extent from reality and could not exactly reflect the climatic conditions pertinent to the samples in question. This is due to differences between the two time models, to the discrepancy of measuring errors and accuracy of the original indirect climatic index values, also accompanied by some random effects, such as migration of carbonates and iron oxides in loess-paleosol.

In the past 0.97 m.y. in Luochuan, there were at least 11 climatic change cycles from warm-humid to dry-cold. The prevailing quasi – 0.08 m.y. oscillation cycles (Fig. 6.10) may represent, more or less, the average duration of the climatic cycles reflected by loess-paleosol assemblages. The quasi – 0.04 m.y. and quasi – 0.02 m.y.

oscillation cycles belong to secondary cycle elements of long-term climatic changes. A dense and uniform sampling from loess-paleosols may be conducive to ascertaining their behavior is the geological records.

2. Implication of Phase Correlation in Oscillation Cycles. According to phase analyses, the order in which oscillation peaks of variables Fe_2O_3/FeO, $CaCO_3$, $-\log K$ and K_d occur can be written as follows (from old to recent):

Quasi−0.08 m.y. oscillation (25×10^3 yr running treatment applied)

$$Fe_2O_3/FeO \xrightarrow{40\times10^3\,\mathrm{yr}} (-\log K) \xrightarrow{\qquad} CaCO_3 \xrightarrow{\qquad} K_d \xrightarrow{40\times10^3\,\mathrm{yr}} Fe_2O_3/FeO$$

80×10^3 yr ——————————————— 0

Quasi−0.04 m.y. oscillation (25×10^3 yr running treatment applied)

$$Fe_2O_3/FeO \xrightarrow{9\times10^3\,\mathrm{yr}} (-\log K) \xrightarrow{3\times10^3\,\mathrm{yr}} K_d \xrightarrow{5\times10^3\,\mathrm{yr}} CaCO_3 \xrightarrow{23\times10^3\,\mathrm{yr}} Fe_2O_3/FeO$$

40×10^3 yr ——————————————— 0

Quasi−0.02 m.y. oscillation (no running treatment applied)

$$Fe_2O_3 \xrightarrow{5\times10^3\,\mathrm{yr}} (-\log K) \xrightarrow{3\times10^3\,\mathrm{yr}} K_d \xrightarrow{2\times10^3\,\mathrm{yr}} CaCO_3 \xrightarrow{12\times10^3\,\mathrm{yr}} Fe_2O_3$$

$22\times10_3$ yr ——————————————— 0

It is evident from the above expression that of the four variables of the sequence, the Fe_2O_3/FeO peak (will low $CaCO_3$ value) and $CaCO_3$ peak (with low Fe_2O_3/FeO value) differ from one another by a half cycle of approximally 0.04 or 0.02, or 0.01 m.y., clearly showing an opposite phase correlation. Fe_2O_3/FeO values indicate the oxidation/reduction conditions under which loess or paleosol was formed, and reflect to a certain extent the variation of temperature as one of the climatic elements. $CaCO_3$ content reflects the leach or accumulation of Ca, indicating, to a certain extent, the humidity conditions under which loess or paleosol was formed. These variables are believed to better indicate the climatic oscillations from warm-humid to dry-cold.

3. Comparison of Climatic Pulsation Cycles Recorded by Deep-Sea Sediments. Using the SIMPLEX age model, Hays et al. (1976) made a frequency-spectral analysis of deep-sea drilling hole RC11−120 and more core E49−18 for three geological variables $\delta^{18}O$, Ts and *C. Davisiana* percentage, and extracted three cycle elements of 87−119×, 37−47× and 21−24×10^3 yr. The former two may plausibly be comparable to the quasi−0.08 and quasi−0.04 m.y. cycles of Luochuan climatic variables. In the climatic variable sequences of 1 and 0.77 m.y. without running treatment, there exists a cycle element of quasi−0.02 (0.022) m.y. comparable to that of 0.021−0.024 m.y. in deep-sea climatic records.

4. Comparison with the Change Cycles of Earth-Orbital Element. The orbit hypothesis of climatic changes and other geological evidence suggest that the radiation-climate system exhibits a lineal response to both the tilt of the earth's axis and precession of equinoxes. The prominent quasi−0.04 and quasi−0.02 m.y. cycle elements of Luochuan loess-paleosol sequence can be compared to the tilt of the earth's axis at about 0.041 m.y. and to the average of precession of equinoxes at about 0.021 m.y. (Hays et al. 1976). The eccentricity of orbit has an approximately 0.093 m.y. cycle which may plausibly be associated with the prevailing quasi – 0.08 m.y. cycle in Luochuan series. For the time being, however, it is difficult to discuss the correlations between them since climate has no simple linear relation with the eccentricity of orbit. The quasi – 0.08, quasi – 0.04 and quasi – 0.02 m.y. oscillation cycles extracted from Luochuan climatic records have a close relation with the long-term changes of the geometric elements of the earth's orbit. However, the relative position changes between the earth and the sun could not completely govern the whole change process of the "interior" climatic system of the earth. It is worthwhile noting that in the vast area of Northwest China the loess deposition, desert expansion and the dry-cold climate of glacial periods are all closely related to the uplift of the Qinghai-Xizang Plateau. The surface-albedo feedback effect and the increase of dust concentration in atmosphere caused by the loess and desert development might all exert a substantial influence on the radiation equilibrium at the mid-latitude, thus to a certain extent affecting the global climate, especially the paleoclimatic changes of northern China.

6.5 Composition of Loess and Loessial Soil and Agricultural Production

6.5.1 Composition and Texture of Loess and Agricultural Production

According to Zhu Xianmo (1981), loessial soils in loess-distributed areas, for example in Shaanxi Province, mainly include the following types: black loam, cinnamon soil, manured loessial soil, yellow cultivated loessial soil and Heizhuang soil (black fertile soil). In addition, there are castanozems and calcic podzolic soil (Peng Lin 1983).

Cultivation-reclamation indices in most areas of the Loess Plateau are generally over 30%, even up to 60 – 80% (Peng Lin 1983). Irresponsible utilization of soil and land resources, for instance unplanned reclamation, denudation, extensive cultivation and ruthless destruction of soil and vegetation are greatly accelerating soil erosion. Therefore the traditional habit of extensive cultivation must be given up; instead a progressive transition must be made from monoagricultural management to a comprehensive farming system, including forestry and animal husbandry.

The Loess Plateau is a strongly eroded area with a total area of $430 \times 10^3\,km^2$. Statistical data show that the annual amount of soil erosion reaches 5 000 – 10 000 ton/km^2 (Gong Shiyang and Xiong Guisu 1979). The amount of silt transported by the Huanghe River averages 1 600 million tons each year. It is of great significance in the realization of China's agricultural modernization to solve the problem of soil erosion on the Loess Plateau. In this paper the authors offer the following proposals for developing agricultural production on the Loess Plateau on the basis of the composition and texture of loess, coupled with the practical experience of the local populace.

1. When attention is paid to water and soil conservation and comprehensive development of farming, forestry and animal husbandry, the composition and textural zonation of loess should be taken into consideration. For example, if terraced fields are built on the sandy loess belt in the northwestern part of the Loess Plateau, the height and gradient of inclination of field banks should be designed somewhat differently from those on the clayey loess belt in the southeastern part; the former should be lower and more gentle, whereas the latter should be higher and steeper.

Moreover, in the construction of dams to create fields, regional variations in the composition and texture of loess should also be taken into consideration. For example, in the sandy loess-covered hilly area a rapid damming technique (hydraulic fill damming) is recommended, but in the loess "Yuan" area where loess or clayey loess is finer and more coherent than sandy loess in the northern part, where the Malan Loess is relatively thin and the more clayey Lishi Loess and Wucheng Loess are extensively exposed in the gullies, the hydraulic-fill damming technique mentioned above is unsuitable.

2. Construction in farming areas should keep pace with the measures of water and soil conservation. The transformation from soil erosion to water and soil conservation should be effected by engineering measures, in conjunction with some local methods such as adjusting soil quality, leveling and deeply ploughing and improving the texture and properties of soils. Three types of landform are typical of the Loess Plateau, namely "Yuan", "Liang" and "Mao". Construction should be carried out in accordance with specific landform conditions.

In the initially cultivated loessial soil, artificial improvement of soil quality can be achieved by leveling and deep ploughing. Furthermore, extensive application of organic fertilizers is important to promote the transformation of sandy textural loessial soil into perfectly granulated structural soil.

3. Attention should be paid to the study of trace elements both in loess and in the upper cultivated soil to enhance their fertility.

4. Attention should also be paid to the influence of clay minerals on the fertility of soil and on their influence on engineering installations.

In addition to engineering measures, biological measures should be fully stressed and strengthened in developing of the Loess Plateau. Reforestation and grassplanting are of great significance in keeping the ecological balance, adjusting the microclimate and preventing soil erosion on the Loess Plateau.

6.5.2 Trace Elements in Loess and Loessial Soil and Agricultural Production

A total of 180 samples of Malan Loess and modern cultivated soil (loessial soil) collected from 16 counties (city) along the middle reaches of the Huanghe River were analyzed for trace elements particularly Zn, Cu, Mn and Mo which are closely related to the growth of crops (Diao Guiyi et al. 1982). We analyzed not only the bulk content of Zn, Cu, Mn and Mo, which are closely related to the growth of crops in the loess, but also the content of these elements present in the form capable of being effectively taken up by crops.

Table 6.10. Contents of trace elements in the Malan Loess from the middle reaches of the Huanghe River (ppm)

Element	Number of samples	Content range		Average content			
		Loess	Soil	Loess	Soil[a]	Crust[b]	Sedimentary rocks
Zn	36	41 – 73	10 – 300	62	50	80	80
Cu	36	14 – 27	2 – 100	23	20	70	57
Mn	36	310 – 676	200 – 3000	534	850	1000	670
Mo	36	0.39 – 0.71	0.5 – 5	0.54	2	2.3	670

[a] After Goldschmidt.
[b] After Vinogradov.

6.5.2.1 The Content and Distribution of Trace Elements in the Malan Loess

Semi-quantitative spectral and partial chemical analyses revealed the following trace elements in the Malan Loess (%): Ti 0.3 – 0.5, Mn 0.3 – 0.75, Sr 0.01 – 0.1, Ba 0.05 – 0.07, Zr 0.01 – 0.03, V 0.007 – 0.01, Cr 0.007 – 0.01, Pb about 0.03, Zn 0.003 – 0.009, Ni 0.002 – 0.004, and Cu 0.001 – 0.003, and minor amounts of Mo, Be, Co, B etc. (about 1 ppm).

In recent years, we have analyzed several trace elements closely related to the growth of crops (Table 6.10), and the results indicate that the content of Zn and Cu in the Malan Loess is lower than the average content of Zn and Cu in the crust and sedimentary rocks, but exceeds the average content of soils. The content of Mn, although lower than the average content of Mn in soil, reaches the average content of Mn in sedimentary rocks. The content of Mo is the lowest, its average value being only one quarter of the average value of soil. Meanwhile, it has been found that the total of trace elements increases from north to south, and Zn, Cu, Mn and Mo also increase in the same direction. Their regional distribution is consistent with the zonal characteristics of the Malan Loess, and is related to the physical geographical environment at the time of Malan Loess accumulation on the Loess Plateau, where the climate is dry in the northwest and moist in the southeast. Although the trace elements in the Malan Loess vary in concentration, there are significant correlations among the different elements. For example, Zn, Cu and Mn show remarkable positive correlations whith each other. In addition, the content ratios of these three elements are relatively stable. This may be related to the fact that the Malan Loess distributed in various areas along the middle reaches of the Huanghe River came from the same provenance.

6.5.2.2 Trace Elements in Loessial Soil in Relation to Agricultural Production

In addition to water, temperature, sunshine, carbon dioxide, and N-, P- and K-fertilizer, trace elements are indispensable to the normal growth of crops. The content of some trace elements in loessial soils in various areas along the middle reaches of the Huanghe River is generally close to that in the parent loess materi-

Table 6.11. The average contents of effective trace elements in loessial soils and parent materials from various localities

Locality	Name	No.	Zn (ppm)			Cu (ppm)			Mn (ppm)			Mo (ppm)		
			A	B	C	A	B	C	A	B	C	A	B	C[a]
Yulin	S.[b]	2	50	0.75	0.015	17	1.55	0.091	405	15.3	0.038	0.50	0.007	0.014
	p.m.	2	51	0.55	0.011	20	1.35	0.068	462	12.5	0.027	0.56	0.007	0.018
Suide	S.	2	59	0.90	0.015	20	1.73	0.087	503	17.4	0.085	0.48	0.007	0.015
	p.m.	2	55	0.90	0.016	20	1.45	0.073	480	18.8	0.039	0.54	0.003	0.006
Tongxin	S.	2	58	0.80	0.014	21	1.30	0.062	500	13.3	0.027	0.49	0.010	0.020
	p.m.	8	60	0.71	0.012	20	0.70	0.035	495	8.1	0.016	0.96	0.122	0.127
Dingbian	S.	4	41	1.25	0.031	14	1.85	0.130	310	29.6	0.096	0.49	0.017	0.035
Jingyuan	S.	2	68	1.03	0.015	23	1.40	0.061	570	17.4	0.031	0.63	0.013	0.021
	p.m.	4	67	0.69	0.010	23	0.70	0.035	567	8.1	0.014	0.89	0.122	0.137
Lanzhou	S.	2	70	1.16	0.017	26	1.70	0.065	550	12.0	0.022	0.56	0.007	0.013
	p.m.	3	65	0.80	0.012	24	0.85	0.035	563	22.3	0.040	1.05	0.043	0.041
Dingxi	S.	2	71	0.80	0.011	22	1.30	0.059	560	29.3	0.052	0.56	0.017	0.023
	p.m.	3	64	0.69	0.011	23	1.15	0.050	566	10.3	0.018	0.56	0.014	0.025
Xifeng	S.	2	58	0.71	0.012	23	1.30	0.057	520	26.2	0.050	0.42	0.017	0.041
	p.m.	4	63	0.52	0.008	23	1.10	0.048	557	9.3	0.017	0.54	0.020	0.037
Lintai	S.	2	62	0.59	0.010	23	1.30	0.057	560	19.8	0.035	0.39	0.23	0.59
	p.m.	4	61	0.55	0.009	23	1.05	0.046	560	10.4	0.019	0.49	0.029	0.059
Yongshou	S.	2	60	0.69	0.012	25	1.30	0.052	570	13.6	0.024	0.42	0.010	0.024
	p.m.	3	63	0.67	0.011	25	1.55	0.062	587	6.8	0.012	0.45	0.007	0.016
Yanan	S.	4	62	0.88	0.014	22	1.74	0.079	526	18.5	0.035	0.68	0.013	0.019
	p.m.	4	64	0.73	0.011	23	1.75	0.076	571	11.9	0.021	0.75	0.010	0.013
Luochuan	S.	4	68	1.04	0.015	24	2.29	0.095	581	51.6	0.089	0.58	0.020	0.035
	p.m.	3	72	0.97	0.014	26	1.57	0.060	616	13.8	0.022	0.49	0.017	0.035
Tongchuan	S.	3.	70	0.88	0.013	26	1.68	0.065	600	26.0	0.042	0.71	0.010	0.014
	p.m.	1	71	0.80	0.011	25	1.70	0.068	535	10.5	0.018	0.46		
Longxi	S.	2	64	0.86	0.014	26	1.10	0.042	530	19.4	0.033	0.39	0.017	0.044
	p.m.	4	62	0.60	0.010	25	1.10	0.044	568	12.9	0.023	0.44	0.022	0.050
Wugong	S.	3	73	0.63	0.009	27	1.53	0.057	676	23.5	0.035	0.68	0.010	0.015
	p.m.	2	74	0.50	0.007	27	0.85	0.032	663	10.8	0.016	0.56	0.013	0.025

[a] (A) Bulk content; (B) effective state; (C) effective coefficient.
[b] S. = soil; p.m. = parent material.

als. The concentration of trace elements in soil cannot be used as a measure of their contribution to crops; what can be effectively absorbed by crops are those trace elements present in the active, effective state. The content of these trace elements in soil is related to the physicochemical properties of soil itself. It can be seen from Table 6.11 that in loessial soils from the middle reaches of the Huanghe River, the contents of Zn, Mn and Mo present in the effective state are all lower than the critical values, and therefore not in sufficient supply for normal crop growth. In order to enhance their content to meet the requirements of normal crop growth, some trace element fertilizers should be spread in accordance with specific conditions. One of the most important crops growing in the loess-distributed areas is maize. As its growth is relatively sensitive to Zn, a Zn-fertilizer should be applied, followed by a Mn- and Mo-fertilizer, depending on the different crops. Application of Cu-fertilizer to crops growing in sandy soil and yellow cultivated loessial soil containing minor effective-state copper may significantly contribute to the increase of agricultural production.

6.5.2.3 Variation of Trace Elements in the Process of Amelioration of Loessial Soil

After many years of cultivation, loessial soils contain as much Zn, Cu and Mn as the parent materials of the Malan Loess. The content of these elements approaches or exceeds the average value of soil. However, Mo is almost one order of magnitude less than the average value of soil. With increasing number of years of cultivation, the soil will become more and more mature, and thus the concentration of Zn, Cu and Mo present in the effective state and substitutional Mn tend to increase. These elements are obviously higher than those in the parent loess materials. For example, the content of effective-state Zn in the parent loess ranges from 1.2 to 1.6 ppm, while its content in the activated layer of loessial soil may be as high as 1.8 – 2.3 ppm after only one year of cultivation. After many years of cultivation, the content of effective-state Zn in loessial soils may reach 2.2 – 2.3 ppm (Wen Qizhong and Yu Suhua 1979). At the beginning of cultivation, effective-state Zn, Cu, Mn and Mo are generally low in loessial soils, reflecting that soils in the newly built fields generally possess a low supply capacity of Zn, Cu, Mn and Mo. At the seedling stage of maize growing in the loess area, the occurrence of "leaf spot" seems due to the lack of zinc. Good results will be achieved by applying trace element Zn-fertilizer in good time. In loessial soils, the content of effective-state Mo is lower than the critical value (0.15 – 0.02 ppm), probably for the growth of leguminous and mustard-family crops. For this reason, proper application of Mo-fertilizer (e.g., ammonium molybdenate) may be beneficial for the growth of crops.

6.6 Geochemical Environment of Loess and Endemic Diseases

Endemic disease refers to a disease which is characterized by certain area and a mass of people and is related to local natural conditions. Thus, the disease is called "unacclimatization" to water and soil by the local people. Diseases of this nature include Keshan disease, Kaschin-beck disease, local fluorine poisoning, goitre, Keding disease, etc. The causes for these local diseases are still unknown, but from the biogeochemical viewpoint, biological and abiotic conditions are very important for the incidence and distribution of the local diseases.

6.6.1 Keshan Disease

6.6.1.1 Distribution of Keshan Disease in China

In 1935, large numbers of patients suffering form myocardial disease died suddenly in Keshan County, Heilongjiang Province. This etiologically undefined dis-

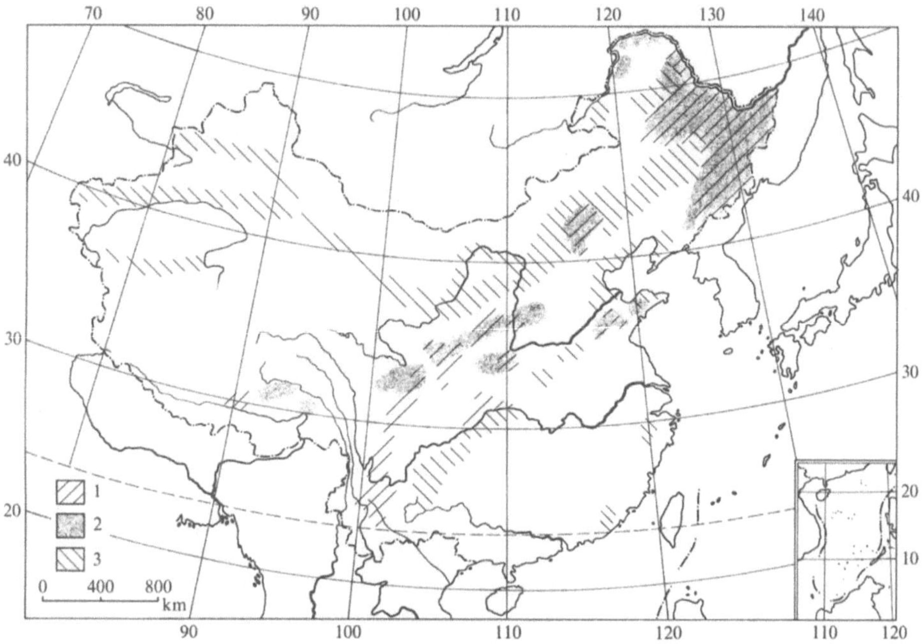

Fig. 6.11. Occurrence areas of Keshan disease, Kaschin-beck disease and local fluorine-poisoning. *1* Keshan disease; *2* Kaschin-beck disease; *3* local fluorine poisoning

ease is called "Keshan disease". Professor Yu Weihan in Harbin Medical University suggested that Keshan disease is a myocardial disease with a chronic course. It mainly impairs the cardiac muscle, but is neither an inflammatory, nor a contagious disease. The epidemic characteristics, in addition to the locality, are frequent occurrence as well as multiannual and multiseasonal (often in winter). It selectively strikes pregnant women and preschool children and shows multiple occurrence in a family. Historical records, in-situ investigation and reports from medical workers indicate that the disease has occurred in some provinces (Fig. 6.11).

Keshan disease in China occurs in areas from Heilongjiang, Hebei, Inner Mongolia and Shaanxi, to Sichuan and Yunnan as a wide zone in northeast-southwest direction (Fig. 6.11). If China is divided into three belts from arid to humid climate, such as northwest inland arid belt, southeast humid belt, and semi-arid and semi-humid belt between them, then Keshan disease is distributed in the middle zone. The disease areas seem to be more or less stable in local extent.

6.6.1.2 Geochemistry of Loess and Keshan Disease

The loess region exhibits a high incidence of Keshan disease. This area includes northern Shaanxi, eastern Gansu, western Henan, and western Shanxi with northern Shaanxi being the most serious (Fig. 6.12).

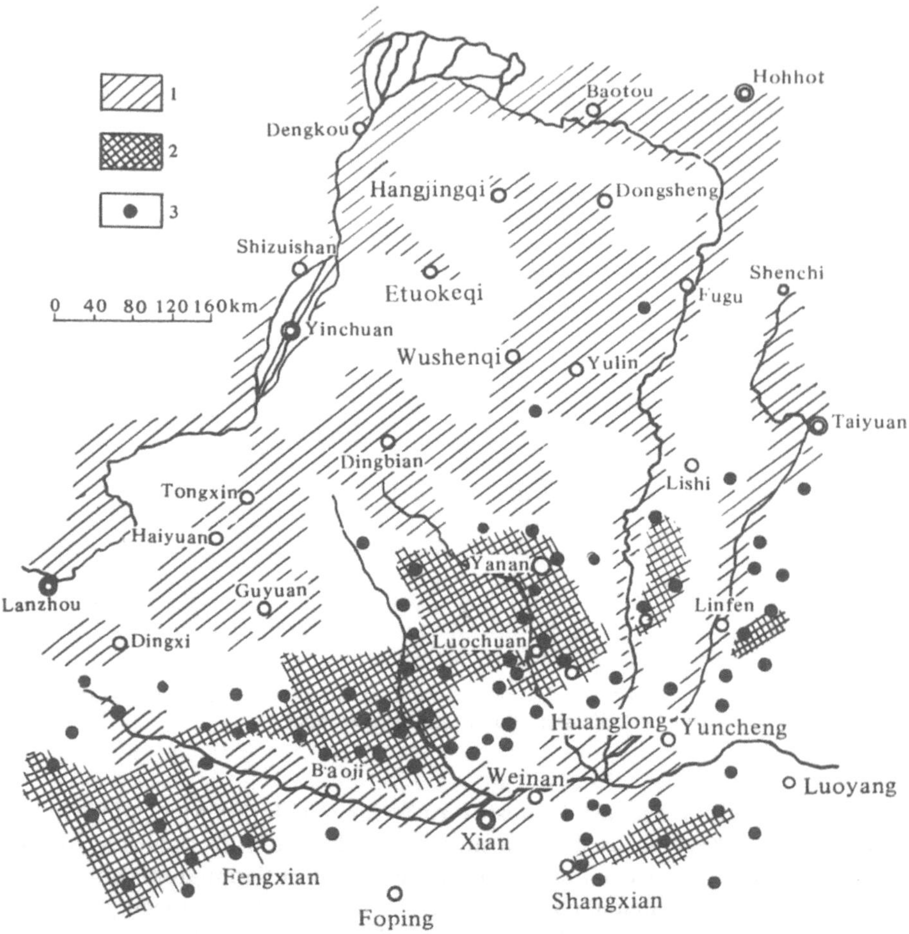

Fig. 6.12. Occurrence of Keshan disease, Kaschin-beck disease and local fluorine poisoning in loess region. *1* Local fluorine poisoning; *2* Keshan disease; *3* Kaschin-beck disease

Geological and Geomorphologic Background. The Loess Plateau is composed mainly of Mesozoic sandstones, with overlying remnant Tertiary Red Clay covered by loess tens to over one hundred meters thick. Ten to twenty or more paleosol layers are developed in the loess layer, and appear to be a series of cinnamon soils. The effect of basement relief and the dissection by running water have brought about the relief of "Yuan", "Liang" and "Mao" in the Loess Plateau (Liu Tungsheng et al. 1964). The Luohe and Jinghe Rivers flow across the northern Shaanxi area. Keshan disease often occurs in the divide zone, but the slight disease and disease-free areas often lie in the lower reaches of the river valleys and the open flat "Yuan". This reflects a correlation of the distribution of Keshan disease in the regions of serious-, slight- to disease-free with locality from upper to lower reaches of some tributaries of the Huanghe River.

The Keshan disease area in Shaanxi Province lies between the north Weihe River valley and the southern Baiyushan mountain, from the Huanghe River in

Table 6.12. SO_4, Ca, Mg, SiO_2 contents in drinking water from Keshan disease areas and disease-free areas in the loess region (mg/l)

Area	Components analyzed					
	SO_4^{2+}	Ca^{2+}	Mg^{2+}	SiO_2	Mineralization (g/l)	Total hardness
Keshan disease	14.22	37.92	16.16	9.30	0.31	8.06
Disease-free	49.96	47.40	20.21	13.92	0.47	7.28

Table 6.13. Ba, Ca, Pb, Zn contents and elements ratios in drinking water from Keshan disease areas and disease-free areas in the loess region

Area	Analyzed composition										
	Ba	Sr	Ca	Pb	Zn	Ca/Sr	Mg/Sr	SO_4/Sr	Ca/Cu	Mg/Ca	SO_4/Cu
Keshan disease	25.40	470.6	0.6	0.5	97.0	37.3	20.9	49.5	32.5	16.1	14.3
Disease-free	253.8	373.0	0.7	0.7	27.2	69.0	49.5	37.6	16.1	20.2	68.5

the east to Ziwuling mountains in the west. The seriously disease-afflicted area lies in the secondary forest area of the Huanglong and Ziwuling mountains.

Quality of Drinking Water. The All-China Joint Etiological Investigation Group on Keshan disease made an analysis and investigation of major elements and trace elements in waters from disease areas (1974). The results indicate that among the major elements in disease areas, Ca, Mg, SO_4 and SiO_2 are lower than those in disease-free areas. Among trace elements such as Cu, Pb, Zn, Sr and Ba, only Pb tends to be lower (Table 6.12 and Table 6.13). Among the element ratios, Ca/Sr, Mg/Sr, SO_4/Cu ratios are lower than those in disease-free areas. However, the Ca/Cu ratio is higher than that in disease-free areas. These ratios reflect a certain difference in contents of chemical elements in water source environments between disease areas and disease-free areas.

Soils. Soils in Keshan disease areas are mainly black loam developed on loess. Black loam is relatively thick, usually 2 m or more and contains a thick humus layer. The whole loam profile exhibits strong $CaCO_3$ reaction. The surface soil layer is rich in nitrogen, phosphorus, potassium and other metallic elements. The effective Mo, total Se and humus contents have been determined. In order to understand the relation between Keshan disease and Mo, Se and humus in soils Table 6.14 shows that effective Mo in disease areas is lower than that in disease-free areas, but humus content in the disease area is higher than that in the disease-free areas.

Food. The main foods in Keshan disease areas of the loess region are wheat and maize, important compositions in food chain, which reflect the natural conditions of water and soil. Results of chemical analyses exhibit irregular variations in the

Table 6.14. Content of total Se, effective Mo and humus in soils of Keshan disease and disease-free areas in the loess region

Area	Analyzed compounds					
	Effective Mo (ppm)		Total Se (ppm)		Humus (%)	
Soil layers	1	2	1	2	1	2
Disease areas	110	96	0.10	0.09	0.17	0.24
Disease-free areas	133	184	0.08	0.08	0.14	0.14

major elements; only element Mo in wheat in the disease area is lower than that in disease-free area while Cu is higher than in disease-free area. Cu and Mo in maize are slightly lower in the disease-free area. Statistical data show that the level of absorbed Cu and Mo in the disease area is higher than that in the disease-free area.

6.6.1.3 Etiological Problem of Keshan Disease

The cause of Keshan disease is a complex problem, which is still unsolved. The absence of one or several chemical elements in water, soil and crops may be the cause of the disease, such as the absence of Mg (after Zhangjiakou Medical College), Se (after Chinese Academy of Medical Sciences, Xian Medical College), and Mo (after Guiyang Institute of Geochemistry, Academia Sinica). From this etiological viewpoint, attention should be paid to the following aspects:

1. Absence of Mo. The element Mo is a direct participant in the metabolism of the cardial muscle in the human body and affects the activity of some important biogenic oxidative-reductive ferments. Determinations of Mo in food from Northeast China region show that Mo content in food (maize) from the disease area is lower than that (in maize and rice) from the disease-free area. The experiment on the application of Mo fertilizer to crops to improve food for the prevention of Keshan disease in Yunnan Province has yielded certain positive results.

2. Absence of Se. Xian Medical College first suggested the absence of Se as a cause of Keshan disease because Se can be effectively used for protecting livestock from disease by muscle necrosis caused by absence of Se. The extent of distribution of Keshan disease is approximately coincident with that of leucomuscle disease of livestock in Shaanxi Province. Thus, medical and geological scientists generally pay attention to the relation between element Se and Keshan disease and the Institute of Hygenics, Chinese Academy of Medical Sciences, and other related departments have had some success in preventing Keshan disease.

6.6.2 Kaschin-Beck Disease

Kaschin-beck disease is an etiologically undefined local disease. The symptoms of the disease are enlargement of finger joints, convulsions, spasms and stiffness in elbow joints, alternative or continuous pain in fingers (toes), wrists, ankles, knee joints. Kaschin-beck disease affects bone development with resulting loss of productivity in the patient. The Kaschin-beck disease area lies in a northeast-southwest direction. In Northeast and Northwest China, the disease areas are, to some extent, coincident with the Keshan disease areas. In Southwest China, the disease has been found only in the Aba district of the Sichuan Province. However, in the Keshan disease area in Yunnan Province, the Kaschin-beck disease has not been found.

6.6.2.1 Environment of Loess and Kaschin-Beck Disease

Geological Background. Kaschin-beck disease occurs in the areas of Yongshou, Chunhua, Linyou and Qianxian Counties on the Loess Plateau, where the thickness of loess reaches tens to over one hundred meters. In areas of geomorphologically eroded and dissected relief, the Kaschin-beck disease occurs more seriously, but does not generally occur on the even surface of "Yuan" (Fig. 6.13).

It is seen from the geomorphic-geologic profile that the Ganjing and Yijing area is weakly eroded and its landform is even, so the occurrence of disease is less frequent. In more intensely eroded areas of Yongshou and Guantou, the occurrence of disease is more frequent. The incidence of the disease in villages varies in the different geomorphic units (Table 6.15).

Quality of Drinking Water. A comparison of the quality of drinking water in disease and disease-free areas shows that mineralization and content of strongly migrating Na^+, Cl^-, SO_4^{2-}, Mg^{2+} and Ca^{2+} and pH in the disease areas are lower than those in disease-free areas. Analysis of trace elements Ba, Sr, Cu, Pb and Zn shows that Cu and Zn in the disease areas appear to be higher than those on the disease-free areas, but Ba, Sr and Pb vary irregularly. According to scientific research of Kaschin disease in Yongshou County (Office of Leading Group for Prevention of Local Disease 1984, unpubl.), other concentrations of the trace

Table 6.15. Relationship between geomorphological unit and distribution of the village where disease occurred[a]

Geomorphological unit	Incidence of disease				
	Free 0	Light 20%	Medium 20–50%	Serious 50%	Sum
"Yuan" surface	0	43	24	0	67
		64.2	35.8		100
Residual "Yuan" slope	0	12	105	51	168
		7.1	62.5	30.5	100
Gully	2	2	0	0	4
	50	50			100

[a] $\frac{\text{Number of the village where disease occurred}}{\text{Percentage}}$.

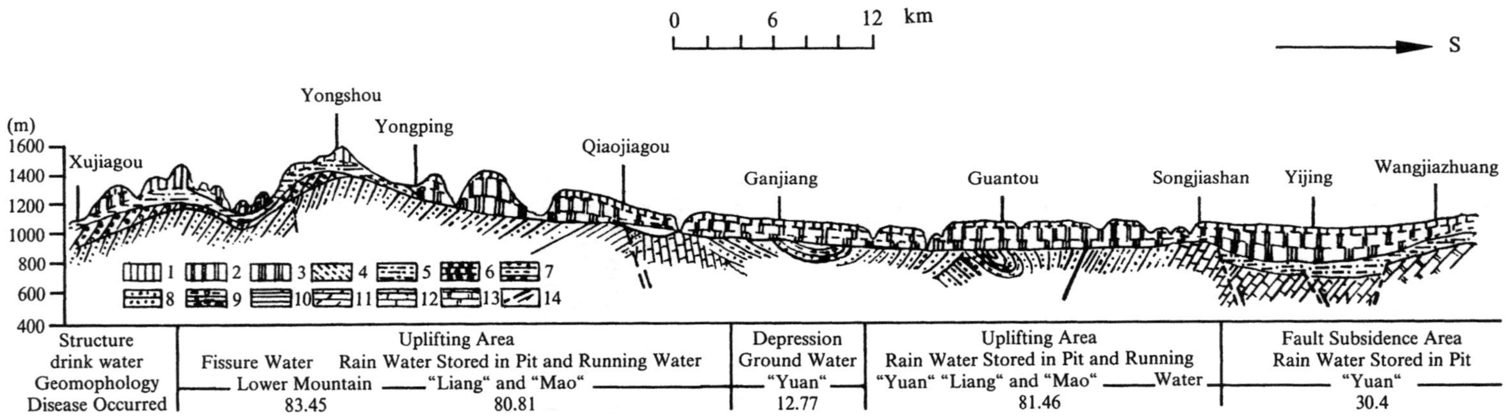

Fig. 6.13. Occurrence of Keshan disease, Kaschin-beck disease and local fluorine poisoning in loess region. *1* Local fluorine-poisoning; *2* Keshan disease; *3* Kaschin-beck disease

Table 6.16. Content of trace elements in drinking water of Yongshou County

Type of drinking water	Elements analyzed								
	Se	Mo	Cu (ppb)	Pb	Zn	Fe	Mn (ppm)	Sr	Cr
Rain water stored in pit	0.12	1.3	4.0	4.1	24.0	2.05	0.15	0.23	0.005
Shallow ground water	0.27	1.1	3.3	3.2	27.2	0.66	0.28	0.57	0.011
Deep ground water	0.19	1.6	7.1	3.1	9.7	0.83	0.03	0.62	0.064
Fissure water	0.18	1.5	4.0	5.3	30.8	0.50	0.49	2.49	0.003
Running water	0.10	0.9	6.0	4.4	16.6	1.65	0.06	0.30	0.006

elements are lower with the exception of higher Fe and Mn content in drinking water in the disease area. This is related to the geochemical environment of Yongshou characterized by leaching, large loss of migrating elements of Se and Mo, and relative concentration of difficulty migrating Fe and Mn (Table 6.16).

Content of Humus. The total content of humus in drinking water in the Kaschin-beck disease area of Yungshou County is 1.06 ppm on average, with a maximum of 5.7 ppm, and a minimum of 0.16 ppm. These values are similar to those in the Kashan disease areas in Northeast China, and 0.5 ppm higher than those in disease-free areas.

The concentration of Se and Zn in maize from the disease areas is less than that from the disease-free areas. Se, Zn and Fe in wheat are less than those in the disease-free areas. A correlation analysis of maize and wheat with the disease suggested that elements Mo, Ba, Fe, Mn and Se are correlated with the occurrence of the disease. For example, Se in wheat and maize has a negative correlation to the occurrence of the disease.

Soils. Parent materials of soils in disease are mostly carbonate-rich loesses. The relation between elements and incidence of the disease shows that incidence of the disease has a negative correlation to water-soluble Se, and a negative correlation to effective Zn.

In summary, we suggest that there is a close relation between the geochemical environment of loess and Kaschin-beck disease. A high incidence of disease occurs under the following conditions: (a) lithological character – older loesses (Lishi Loess and Wuncheng Loess); (b) geomorphology – erosion mounts and hills, residual "Yuan" and slopes of even "Liang"; (c) water source – surface and shallow waters; (d) soils – red clay, two-colored soil, clayey black loam.

6.6.2.2 Etiological Problem of Kaschin-Beck Disease

Kaschin-beck disease is one of the serious local diseases in the loess region. In 1935–1938, A. L. Vinogradov suggested that the imbalance of Ca/Sr ratio can induce the disease to occur. Experiments on animals with strontium nitrate made by Jilin Medical University suggested that simple Sr excess, or more Sr and less

Ca produced rickets, but not Kaschin-beck disease. Comparative analysis of water qualities in disease areas shows that contents of SO_4, K, F, B and I are evidently lower than those in disease-free areas. The investigation made in the area of Yongshou, Cuanhua and Quianxian Counties by the Institute of Geochemistry, Academia Sinica, indicates that S may be an important element for the disease. The Institute for Kashan Disease of Harbin Medical University suggested that Kaschin-beck disease is caused by the toxin of sickle bacteria.

6.6.3 Disease by Local Fluorine Poisoning

Vital signs of local fluorine poisoning are mainly spot enamel formed on teeth and deformed skeleton. The spot enamel on teeth is spots and stripes pale yellow or brown-black in color. As the luster darkens, teeth become destroyed. The deformation of skeleton is the result of bone sclerosis. However, for few patients, loosening and calcification at points of adhering muscle and ligaments, bending and deformation of elbow joints and stiff in lumbar vertibra have been found.

6.6.3.1 Local Disease by Fluorine Poisoning in China

Disease by fluorine poisoning in China has been found mainly in the following four areas (Liu Tungsheng et al. 1980b): (a) Zhaodong, Zhaochou and Zhaoyuan area in Heilungjian Province; (b) along the North China coast; (c) Hubei, Guizhou, Sichuan, southern Shaanxi, northeastern Yunnan Provinces; (d) some scattered areas with local high fluorine poisoning, such as hot spring areas and small arid basins.

6.6.3.2 Geochemistry of Fluorine in Loess Region and Disease by Fluorine Poisoning

In the loess region disease by fluorine poisoning frequently occurs. Loesses contain carbonates, and display alkaline reaction. A study on material composition of loesses (Liu Tungsheng et al. 1966) suggested that the loesses and paleosols contain fluorine-bearing minerals, such as mica, amphibole, tourmaline, apatite Data analysis made by Chen Qingmu and Yuan Ziyun reveal that content of fluorine in Malan Loess in Luochuan area, Shaanxi Province, is 490 – 520 ppm, 550 ppm on an average in older Lishi Loess, 560 ppm on an average in paleosol and 285 – 500 ppm in nodular layer of loess (Liu Tungsheng et al. 1980). Therefore, loess sediments themselves contain a certain amount of fluorine. Men who live in the environment with different loess layers, can be more or less affected by fluorine poisoning. Fluorine comes partly from loess, and partly from fluvial-lacustrine sediments (aquifers) in the loess region. But the areas with most serious fluorine poisoning are in the arid inland belt north of the Loess Plateau, such as the Yulin-Dingbian zone, and in the arid and semi-arid valley basins of the Hunaghe River, Weihe River, Jinghe River and Luohe River.

6.6.4 Other Local Disease in Loess Region

In loess regions there are other common local diseases, such as local goiter, Keding disease and others. Local goiter is called by local resident in Shaanxi Province "yingguagua" (a tumor of the neck), and enlarged neck disease in Northeast China. The symptoms of Keding disease are cretinism, deaf-mutism, low function of thyroid and convulsional nerve muscle in varying degrees. It occurs only in areas with a long history of local goiters. These two kinds of disease commonly occur in the loess region, and often occur repeatedly or alternately with Keshan disease and Kaschin-beck disease.

6.6.5 Local Cancer of the Esophagus

Local cancer of the esophagus is a disease characterized by a blockage of the the esophagus. The cause of the disease is still unknown. On a global scale the disease occurs mainly along the coast of Caspian Sea of northern part of Iran, Middle Asia, Mongolia, Siberia and the northern part of China (Fig. 6.14). The areas with high occurrence of cancer are concentrated in North China, especially in the loess region. It can be seen (Fig. 6.15) that the highest occurrences of cancer lie in certain areas. Including the northwestern part of Uygur Autonomous Region of Xinjiang, and Yanting district of northwestern Sichuan, the Yangzhong of Jiangsu and Nanao of Guangdong Provinces, and the juncture Hebei, Henan and Shanxi Provinces, i.e. along the southeastern border of the Loess Plateau (Fig. 6.15).

Fig. 6.14. Map showing the belts with cancer of the esophagus (After Kmet and Mahboubi 1972)

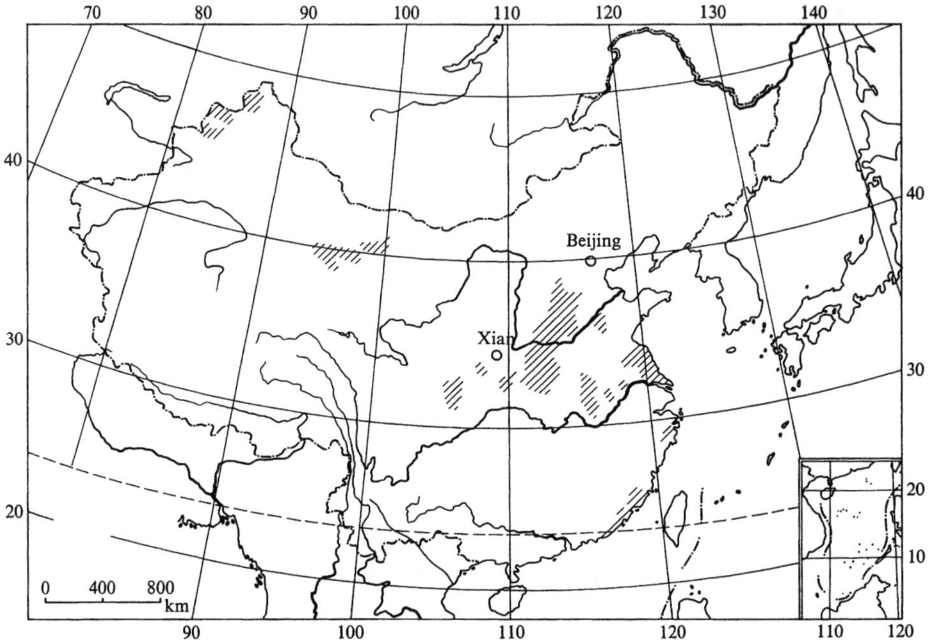

Fig. 6.15. Distribution of zones with high occurrence of cancer of the esophagus. (After the Office for Prevention from Tumour, Ministry of Public Health)

6.6.5.1 Environmental Background of Zones with High Occurrence of Cancer of the Esophagus

Relationship between grain size of loess and occurrence of cancer of the esophagus.

On the Loess Plateau, the mortality rate from cancer of the esophagus is less than 10/100000 persons in a zone from central and eastern Gansu to Dingbian, Jingbian, Wuqi of northern Shaanxi. It increases to 10–20/100000 persons in a zone from Yanan (excluding Huanglong), Tongchuan, Xianyang, Weinan districts of Shaanxi Province, and Luliang, Yuncheng districts of Shanxi Province. A further increase to 20–40/100000 persons occurs to the southeast, and up to 40–80/100000 persons in the eastern Shanxi area, and even exceeds 80/100000 persons in individual areas such as in Yangcheng. The mortality from cancer of the esophagus gradually increases from northwest to southeast with decreasing grain size from coarse in northwest to fine grains in southeast (Fig. 6.16).

This correlation reflects some difference in the properties of the loess. For example, increasing content of clay may affect the biogeochemical cycle, and cause the food chain to change, resulting in high incidences of cancer.

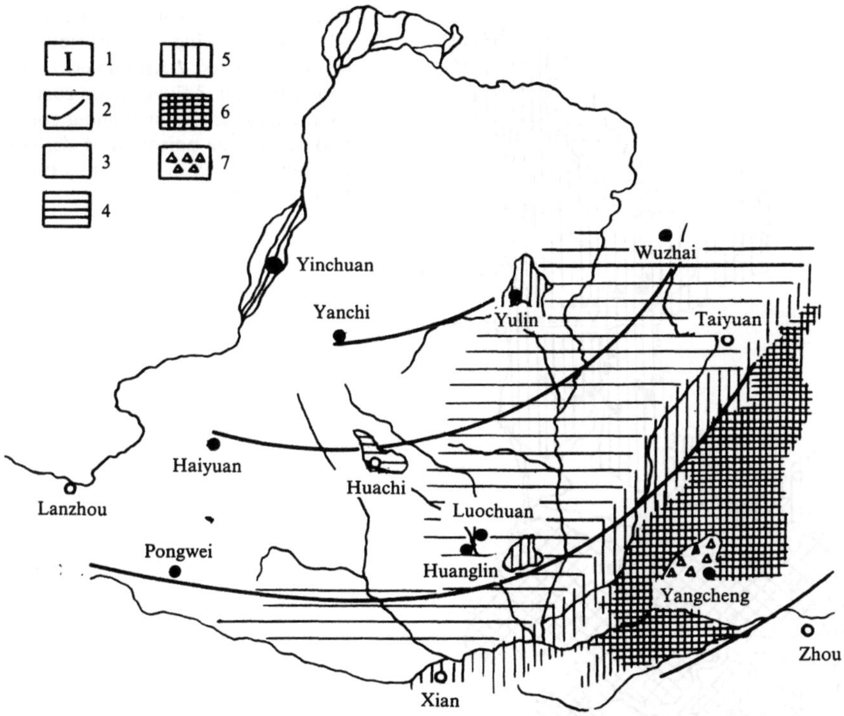

Fig. 6.16. Correlation of mortality by cancer of the esophagus to zonation of loess grain size in loess region along the middle reaches of the Huanghe River. *1* Zonation of loess grain size: *I* sandy loess; *II* loess; *III* clayey loess; *2* boundary between zones of loess grain size; *3* mortality by cancer of the esophagus 10/100000 persons; *4* mortality 10–20/100000 persons; *5* mortality 20/100000 persons; *7* mortality more than 80/100000 persons

6.6.6 Local Diseases in Loess Region and Geological Investigation

Local diseases, such as Keshan disease, Kaschin-beck disease and disease by fluorine poisoning in loess regions show zonal distribution and a close relation to the local environmental conditions. For example, in Shaanxi Province (Fig. 6.17), several relationships occur such as: (a) In the zone from Yulin to Dingbian to the north of Yanan (at the southern border of inland arid desert), the fluorine poisoning is predominant; (b) In the northern Shaanxi loess plateau, (semi-arid and semi-humid area with loess sediments), Keshan disease and Kaschin-beck disease is predominant, sometimes local fluorine poisoning occurs, and commonly light local goiter occurs; (c) On fluvial plains, fluorine poisoning occurs; (d) In southern Shaanxi (semi-humid and humid area), local goiter is predominant, and Kaschin-beck disease and Keshan disease also occur.

Thorough study of environmental geology and material composition of loess is of great importance for identifying the causes of local diseases. Keshan disease and Kaschin-beck disease are the etiologically undefined local diseases, but local goiter and local fluorine poisoning appear to be etiologically defined, and have

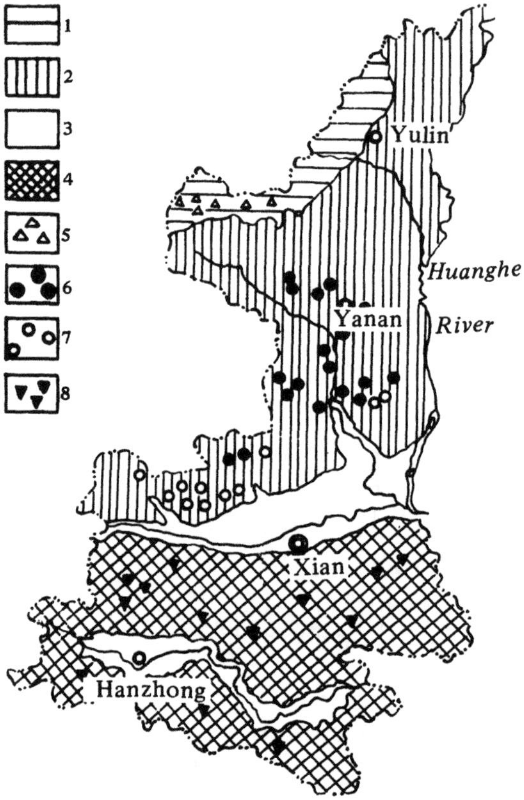

Fig. 6.17. Geomorphological zoning of Shaanxi Province and distribution of local diseases. *1* Arid desert zone; *2* loess plateau; *3* fluvial plain zone; *4* southern Shaanxi mountainous zone; *5* local fluorine poisoning; *6* Keshan disease; *7* Kaschin-beck disease; *8* local goiter

a more stable relation to the regional environment. Thus, the study of natural conditions of regional environment for the occurrence of local fluorine poisoning and goiter is also of great importance for identification of the etiology of Keshan disease and Kaschin-beck disease.

In the study of local disease, attention should be paid to the whole process of interaction and effects of natural factors from a comprehensive viewpoint. The internal and mutual relations between various materials must be considered in the analysis of various chemical compounds.

6.7 Property of Engineering Geology of Loess Deposits and Evaluation of Regional Stability

In this section we will discuss the property of engineering geology of loess deposits (including loess and secondary loess), the characteristics of regional distribution of related factors, and the engineering geological division in the area of middle reaches of the Huanghe River. As the engineering geological division can provide a geological basis for siting, projecting, construction and safe operation of engineeering buildings, the division itself should involve the evaluation of regional tectonic stability. It will have to be concerned with the crustal tectonic activity and the evolution of recent geological environment in various regions (segments). Many studies show that buildings in the middle reaches of the Huanghe River are endangered and extensively destroyed by rapid geological events (such as strong earthquakes, landslide, etc.) and slow geological events (such as sharp erosion, siltation, aseismic cracking, etc.). The frequency and intensity of these rapid and slow geological events depend mainly on the recent crustal tectonic movement and their joint effects with the bioclimatic environment. In addition, this section will also discuss some principles and the basis for an engineering geological division of the middle reaches of the Huanghe River and evaluation of the regional engineering stability of each region (segment).

6.7.1 Characteristics of Regional Distribution of Physico-Mechanical Properties of Loess

The engineering-geological properties of loess-like soil often refer to a sum of physical and mechanical properties, collapsing and other properties of these sediments. According to Zhang Zonghu et al. (1959), Northwest China Institute of Hydrological Science (1959, unpubl.), Institute of Hydrogeology and Engineering Geology, Ministry of Geology (1959), Qinghai-Gansu Comprehensive Expedition, Academia Sinica (1962), Lin Zaiguan and Liang Weiming (1982), Zheng Yanwu (1982), Zhai Lisheng (1983), and Hydrogeological Team No. 2 of Bureau of Geology and Mineral Resources of Shaanxi Province[12], a vast amount measurements of various physical properties (such as dry bulk density γ_d, void ratio ε, plastic limit I_w, liquid limit q_w, coefficient of compressibility a, friction angles c, cohesion ϕ, etc.), collapsing indices (such as relative collapsing coefficient δ_s, collapsing volume Δzs, etc.) of loess-like soil of the middle reaches of the Huanghe River permit us to summarize the following aspects.

Indices of various physico-mechanical properties of loess-like soil are interrelated, and generally reflect the material composition, structure and texture of the sediments. They clearly represent the relationship of dry bulk density to void ratio, dry bulk density or void ratio to relative collapsing coefficient, and dry bulk density to grain size composition, $CaCO_3$ content, microstructure types, etc. In addition they also represent variations in physico-mechanical properties and material composition with depth of the occurrence of soil layers.

The relation of bulk density (γ_d) to void ratio (ε) of loess and secondary loess for different geomorphologic units of the middle reaches of the Huanghe River is shown in Fig. 6.18, and listed in Table 6.17. It is readily seen that this is a negative linear relation between them. As dry bulk density (γ_d) is larger, the void ratio (ε) is less, and vice versa.

The Heimugou loess profile near Potou is a typical one on the loess Plateau. Dry bulk density, grain size composition (for details, Chap. 5, Sect. I), $CaCO_3$

[12] Hydrogeological Team No. 2 of Bureau of Geology and Mineral Resources of Shaanxi Province, 1982. Report on the regional engineering geological investigation in area of the middle reaches of the Hunaghe River (unpubl.).

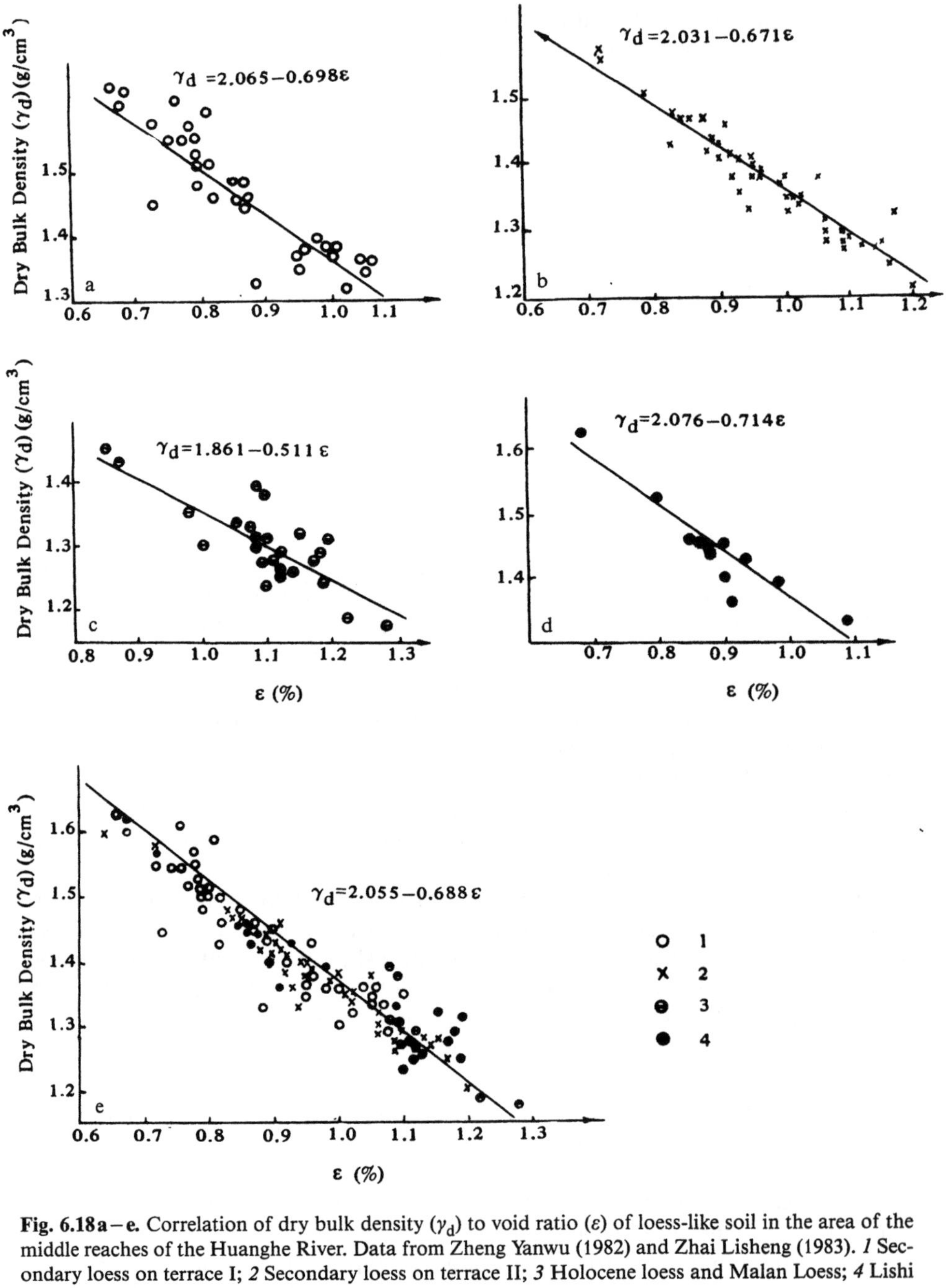

Fig. 6.18a–e. Correlation of dry bulk density (γ_d) to void ratio (ε) of loess-like soil in the area of the middle reaches of the Huanghe River. Data from Zheng Yanwu (1982) and Zhai Lisheng (1983). *1* Secondary loess on terrace I; *2* Secondary loess on terrace II; *3* Holocene loess and Malan Loess; *4* Lishi Loess and Wucheng Loess

Table 6.17. Linear correlation of bulk density γ_d to void ratio ε of loess in the area of the middle area of the middle reaches of Huanghe River

Morphology	Material	Linear regression equation	Correlation coefficient	Significanty
Terrace I	Secondary loess	$\gamma_d = 2.065\varepsilon - 0.699$	0.872	fine
Terrace II	Secondary loess and loess	$\gamma_d = 2.031\varepsilon - 0.671$	0.928	fine
"Yuan"	Holocene loess and Malan Loess	$\gamma_d = 1.861\varepsilon - 0.511$	0.773	better
	Lishi Loess and Wucheng Loess	$\gamma_d = 2.076\varepsilon - 0.714$	0.928	fine
Total		$\gamma_d = 2.055\varepsilon - 0.688$	0.891	fine

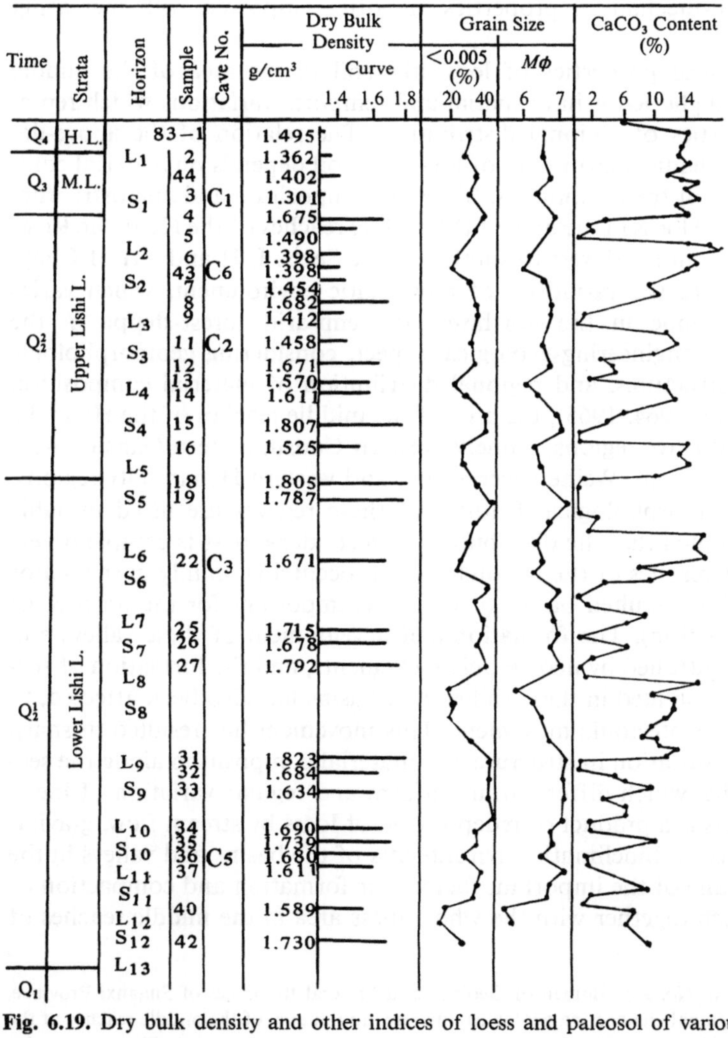

Fig. 6.19. Dry bulk density and other indices of loess and paleosol of various horizons along the Heimugou profile in Luochuan County

content (Lu Yanchou 1981) and other indices show regular, synchronous variation (Fig. 6.19) of loess and paleosol at different horizons along the profile, and their microstructure types show also their corresponding variation (for details, see Sect. VII of Chap. 5). Generally, as the dry bulk density of the horizons increases, the grains become finer, $CaCO_3$ content is lower and texture is more compact. In addition, dry bulk density increases approximately with depth, excluding upper and lower sandy loess layers. Study of physico-mechanical properties of different loess layers in Luochuan profile by the Hydrogeological Team No. 2 of Bureau of Geology and Mineral Resources of Shaanxi Province suggests[13] that dry bulk density of loess increase with depth of occurrence, void ratio and porosity decrease, and the relative collapsing coefficient also decreases. Physico-mechanical properties of loess-like soil at various sites along the middle reaches of the Huanghe River vary similarly with depth (Zheng Yanwu 1982; Zhai Lisheng 1983). The physico-mechanical properties are often improved with increasing depth.

Physico-mechanical properties of loess-like soil in the area of the middle reaches of the Huanghe River have spatial and temporal variations which represent the characteristics of regional distribution. The relation of the above-described physico-mechanical properties of loess-like soil depends on material composition and texture of their sediments. Material composition and microstructure synchronous with the loess in the area of the middle reaches of the Huanghe River change regionally from northwest to southeast (see Sects. I, II and VII of Chap. 5), and correspond to the variations of bioclimatic environment, which varies from arid desert-steppe in the northwest to semi-arid forest-steppe in the southeast. From the engineering-geological aspect, considering geomorphologic and stratigraphic structures, and regional distribution of material composition (Liu Tungsheng et al. 1964, 1965), the area of the middle reaches of the Huanghe River, is divided into five regions, namely, western Gansu, eastern Gansu-northern Shaanxi, Shanxi basin, Weihe River Valley, and western Henan hilly regions. Geological and geomorphological features of these regions are listed in Table 6.18. It should be noted that the distribution of secondary loess is comparatively local in the middle reaches of the Huanghe River, occurring mainly in the major river valleys and down-faulted basins (it is just as important for the practice of engineering construction). The formation and development of these valleys and basins have been controlled by neotectonic movements, and the formation of secondary loesses accumulated in these valleys and basins has also been affected, to various extents, by neotectonic movement. This movement has resulted in sharp lateral and upward variation in structure and material composition along the secondary loess profile, which differs from uniform and regular variation of loess. The secondary loess is a product of redeposition of loess by stream flow, gravitational effect and recent bioclimatic environments of the basins and valleys in the studied region – one of the important factors for formation and compaction of loess-like soil, which together with the whole loess area in the middle reaches of

[13] Hydrogeological Team No. 2 of Bureau of Geology and Mineral Resources of Shaanxi Province, 1982. Report on the regional engineering geological investigation in area of the middle reaches of the Huanghe River (unpubl.).

Table 6.18. Geologic-geomorphological characteristics of loesses in middle reaches of the Huanghe River

Region	Distribution	Stratigraphic structure	Geomorphologic feature	Grain size composition (%)			
				>0.05	0.05 – 0.005	<0.005 (mm)	
Western Gansu	Loess is widely distributed as a continuous cover with thickness less than 100 m.	Since Mid-Pleistocene loess is accumulated. Brown paleosols less developed in Lishi L.	Ridge and gorge are predominant, valleys and gorges densely cut it.	11.5 – 38.5	45.5 – 67.8	11.5 – 33.0	$Q_{3\text{-}4}$
				9.8 – 23.5	56.5 – 65.5	20.0 25.8	$Q_{1\text{-}2}$
Eastern Gansu-Northern Shaanxi	Loess is widely distributed as a continuous cover. Malan L. is 10 – 20 m thick.	Various loess can be compared stratigraphically. Black loam is found within the Holocene L. and Malan L.	Typical "Yuan", valleys and gorges are developed. Drainage system has north-south trend in E. Gansu, and dentrictic in N. Shaanix.	11.40 – 62.00	29.60 – 67.3	7.3 – 29.6	$Q_{3\text{-}4}$
				10.30 – 40.98	44.2 – 65.0	15.0 – 38.4	$Q_{1\text{-}2}$
Weihe R. valley	Loess covers discontinuously on piedmont, plarform and terraces.	Loess developed on platform, upper Lishi L. accumulated on terraces.	Platform (intensive cut) and terraces (less cut) are predominant.				
Shanxi basin	Loess is scatteredly distributed in intermontane basins, sometimes on terraces.	Various loess is accumulated. But on terraces only Malan L. and Lishi L.	Ridge-gorges and terraces are predominant. Erosion is weak.	11.9 – 59.50	24.60 – 66.40	11.70 – 33.00	$Q_{3\text{-}4}$
				9.20 – 40.00	37.80 – 67.60	9.20 – 43.6	$Q_{1\text{-}2}$
Western Henan hills	Lishi L. and Malan L. scatteredly distributed in intermontane basins and on terraces.	3 – 12 brown paleosols are found in Lishi L., and Malan L. is locally found.	Ridge-forges are predominant. Terraces are fully developed, and "Yuan" are local.				

the Huanghe River represent their regional variation from northwest to southeast. Therefore, the physico-mechanical properties of secondary loess in the whole area studied represent, to a certain extent, the same characteristics as that of loess (Table 6.18). Thus, the valleys and basins filled with secondary loess in different loess regions can be considered as subregions or zones in the corresponding regions.

Statistical data of physico-mechanical properties of loess and secondary loess of different times in different geomorphologic units of various loess regions along the middle reaches of the Huanghe River are listed in Table 6.18. The data reflect the spatial and temporal variation of physico-mechanical properties of loess and secondary loess. For the Holocene loess in western Gansu, Malan Loess and secondary loess in the river valleys, their dry bulk density (γ_d) is relatively low, and void ratio (ε), relative collapsing coefficient (δ_s) and coefficient of compressibility (a) high. These values are less for eastern Gansu-northern Shaanxi region, but for loess-like soil in the western Henan hilly region, dry bulk density (γ_d) is large, the void ratio (ε) is less, relative collapsing coefficient (δ_s) and coefficient of compressibility (a) are significantly lower than that in other regions. The characteristics of regional distribution are an important factor for evaluation of regional stability and engineering-geological division of loess-like sediments.

6.7.2 Collapsing Loess and Its Regional Distribution

Collapsing usually refers to rapid subsidence, compaction and lateral rheological behavior of loess-like soil waterlogged at the loading of overburden. It is a specific mechanical or construction property of loess-like soil (especially loess-like soil formed in the Malan stage). Loess-like soil with this property is named "collapsing loess". It is classified into two types, "gravity-collapsing loess" and "nongravity-collapsing loess" (Huang Qiang 1963; Zhai Lisheng 1983). Malan Loess, Holocene loess and secondary loess in the area of the middle reaches of the Huanghe River often show significant collapsing, but the older and deeper buried Lishi Loess and Wucheng Loess usually do not show collapsing. However, some older and deeper buried loess layers may show obvious collapsing under greater soakage pressure.

In addition to consideration of geomorphologic and stratigraphic distributions of the loess-like soil during evaluation of their settlement characteristics and regional distribution, three parameters are also considered: (1) relative collapsing coefficient (δ_s), that is the collapsing amount of unit thickness of soil layer under the loading of 2 kg/cm^2. This parameter is used for identification intensity of collapsing of various soil layers. (2) Thickness of collapsing soil layer, i.e., the thickness of a soil layer with coefficient $\delta_s = 0.015$. It is of importance for construction measures in an interval of depth under a foundation affected by collapsing. (3) Initial pressure for collapsing reflects the sensitivity of a soil layer to collapsing. Loess with a value less than that of saturated overburden and gravity-loading, is called "gravity collapsing loess".

The interpretation of the current data (Fig. 6.20, Tables 6.19 and 6.20) indicates that relative collapsing coefficient (δ_s) of loess-like soil in the area of the

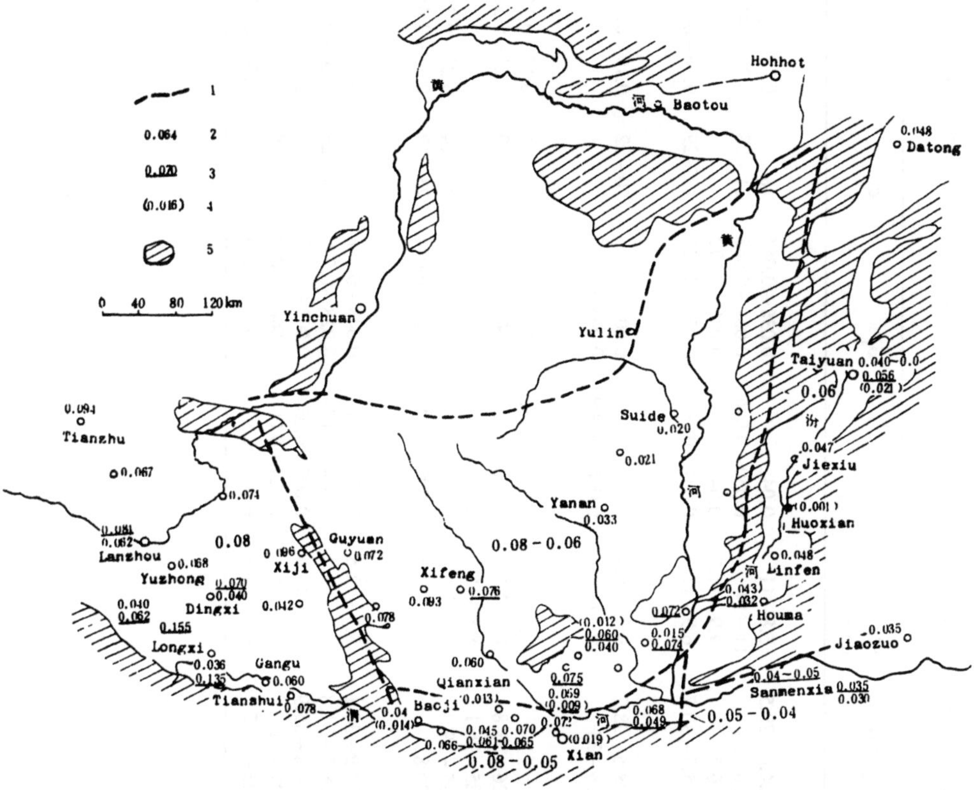

Fig. 6.20. Distribution of δ_s value of surfacial (to depth of 10–15 m) loess-like soil in area of the middle reaches of the Huanghe River. (Data from Zhai Lisheng 1983; Zheng Yanwu 1982). *1* Regional boundary; *2* secondary loess; *3* Malan Loess; *4* Wucheng Loess-Lishi Loess; *5* basement rock

middle reaches of the Huanghe River varies in space and time. The value (δ_s) of synchronous loessial sediments tends to gradually decrease from northwest to southeast. For example, δ_s of Malan Loess is usually larger than 0.08–0.07 in western Gansu, 0.08–0.06 in the eastern Gansu-western Shaanxi region, but less than 0.05–0.04 in the western Henan region. In all the regions, δ_s of the Holocene loess and Malan Loess is significantly larger than that of Lishi Loess and Wucheng Loess, and collapsing decreases with increasing age of deposition. The thickness of collapsing loesses shows a tendency towards the same spatial variation mentioned above (Table 6.20). The thickness of collapsing loess decreases from 27 m in western Gansu, and to 7–12 m in western Henan. The thickness of secondary loess is 16 m in western Gansu, but decreases to 2–10 m and 4–8 m respectively in Shanxi and western Henan. Data on initial pressure for collapsing are relatively scale. The data published by Zheng Yanwu (1982) indicate that the pressure value shows a similar trend of variation. For example, the initial pressure for collapsing of secondary loess is 0.35 kg/cm^2 in the Lanzhou region, 1.0 kg/cm^2 in Guanzhong region, 1.16 kg/cm^2 in the Sanmenxia region. It increases gradually from northwest to southeast.

Table 6.19. Data on physico-mechanical properties and collapsing of loess and secondary loess of various times. (Statistical data are cited from Zheng Yanwu 1982; Zhai Lisheng 1983)

Regions	γ_d (g/cm^3)	ε (%)	I_w (%)	q_w (%)	Plastic index	a (kg/cm^2)	δ_s	c	ϕ
Holocene L. and Malan L.									
W Henan hills	1.32 – 1.45	0.85 – 1.15	1.80 – 18.6	24.0 – 29.5	5.8 – 11	0.04	0.036 – 0.054	18	0.27
Weihe valley	1.24 – 1.33	1.0 – 1.19	17 – 20	30 – 34	11 – 14	0.030 – 0.046	0.061 – 0.088	21 – 28	0.2 – 0.44
E Gansu-N Shaanxi	1.28 – 1.33	1.03 – 1.10	18.5	30.2	9.9 – 11.7	0.042	0.076 – 0.079		
W Gansu	1.18 – 1.31	1.12 – 1.28	16.5 – 20.0	25.3 – 30.7	6.9 – 11.4	0.041 – 0.105	0.062 – 0.155	19 – 29	0.16 – 040
Lishi L. and Wucheng L.									
W Henan hills	1.33 – 1.40	0.9 – 1.09	17.8 – 18.8	28.5 – 28.7	9.9 – 10.9		0.002	20 – 22	0.19 – 0.42
Weihe valley	1.36 – 1.46	0.85 – 0.93	17 – 20	18 – 33	10 – 14	0.006 – 0.009	0.003 – 0.016		
Secondary loess on terrace I									
Shanxi basin	1.48 – 1.62	0.68 – 0.86			7.2 – 16.7	0.017 – 0.032	0.003 – 0.015	21 – 31	0.13 – 0.35
W Henan hills	1.46 – 1.55	0.76 – 0.86	16 – 20	26 – 33.8	10 – 13.7	0.030 – 0.033	0.002 – 0.006	20 – 29	0.21 – 0.50
Weihe valley	1.36 – 1.63	0.66 – 1.04	17 – 20	25 – 26.1	8.8 – 15	0.024 – 0.068	0.002 – 0.040		
E Gansu-N Shaanxi	1.32 – 1.34	0.89 – 1.09	14 – 19	25 – 29	9 – 12.5	0.006 – 0.066	0.029 – 0.044	23 – 24	0.23 – 0.71
W Gansu	1.33 – 1.45	0.9 – 1.06	17.6 – 19.4	30.8	8.5 – 13.5	0.007 – 0.045	0.032 – 0.044	23 – 32	0.25 – 0.45
Secondary loess on terrace II									
Shanxi basin	1.33 – 1.42	0.11 – 1.0	15.9 – 17.6	24.9 – 30.1	8.5 – 12.5	0.020 – 0.060	0.043 – 0.048	21 – 32	0.19 – 0.36
W Henan hills	1.38 – 1.47	0.83 – 0.96	17.2 – 20.0	28.2 – 31.8	9.5 – 13.2	0.014 – 0.037	0.028 – 0.039	17 – 23	0.26 – 0.56
Weihe valley	1.30 – 1.60	0.91 – 1.12	15.3 – 20.8	26.8 – 35.9	9.4 – 15.4	0.007 – 0.03	0.003 – 0.072	23	0.33
W Henan-N Shaanxi	1.21 – 1.28	1.09 – 1.20	18.5 – 20	27 – 31	7.6 – 11.5	0.024 – 0.063	0.060 – 0.093	22 – 28	0.09 – 0.25
W Gansu	1.25 – 1.41	9.90 – 1.17	16.3 – 19.2	23 – 31.5	6.5 – 21.3	0.017 – 0.042	0.050 – 0.094	19 – 24	0.34 – 0.67

Table 6.20. Thickness (m) of loess and collapsing loess layers in area of the middle reaches of the Huanghe River. (Statistical data are cited from Zheng Yanwu 1982; Zhai Lisheng 1983)

Region	Loess		Collapsing loess		Thickness (m) of collapsing loess at typical location	
	Plateau or piedmont platform	Terraces I and II (secondary)	Plateau or piedmont platform	Terraces I and II (secondary)	Plateau and tablelands	Terraces I and II (secondary)
Western Gansu	20 – 60	5 – 20	10 – 27	4 – 6	Lanzhou (27), Xining (10 – 25), Lintao (3), Dingxi (20)	I. Xining (2 – 4.5), Pinganzhou (2.5 – 4.5), Lanzhou (2 – 5), Tianshui (5), Jingning (6) II. Xining (4 – 15), Lanzhou (5 – 16), Tianxhui (12.5), Jingyuan (8.6), Gangu (10)
Eastern Gansu-Northern Shaanxi	50 – 170	5 – 30	9 – 20	4 – 8	Guyuan (9 – 10), Xifengzhen (9 – 10)	I. Suide (3), Zichang (1.5 – 6.5), Yan'an (4.5), Pingliang (6), Binxian (6 – 9.8), Zhenyuan (4) II. Yan'an (4.5), Pinglian (6 – 9.8), Zhenyuan (4)
Weihe valley	50 – 100	5 – 20	6 – 14	4 – 11	Xianyang (9), Qianxian (5 – 14), Hancheng (10), Pucheng (6 – 13), Guozhen (9), Liquan (5 – 12), Fufeng (8), Yaoxian (15), Tongchuan (5 – 11)	I. Baoji (1.5 – 3), Chaijiapo (2 – 2.5), Zhouzhi (2.5 – 3), Xianyang (1.5 – 3), Tongchuan (3 – 5) II. Baoji (6 – 11), Xian (12), Wugong (4 – 5), Pujizhen (4 – 8), Weinan (11), Tongchuan (6)
Shanxi	30 – 100	8 – 15	5 – 17	2 – 10	Taiyuan (17), Houma (10), Houxian (15)	I. Taiyuan (2 – 10), Linfen (8 – 9), Houma (6) II. Taiyuan (5 – 8), Datong (5.8), Yuci (8.5), Jiexiu (6), Linfen (8 – 9), Houma (6)

Synthesizing the above three parameters shows that in collapsing of loess-like soil an area of the middle reaches of the Huanghe River decreases from northwest to southeast. This is coincident with the regional distribution of material composition and microstructure, and implies the reason for collapsing. Zhu Haizhi, Lu Yanchou and others have discussed the question of collapsing in relation to material composition and microstructure of loess-like soil, and emphasized the relationship of microstructure and collapsing (Liu Tungsheng et al. 1966). Collapsing of loess-like soil is a process during which the structure of ground mass abruptly changes because of disrupture and losses of stability under the effect of water loading. Thus, the initial microstructure type of ground mass has a decisive effect of collapsing. Study of the microstructure of loess-like soil conducted by Gao Guorui (1981) also verified this suggestion. Moreover, the formation of microstructure of loess and micromorphology of paleosol depends on the bioclimatic environment in which loess and paleosol have been formed, as well as epigenetic diagenesis (Lu Yanchou and An Zhisheng 1979; An Zhisheng and Wei Lanying 1979, 1980; see Sect. I of Chap. 5). For the Holocene and Late Pleistocene loess-like soils undistributed by sharp diagenesis, the distribution of their microstructure essentially resulted from the regional change of bioclimatic environment. Therefore, the regional distribution of collapsing of loess-like soil in the area of the middle reaches of the Huanghe River has been controlled to a large extent by bioclimatic environment.

6.7.3 Engineering-Geological Division and Regional Stability

The previous studies indicate that engineering geological division of loess area has been carried out on the basis of the following two principles:

1. From viewpoint of engineering geological genesis, the division has been carried out by using the index of construction properties of loess in small areas for large-scale division.
2. The division has been carried out on the basis of natural conditions for formation of loess in connection with some construction properties, such as that made by Zhai Lisheng (1983). Other division of loess areas of the middle reaches of the Huanghe River have been made by Lin Zaiguan and Liang Weiming (1982) on the basis of thickness, structure and collapsing of loess-like soil. Moreover, Zhang Zonghu et al. (1959) proposed division of the Loess Plateau.

The division on the basis of the above two principles has been made with little consideration of changes of geological environment that can affect engineering buildings. In practice, the cases of engineering buildings affected by changes in geological environment are numerous. For example, the occurrence and development of ground cracks, found in Xian City since 1978, have severely endangered a number of buildings (Liu Jingwen and Zhang Jiaming 1982). Heimugou headward erosion in Luochuan County, and new gully streams cut the basements of municipal walls, which had been intact up to 1958, and endangered the normal development of the city. In August, 1955, the Wolongsi landslide in Baoji County

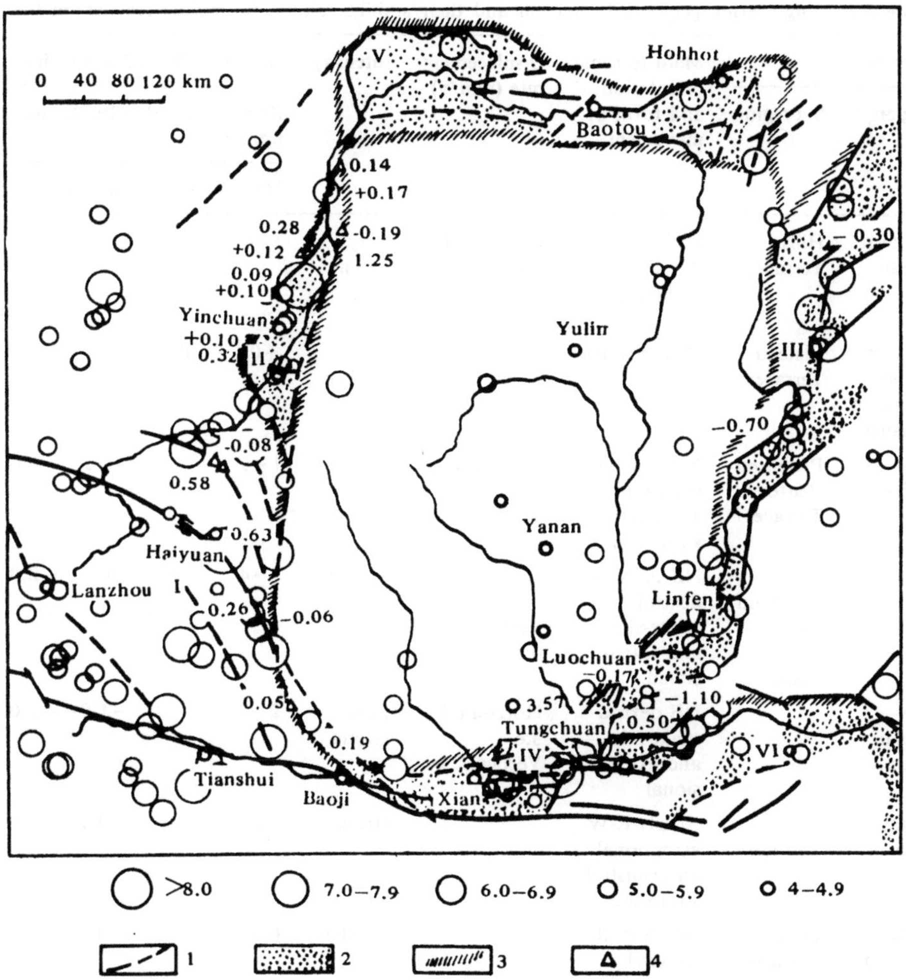

Fig. 6.21. Map showing neotectonic regions, major active faults and strong earthquake epidenters along the middle reaches of the Huanghe River. *1* Active faults; *2* down-faulted basins; *3* boundary between regions; *4* geodesic short baseline measurements across faults. + indicates the uplifted block; – indicates the subsided block

caused the Longhai railway to move southward 110 m[14]. The rapid siltation in the Sanmenxia Reservoir seriously affects not only the normal operation of the reservoir, but also endangers the buildings in some areas of the Guanzhong region. The Haiyuan earthquake with M = 8.5 in 1920 destroyed a mass of cave dwellings and civil buildings and killed at least 230000 people (Historical Research Group of Seismological Commission, Academia Sinica 1956). In addition the earthquake caused 650 loess collapses in Xiji County and formed 41

[14] Shi Chuanyi, Liu Guangdai (1983) An outline of Shaile landslide in Dongxiang County, Gansu Province.

Table 6.21. Engineering geological Division and their stability along the middle reaches on the

Division		Main tectonic orientation and form of movement	Displacement rate (mm/y)	Intensity	Active faults and seismicity	
					Recent ground deformation (mm/y)	
Region	Subregion				Horizontal	Vertical
Western Gansu	Hills Terraces	NW over-thrust-strike-slip	Horizontal 3 – 9	Very strong	3 – 6	+ 1 – + 3
Eastern Gansu-Northern Shaanxi	Plateau Terraces	Slow relative uplift		Very weaker	+ 0.1 – + 0.5	
Weihe valley	Piedmont Platform Terraces	NEE or NE extensional and dextral strike-slip	Horizontal 4.3 Vertical 1 – 2	Strong	1 – 3	– 0.5 – – 2.0
Shanxi basins	Hills Highland Down-faulted basin	NE or NNE dextral strike-slip and exten-sional	Horizontal 1 – 3 Vertical 0.5 – 1	Strong	1 – 4	– 0.5 – – 2.0
Yin-chuan basin		NE dextral strike-slip and exten-sional	Horizontal 3.6 Vertical 2.0	More strong	2 – 4	– 1.0 – – 3.0
Hetao area	Hills Terraces	Nearly E-W extensional and sinistral strike-slip		strong	0.3 – 2	– 0.5 – – 2.0
Hills in western Henan	Hills Terraces	NE dextral strike-slip and over-thrust		Weaker	0.5	– 0.5 – + 1.0

dammed lakes (Zhu Haizhi 1982). Similar destruction took place during the Huaxian earthquake (M = 8) in 1556, the Linfen earthquake (M = 8) in 1695, and Zhaocheng-Hongdong earthquake (M = 8) in 1303. Examples of the hazards are numerous. Therefore, the engineering geological division should provide geological constraint for the reasonable distribution, projecting, construction and safe operation of massive buildings, but during division it is necessary to consider the differences in probability and intensity of the rapid geological events (such as earthquakes, landslides, etc.) and slow geological events caused by the differences of crustal tectonic movement and erosion-sedimentation in various locations, and the prediction of recurrence frequency and intensity of these events is an important principle.

Huanghe River seismicity

Historic strong earthquake: Intensity	Historic strong earthquake: M_{max}	Recent moderate and small earthquake	Seismic intensity: General	Seismic intensity: Maximum	Estimation of regional engineering stability: Major geological events which can affect engineering buildings	Estimation of regional engineering stability: Principal category
High and many	8.5	Many	7–8	11	Large earthquakes, fault displacement; intense settlement, and gravity settlement, slumping, erosion and siltation; mountain mud flow.	VII–X
Low and rare	5–5.5	Rare	6	8	Strong erosion and siltation, slumping, and collapsing.	III–V
High and less	8.0	Less	7–9	11	Collapsing, slumping, large earthquakes and fault creep; siltation and salting.	VI–IX
High and less	8.0	Many	7–9	11	Strong earthquakes and fault displacement, siltation and salting; mountain mud flow; possible settlement.	VI–IX
High and many	8.0	Many	7–9	10	Strong earthquakes and fault displacement, siltation and salting, mountain and flow; possible settlement.	V–IX
Low and less	6.3	Less	6–7	8	Earthquakes and fault displacement, mountain flood and mud flow, siltation and salting.	IV–VI
Low and less	6	Less	6–7	8	Erosion and washing, siltation, slumping; creep, possible settlement.	III–VI

Synthesizing the historical records of earthquakes and their hazards in the period of 2000 and more years (Historical Research Group of Seismological Commission, Academia Sinica 1956; Zhu Haizhi et al. 1982), data on recent earthquakes and distribution of active faults in China (Institute of Geology, State Seismological Bureau 1979), data magnitudes and average ratios (Ding Guoyu and Lu Yanchou 1984) of fault displacements, recent crustal deformation (Chen Qingxuan et al. 1980), geodesic short baseline measurements across faults (Liu Guangxun et al. 1982) and neotectonic data (active faults and seismicity, Fig. 6.22, Table 6.21), we can find that the levels of neotectonic and recent tectonic movements are comparatively high, but are different in various locations of the middle reaches of the Huanghe River. Thus the middle reaches of the Huanghe

River can be divided into seven units – regions, namely western Gansu, Yinchuan basin, Hetao basin, Weihe River valley, Shanxi down-faulted basin, Ordos Plateau (eastern Gansu-northern Shaanxi) and western Henan. Basic data on seismicity and fault movement and the related parameters are given in Fig. 6.21 and Table 6.21. In general, faulting and seismicity are high in western Gansu, and lower in Yinchuan basin, Shanxi down-faulted basin and Weihe River valley, and very weak in eastern Gansu–northern Shaanxi. The later region lies in an area of bulk uplifting, and no significant active faulting and earthquakes ($M_s > 6$) occurred.

In summary, the area of the middle reaches of the Huanghe River, where loess is distributed, has been preliminarily divided into seven engineering geological regions and several subregions (Table 6.21) in accordance with the regional distribution of loess-like soils (and their stratigraphical, and lithological characteristics, physico-mechanical properties and collapsing), regional geomorphologic, neotectonic and recent tectonic activities. A relative quantitative evaluation of engineering regional stability in various regions is made, estimation of stability to unstability has been defined into ten grades of I–X, and the possible unexpected geological events, including rapid and slow geological events, have been indicated. The classification of stability is schematic, and remains to be further studied. The work may serve only as a reference for practice.

6.8 Environment and Erosion of Loesses

6.8.1 Circumstance of the Erosion of Loesses on the Loess Plateau

Loess Plateau covers an area of 580000 km^2, in which loess are continuously distributed in an area of 274000 km^2 (Liu Tungsheng et al. 1964). In the loess regions, soil loss is severe, and the erosion moduli is 200–30000 tons/$km^2 \cdot yr$[15]. This corresponds to the surficial scouring depth of 0.01–2.0 cm/yr. It has been known from the moduli charts of sediment load and coarse-particle (>0.05 mm) plotted by the Huanghe River Water Conservancy Committee of the Ministry of Water Conservancy and Electric Power in 1956–1963 and 1965–1977 (Fig. 6.22 and 6.23) that the area where soil loss is most severe mainly lies west of Luliang Mts., east of Taohe River, south of the Great Wall and north of Weihe River on the Loess Plateau. Three intense erosion regions, where moduli of sediment load is larger than 10000 tons/$km^2 \cdot yr$, are region I with Shanxi–Shaanxi tributaries of Huanghe River from Hekouzhen to Wubu County, region II the source region of the Wuding River, Jinghe River and Luohe River, and region III the river source area of the Liupan Mountains[16].

---

[15] Huanghe River Water Conservancy Committee of the Ministry of Water Conservancy and Electric Power (1965) The charts of moduli of sediment load and coarse particle plotted, 1956–1963.

[16] The Research Institute of Yellow River Conservancy Commission et al. (1978) The effects of water and sands from different areas along the Huanghe River basin on the siltation of lower reaches of the Huanghe River.

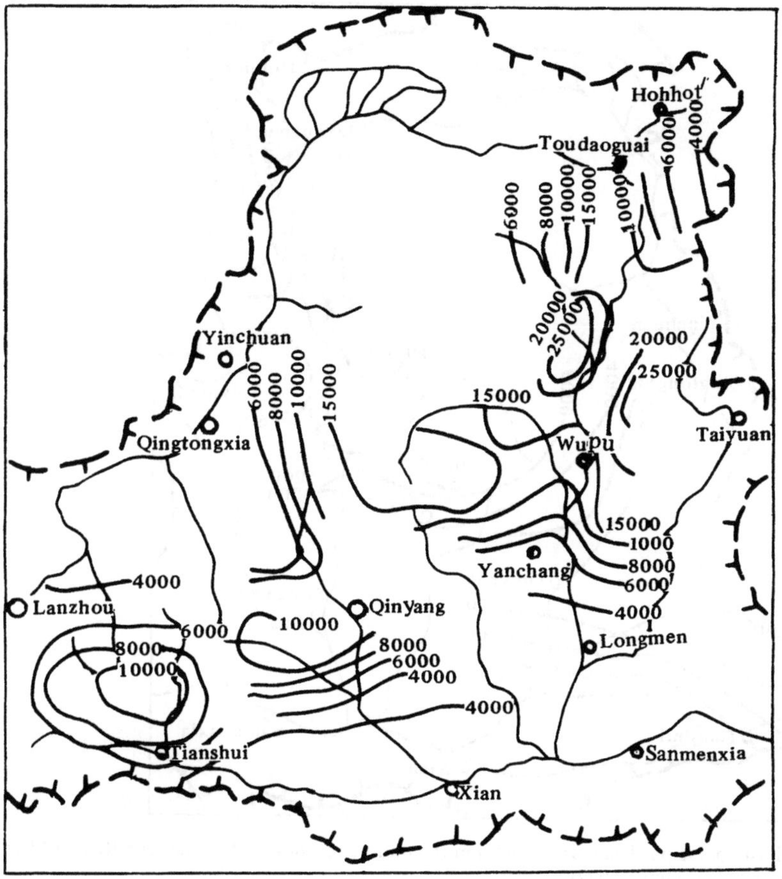

Fig. 6.22. Moduli of sediment load in the middle reaches of the Huanghe River (1965–1977). (After Gong Shiyang and Xiong Guisu 1979)

About 1.6 billion tons of sediment load flow into the Huanghe River each year. Approximately one fourth of this sediment flows in the lower reaches of the channel and causes great disasters. Of the 400 million tons of material in the lower reaches, nearly 69% consists of coarse particle, with a diameter over 0.05 mm. Thus, investigation of the source of the coarse particle is an important for revealing regularity of loess erosion. Hydrological data indicate that coarse particles make up to 43% of the particles in the Huanghe River. Coarse particles in the upper reaches of the Huanghe River from the town of Hekouzhen make up only 5.4% of the particles, and the total amount of particles in the channel of the upper reaches is also small. Coarse particles of the Fenhe River make up only 1.13% of that found in the Huanghe River. This indicates that regions I and II are the areas of most intense erosion of loess, and also the source regions for coarse particles.

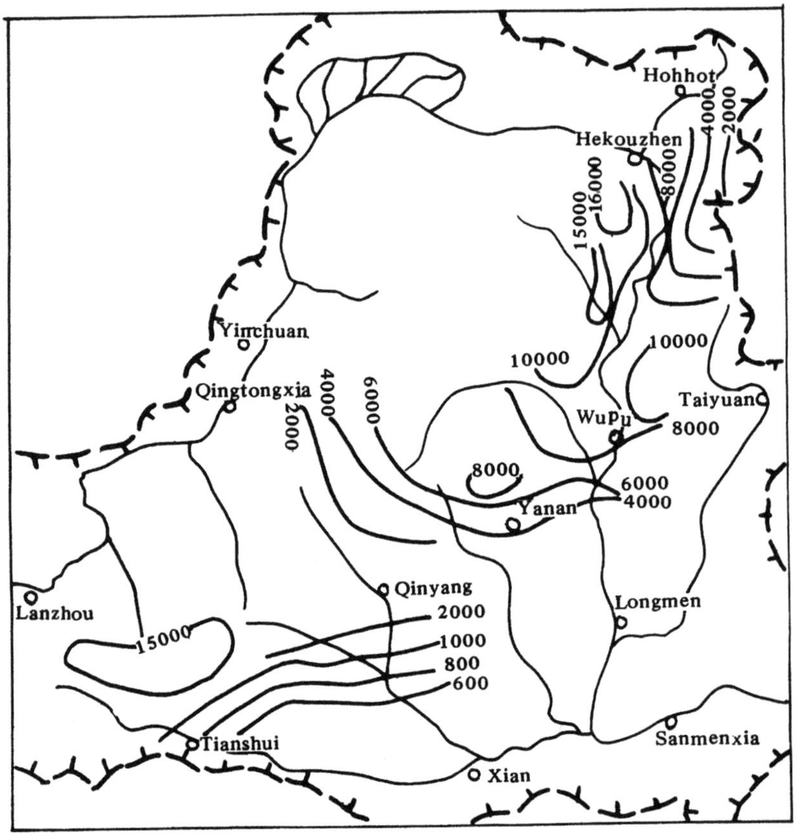

Fig. 6.23. Moduli of coarse particle (1965–1977) in the middle reaches of the Huanghe River. (After Gong Shiyang and Xiong Guisu 1979)

6.8.2 Geological Environment and Erosion of Loess

The Paleozoic and Mesozoic sandstones, shales, sandy siltstones are mainly covered by loess, and exposed only at the bottoms of recent valleys and gullies. Thus, the Late Tertiary Red Clay, the Quaternary loess and sand (especially loess), are the important sources for the sediment load carried by flows into the Huanghe River. Therefore, the material composition, stratigraphic structure and history of loess erosion are the important factors affecting the degree of recent erosion of the loesses.

6.8.2.1 Loess Belts and Loess Erosion

Loesses in the area of the middle reaches of the Hunaghe River display a change in grain size form coarse in the northwest to fine in the southeast, and the area covered by loesses is conventionally divided into three belts, sandy loess belt, loess belt and clayey loess belt (Liu Tungsheng et al. 1964). Loesses in every belt have a different capability for resistance against erosion in relation to their different

Table 6.22. Material composition, microstructure, coherence and erosional amount of Malan Loess in different belts

Loess belt	Microstructure[a]		Grain size composition[b]			Coherence[c]	Instant Erosion maximum	
	Cementation	Interparticle porosity (%)	Medium (mm)	Fine-grain (%)	Clay-grain (%)		Moduli (T/km^2)	Sand amount (kg/m^3)
Belt I	Contact	6.15 – 4.95	0.026 – 0.076	23.6 – 72.4	7.0 – 20.7	0.128	15000	1000
Belt II	basement	4.95 – 2.74	0.016 – 0.032	11.1 – 31.5	8.1 – 30.4	0.06 – 0.608	2000	600
Belt III	basement	2.50	0.018 – 0.027	11.4 – 21.9	18.0 – 30.4	0.12 – 0.552	2000	600

microstructure and coherence (Table 6.22). It can be seen (Fig. 6.22 and 6.23) that Regions I and II, where the most intense erosion occurs, are located in the sandy loess belt, and the erosion decreases from sandy-loess belt to clayey-loess belt.

The northwest border of the sandy-loess belt is severely desertified. This is a cause for the increasing amount of coarse particles in the Huanghe River. For example, in the Huangpuchuan River the amount of coarse particle greater than 0.05 mm increased 5% in 1974 since 1960s[17].

6.8.2.2 Stratigraphic Structure and Erosion of Loesses

The Huanghe River and its tributaries deeply dissect the loess and its underlying basement rocks on their walls, forming gullies and gorges at depths of tens to hundreds of meters. The multilayer structure of loess largely affects the intensity of loess erosion.

The stratigraphic structure of loess represents multiple superimposition of loess, paleosol and calcic nodules in their sediments, and sometimes sand layers are intercalated in the sandy loess belt. This multilayer structure displays different resistance to the erosion in the form of "stable" and "unstable" types due to different combination of these layers. For example, many paleosol layers and calcic nodules are usually more sensitive than loess to erosion. In particular, the hard and dense calcic nodules are important supporting layers for the loess above them. This allows the loess to remain on the steep walls and tunnels to be stable.

Sand interlayers in loesses are loose and easy to erode. But paleosol containing more clay, Red Clay, weathered mudstone and shales are good water-resisting layers. They form the beds for landslides, and cause the steep loess walls to be unstable. The underlying hard basement rocks for loess have different effets on erosion in relation to their occurrence at different geomorphic positions. The basement rocks on the side-walls of valleys in the "Yuan" area can support thick loess on top of them, and maintain the steep loess walls

[17] The Planing Office of Yellow River Conservancy Commission (1965) The charts of moduli of sediment load and coarse particle in 1956 – 1963, (unpubl.).

stable. However, when the heads of gullies and valleys deeply cut the basement rocks on the watershed divide, they become catchment regions where thin loess layers can be carried by flows and the exposure of basement rocks increases the unstability of loess.

6.8.2.3 Fossil Erosion of Loesses

The accumulation of thick loesses started 2.4 m.y. ago and lasted the whole Quaternary period, during which the climate changed frequently. Thus, the ancient loess plateau recorded the evidence of fossil erosions in the process of its formation and evolution. The fossil erosions here mentioned are mainly the product of natural erosions. But owing to human activity the artificial effect may increase the natural erosion – so-called man-made erosion, and the recent erosion takes place on the basis of fossil erosion. The indicators showing sedimentary intermitances of loesses are mainly: (1) paleosol (buried soil layers), (2) denutational surface (disconformity), and (3) interlayers of sediments of other origin (excluding aeolian sediment) (Liu Tungsheng et al. 1965). Analysis of the characteristics of fossil erosion of loess permits us to find the following regularities.

1. Cyclicity of Erosion-Denudation and Sedimentation of Loess. The change of environment of the Loess Plateau is characterized by an alternation of arid steppe, steppe and forest steppe environments (Lu Yanchou and An Zhisheng 1979). This reflects a sequence of change in dry-cold and warm-wet climates. The dry-cold period is a period of loess sedimentation, during this period the denudation and erosion of loess are relatively weak. The warm-wet period is a period of slow accumulation of loess and development of paleosol, during this period the rainfall increases, and the relative denudation-erosion could be more intense than that in dry-cold period.

The recent meteorological and hydrological data also indicate that in a dry period, the erosion of soil (loess) caused by precipitation tends to decrease. It can be seen from the change of water and sand of the Shanxian County during 1920–1969 (Table 6.23) that from 1920 to 1929 the rainfall and mean runoff are less and the mean amount of sediment load is only 11.9 million tons. From these

Table 6.23. Change of water and sediment load in various periods in the Shanxian County , Henan Province[a]

Time interval	Mean annual rainfall amount (mm)	Mean annual runoff amount ($\times 10^9$ m^3)	Mean annual account of sediment load ($\times 10^9$ ft)
1920–1929	376	32.5	1.19
1930–1939	440	48.9	1.76
1940–1949	456	54.0	1.71
1950–1959	454	52.2	1.86
1960–1969	410	59.1	1.88

[a] Data are cited from the Science and Technology Office and Hydrological Division of the Huanghe Hydrological Commission.

data it is possible to postulate that during the period of formation of the paleosol the more intense denudation – erosion took place. Spectral analysis of the Luochuan loess-paleosol sequence suggests that the changes of loess-paleosol sequence take place in some cycles of approximately 80 thousand, 40 and 20 thousand years (see Sect. IV, Chap. 6). This cyclicity may have certain implications for the study of cyclicity of denudation and erosion of loesses. The existence of two or three black loam layers in the Holocene loess indicates also that there may be 2–3 fossil erosional surfaces.

2. The Ages of Main Major Erosional Periods. Results of the paleomagnetic and thermoluminescence determinations enable us to determine the ages of some important denudational surfaces (unconformities), and hence to determine the major erosional periods as follows:

1. 2.4 m.y. ago there was an erosional period corresponding to a denudational surface between Wucheng Loess and Red Clay.
2. Approximately 1.48 m.y. ago, there was an erosional period corresponding to the denudational surface between Wucheng Loess and Lishi Loess.
3. A denudational surface 0.6–0.7 m.y. ago corresponds to the erosional period before the formation of the fifth paleosol.
4. A denudational surface between the upper and lower parts of Lishi Loess 0.5 m.y. ago corresponds to the formation period of the fifth paleosol.
5. 0.3 m.y. ago, there was an erosional period corresponding to the denudation surface of the third paleosol.
6. 0.2 m.y. ago, there was an erosional period corresponding to the denudation surface of the second palesol. On the slopes of river valleys, the second loess after this erosional period often unconformably overlies the older loess.
7. 0.1 m.y. ago, there was an erosional period corresponding to the denudation surface between the Malan Loess and Lishi Loess.

50000 yr ago, there was an erosional period corresponding to the denudation surface of Salawusu period. The sediments formed in this erosional period are comparable with the sediments on the second terrace at 20 meters from the bed of the Wuding River (Yuan Sixun et al. 1983).

3. Paleorelief and Erosion of Loess. The occurrence of erosional surfaces which are one of the characteristics of paleosols are controlled, to a certain extent, by a paleorelief formed at the end of the Tertiary. This paleorelief has largely affected the accumulation and subsequent erosion of loess. For example, the Baiyu Mountain watershed divide represents the "Liang" and "Mao" hilly relief with great density of gullies, and erosion modulum is over 10000 tons/km^2·yr. For another example, the Dongzhi "Yuan" is an open basin, with an underlying plane paleorelief. The thickness of the loess is great with deep gullies and moderate erosion of 4000–10000 tons/km^2·yr. However, the area of Yongshou, Chunhua and Yaoxian Counties is a monoclinal plain. The erosion modulum is usually 2000 tons/km^2·yr with the covering loess of the fossil river terraces, such as on the Weihe River terraces, less eroded.

6.8.3 The Gully System and the Mosaic Environment of Loess

1. The Period of Gully System Formation. The gully system formed by the recent erosion on the Loess Plateau is inherited from the old gully system. The development of the dentrital gully system caused the loess to be eroded more rapidly. Liu Tungsheng et al. (1966) suggested that the late Pleistocene Salawusu formation is distributed on the first-order tributary of the Huanghe River (Weihe River), the second-order tributary (Jinghe River) and the third-order tributary (Zhaojiachuan). It follows that some fourth-order tributaries have been formed as early as 50000 yr ago. In the Zaocigou gully (west of Luochuan City), comparable to the fourth-order tributary of the Huanghe River, Malan Loess overlies the earlier loess. The denudation surface before the Malan Stage was formed 0.1 m.y. ago. These data permit us to date the age of formation of this tributary gully at least 10×10^4 yr ago. The head of this tributary has cut into the archeological relics of color-pottery culture dated at $5-6\times10^3$ yr ago. This indicates that the heads of gullies extended forward approximately 200 m during the last $5-6\times10^3$ yr, (i.e., $3-4$ cm per year on an average). Therefore, the tributary gully-system from the first to the fourth and the fifth orders was formed earlier than the first human activity, and increasingly extended under natural influence.

The erosional depth of the valleys and gullies can be estimated from paleosols and sediments on the terraces. For example, on the Xujiahe Village in Luochuan County, the fifth paleosol is located at 80 m above the bed of the Luohe River. This indicates that before the formation of the paleosol, the Luohe Rivers was deeply cut. 0.05 m.y. ago, Salawusu sediments accumulated to the height of recent beds of the fourth and fifth order tributaries of the Huanghe River. This phenomenon shows that the depth of dissection of these tributary gullies had reached the recent beds at least 0.05 m.y. ago.

2. Small Tributaries Formed by Natural Erosion. It can be seen from satellite images, aerophotographs and the topographic map on a scale 1:50000 that the fourth and the fifth tributaries of the Huanghe River are distributed in the form of an asymmetric plume. Gullies and valleys on the ubac (shade slope) are denser, and those on the adret (sun slope) and shorter and rare. This may be related to the great moisture on the ubac, high water content in the soil, and long freezing intervals which led to the nodular structure of soil. Thus the surfacial soil on ubac is looser than that on adret. The loess is more intensely eroded, and gullies are denser and longer on ubac than on adret. The asymmetric plume pattern of gullies is a consequence of natural erosion.

3. Gully System and Mosaic Environment of Loesses. The ecological environment on the Loess Plateau represents desert steppe, and forest steppe respectively from northwest to southeast. Because of the scattered distribution of exposed rocks on the mountainous area and the formation of deep gullies by erosion, different conditions existed for the growth of plants within zonal ecological environment, and hence a mosaic stereoscopic loessy ecological environment has been formed. For example, in the Laoshan Mt, Ziwuling Mt, Huanglong Mt and other areas, secondary forest on the thin loess and basement rock and grass cover are distributed in the form of islands. Deeply cut gullies exposed the ground water, and thus, the bottom parts of gullies are moister and suitable for the growth of woody plants, and hence a forest or forest steppe environment has been formed. If the moisture is abundant, a meadow environment might be formed. Gullies dentrically dissected the forest steppe and steppe environment. The environmental system also has appeared in the past geological periods, as suggested by results of palynological analysis. In Potou Village of Luochuan County only a few spore and pollen grains are found in Lishi Loess with ages of 0.92 – 0.41 m.y., indicating a steppe environment. However, not far from Potou at Nianzigou the assemblage of the loess under the fifth paleosol is dominated by spore and pollen of woody plants, indicating a forest steppe environment.

Because the area covered by the gullies (including slopes at 20° and more) occupies 60% surface or more of the "Yuan" area, the structural characteristics of mosaic ecological environment must be considered in the planning of land use of the Loess Plateau. This is especially true for specific microregional environmental conditions, where different measures must be taken for planning the reformation.

6.8.4 Geological Environment and Some Problems of Water and Soil Conservation

1. The sandy-loess belt is an important area for conservation of water and soil. The northern part of this belt must be preserved from desertification. It is necessary to determine the effect of basement rocks in this area as a source of coarse particle in the Huanghe River.
2. The natural disasters on the Loess Plateau such as storms, floods, droughts, windstorms and others are frequent. Therefore, in estabilishing a modern vegetation environment we cannot establish an ecological with only single species of plants. Though the spreading of single species of plants can play a significant role, it is difficult to save them from natural disasters. Thus efforts must be taken to establish experimental fields for planting trees and grass, introducing and culturing new species of plants in order to establish multispecies and multilayer ecological environment.
3. Gullies spread length and breadth on the Loess Plateau. Considering the natural ecological environment mosaic with mountain hills, loess "Yuan", "Liang", "Mao" and gullies, a large effort must be taken to reform the gullies and conserve the slope lands.
4. Engineering and biological measures are two methods for conservation of water and soil, especially for the recently eroded gullies. Building dams for the formation of soil, especially for the recently eroded gullies. Building dams for the formation of silted lands is also an important measure. It is beneficial to raise the basic level for erosion section by section from the upper to lower reaches of the river, reduce upstream erosion rates and prevent soil loss. Meanwhile, the building of dams for the formation of silted lands is an important measure for establishing the natural ecological system on mountain hills, "Yuan", "Mao" and gullies. This will allow a reasonable stereoscopic natural ecological system to form on the Loess Plateau.

5. In the process of transformation of nature, attention must be paid to the side effects of the transformation measures. For example, in areas where evaporation is larger than precipitation, the silted lands in gullies becomes saltier, and hence reduces the production of the silted lands.

CHAPTER 7

Methods in Loess Research

7.1 The Principles of Stratigraphic Classification and Nomenclature of Loess Sequence

It is worth noting that the theory of stratigraphy during the past 30 years has achieved remarkable success. After the 1950s, the theory of stratigraphic classification has gradually turned from the traditional unified concept to the multiple concept of modern stratigraphic classification, which emphasizes that the characteristics on which classification are based are many-sided. The characteristics are grouped under headings; one is the "observable" or "tangible" and objective properties of matter, and another is the subjective "interpretation" or "inferential" abstract attributes. The stratigraphic position of change in one property does not necessarily coincide with that of another one. Therefore, there are many kinds of characteristics and attributes of rock formation, and there are many kinds of stratigraphic classification and stratigraphic unit as well. The purpose, scope, content and method of stratigraphy are determined by the concept of stratigraphic classification. In the study of the stratigraphy of loess sequence, attention should be paid to the modern theories and concepts of stratigraphic classification. Because the concepts of stratigraphic classification are frequently reflected in stratigraphy codes or guides, it is appropriate first to introduce the status quo of the Quaternary stratigraphic codes.

7.1.1 The Codes for Quaternary Stratigraphy

The Quaternary, as the latest system of the earth history, is the one with certain differences in characteristics from the others, and often involves certain particular questions concerning the stratigraphic classification. It is necessary to formulate a special code for the Quaternary. Now, there is not a formal international stratigraphic code for the Quaternary, but there are some national or continental codes for the Quaternary stratigraphic classification and nomenclature. These include:

1. The "Key to the Interpretation and Nomenclature of Quaternary Stratigraphy" (first and provisional edition) by INQUA Commission of Stratigraphy (1969), compiled by G. W. Luettig, R. Paepe, R. C. West, W. H. Zagwijn.
2. The "International Stratigraphic Guide" by ISSC of IUGS Commission of Stratigraphy, edited by H. D. Hedberg in 1976.

3. An informal stratigraphic code "Report on Stratigraphic Nomenclature in the New Zealand Quaternary" by the National Committee for Quaternary Research, the Royal Society of New Zealand, edited by M. Gage, K.B. Lewis, J.V. Eade, and R.H. Herzer (1977).
4. The "Stratigraphic Code of the USSR" by Interdepartmental Stratigraphic Committee of the USSR, compiled by A.I. Zhamoida, O.P. Kovalevsky, A.I. Moissejeva, V.I. Yarkin.
5. The "Strategraphic Guide of China and Explanation of Stratigraphic Guide of China" by the All-China Stratigraphic Commission, Compiled by special revision group led by T.H. Yin, 1982.
6. The "North American Stratigraphic Code" by North American Commission on Stratigraphic Nomenclature, 1983.

The previous six codes or guides, except the code of the USSR, all belong to the multiple concepts of stratigraphic classification, and these are the important references for working out the Quaternary stratigraphic code of China.

7.1.2 The Status Quo of Loess Stratigraphic Classification in China

Loess stratigraphy has been studied for over 100 years F. v. Richthofen (1882) first gave loess deposits of China stratigraphic significance, and suggested that loess located in central China corresponded to that located in the Rhine Valley. L.F. Yih (1920) introduced the loess deposits of the Zhaitang Basin, Beijing, as reported in the manuscript of Andersson (1939). Barbour (1930) first used the name of Malan Loess, sometimes called the Malan Loess Stage of Accumulation, Malan Stage Loess, or Malan Loess Association. Teilhard de Chardin and Young (1930) confused the Malan Loess with the True Loess in Shanxi and Shaanxi Provinces. Liu Tungsheng and Chang Tsunghu (1962) designated the loess stratigraphic units in descending order as Malan "Huangtu" (Loess), Lishi "Huangtu" (Loess) and Wucheng "Huangtu" (Loess). From then on, the loess lithostratigraphic sequence of North China was subdivided by rock characteristics into units of rank of formation. This is an important step in the loess stratigraphic classification, because this has established an objective material concept. Unfortunately, the concepts of stratigraphic classification and nomenclature of the Quaternary in China have been frequently influenced by the traditional Soviet concepts of stratigraphic classification and nomenclature.

7.1.3 Loess Stratigraphic Classification and Nomenclature

Loess is not a soil in the concept of pedology, but commonly refers to a loose, yellowish accumulation. In geology, loess is a widespread, homogeneous, commonly nonstratified, porous, friable, slightly coherent, usually highly calcareous, fine-grained blanket deposit, consisting predominantly of fine sand. Loess is generally buff to light yellow or yellowish brown, often contains shells, bones and teeth of mammals, and is traversed by networks of small narrow vertical tubes left

Table 7.1. The history of classification of the loess stratigraphic sequence of China

v. Richthofen (1882)	Andersson (1920)	Barbour (1930)	Teilhard and Young (1930)	Liu Tungsheng (1959)	Zhang Zhonghu et al. (1959)	Kes' (1959)	Yan Zhen (1960)	Liu and Chang (1962)
	Redepos-		True		4th			
	ited		loess	New	Fom.	Yellow		Malan
	loess		(Malan	loess	3rd	loess	New loess	“Huangtu”
			Loess)		Fom.			(Loess)
Loess		Malan						
	Original	loess				Loess		
	loess				2nd	with		Lishi
							Old loess	“Huangtu”
			Reddish	Old	Fom.	buried		(Loess)
			clay	loess		soils		
								Wucheng
					1st	Red	Paleoloess	“Huangtu”
					Fom.	loess		(Loess)
			Hipparion red clay					

by successive generations of grass roots, which allow the loess to stand in steep or nearly vertical faces. Consequently, there are numerous buried soil beds occurring in loess sequence, so loess stratigraphic classification should belong to the category of lithostratigraphic classification and nomenclature.

The loess stratigraphic classification is based upon the organization of loess into units which are characterized by specific properties distinguishing it from adjacent units, on the basis of the lithologic character and change, and in the light of the types and change of buried soil intercalated in it.

The loess stratigraphic unit is a three-dimensional body, consisting of loess with or without intercalated buried soil beds, and spatially traceable. It is basic unit of mapping in the regional geology of the Quaternary. The boundaries between loess stratigraphic units are often diachranous.

The term of unit of loess stratigraphy is “loess”, which corresponds to the “formation” of the lithostratigraphic unit. The loess stratigraphic unit should be named after the rules of nomenclature procedure of “Stratigraphic Guide of China and Explanation of Stratigraphic Guide of China”. The name of a loess stratigraphic unit should be coined from an appropriate local geographic name, combined with the term “loess”. The name of a formally named loess stratigraphic unit should always be capitalized, for example, Wucheng Loess.

The study of a loess stratigraphic sequence often involves many categories of stratigraphic classification, such as pedostratigraphic classification, magnetostratigraphic classification, biostratigraphic classification and chronostratigraphic classification, so the stratigraphic classification of loess sequence is also a multiple classification.

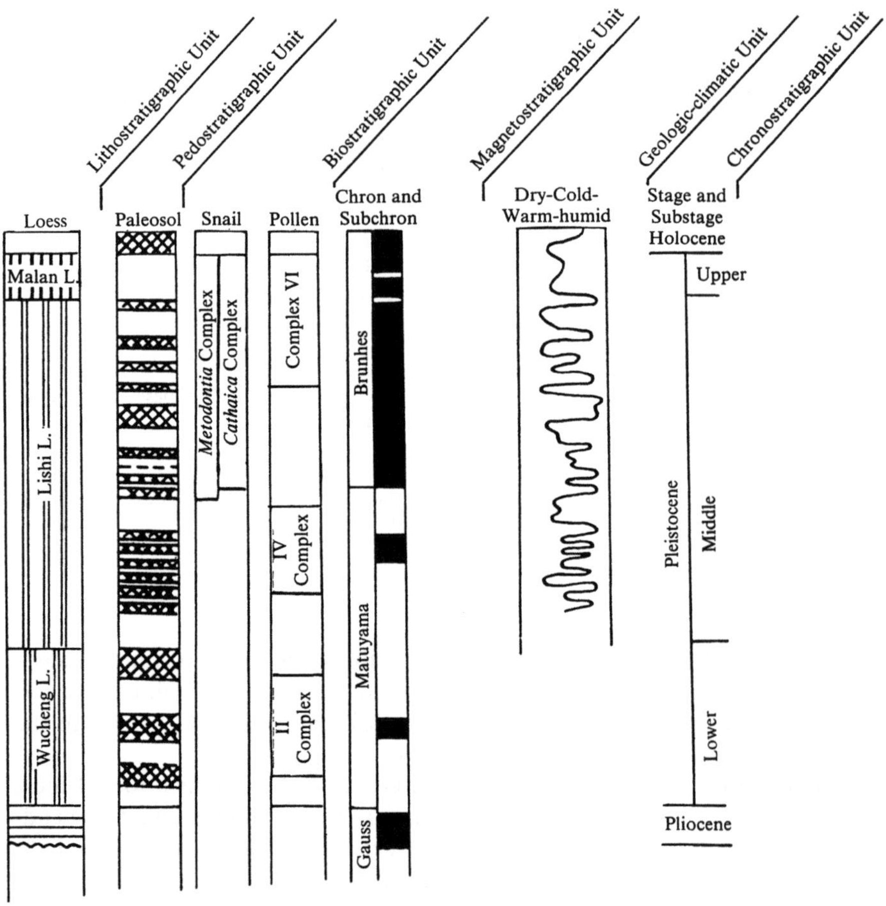

Fig. 7.1. Multiple classification of loess stratigraphic sequence

7.2 Thermoluminescent Dating of Loess

A thermoluminescence (TL) growth curve, determined from quartz which is 1 – 8 μm in size, can be used to obtain a formation age of quartz. The characteristics of the crystal lattice of quartz are beneficial to dating the formation age by using the TL technique. The use of this technique for dating the loess and what is the key to loess dating are presented in this section.

7.2.1 Methods for Determining the Age of Loess

The essential basis for using thermoluminescent dating of loess is that the loess is eolian. The quartz in loess when floating in the air or deposited on the earth is exposed to sunlight during the transportation by wind, and most of the originally stored energy in the crystal of quartz. Only this kind of loess can be used for thermoluminescent dating. The procedures for dating are as follows.

Samples collecting and composition analysis.

Table 7.2. The distribution of annual dose (the contents of uranium and thorium are 1 ppm in loess respectively)

	$\times 10^{-5}$ Gy	$\times 10^{-5}$ Gy	$\times 10^{-5}$ Gy
Thorium series	74.0	2.90	5.00
Thoron and its daughters	43.0	1.85	3.17
Fractions of thoron and its daughters	0.581	0.638	0.634
Uranium series	278.3	14.64	12.68
Radon and its daughters	152.1	8.71	11.99
Fractions of radon and its daughters	0.547	0.595	0.946
Potassium		68.2	20.5

The latitude and altitude of the selected location were recorded. The field dosimeter was used to measure the radiation intensity of the area. Abnormal radiation environment was avoided as far as possible. The surface layer of the loess 5 – 10 cm thick was stripped off and a block of fresh loess 1000 cm^3 in size was taken, put it in a plastic bag as fast as possible, wrapped with black paper, and labeled. The samples were not exposed to any light sources, high temperature heating and irradiation during transportation and storage. After the samples were put into the laboratory, the content of water was measured and the contents of uranium, thorium and potassium determined, to define the radiation dosage of the sample and the annual radiation dose rate. Table 7.2 illustrates some available data for calculating the dosage.

Three kinds of correction should be considered.

1. The Correction for the Percentage of Water Content of the Sample. Because the water content could reduce the radiation dosage, the absorption due to the water should be ruled out to obtain the real radiation dosage. The procedure for determining the water content was that one piece of the sample was taken, and its weight (W′) determined. The sample was then put in the drying box at a temperature of 105 °C for one hour, and the dry sample was weighed (W). The water content of the sample was then calculated as:

$$R = \frac{W' - W}{W'} .$$

The formulas for the correction of different kinds of radiation are:

$$D_\alpha = \frac{D'\alpha}{1 + 1.5\,R} ,$$

$$D_\beta = \frac{D'\beta}{1 + 1.25\,R} ,$$

$$D_\gamma = \frac{D'\gamma}{1 + 1.14\,R} ,$$

where $D'\alpha$, $D'\beta$, $D'\gamma$, and $D\alpha$, $D\beta$, $D\gamma$ are the radiation doses before and after the correction of water content, respectively.

2. The Correction of Radon Escape. The parent nuclei in three natural radioactive series are the isotopes of uranium and thorium. One kind of gas formation during the decay chain is called the emanation element as it escapes easily and diffuses in the loess. After the diffusion, radon and its daughter nuclei which are produced through the procedure of decay will be transferred from one place to another. Thus, it is an entire radiation series and will be destroyed. If we do not consider the effects from the diffusion of the emanation, we will get the incorrect result from evaluating the radiation dose from the entire radiation series. After the content of uranium and thorium is determined the diffusive quantities of the emanation should be determined. The corrected results could be obtained during estimating the radiation dose. Table 7.2 gives the dose fraction of the radon isotopes and their daughters in the whole radioactive series. The quantity of the emanative excape is closely related to the nature of the sample and its environment. So, each sample should be measured one by one. The determination of the escaping fraction of radon and thorium emanation is to measure the a-particle counts of a piece of loess sample, first to record that belonging to the total radioactive series a_0, and then to record that belonging to the emanation and its daughters a_1. The escaping fraction can be calculated by the following formula,

$$X = \frac{2a_1}{a_0} .$$

For most of the loess samples $a_1 \doteq 0.5\, a_0$ or $X \doteq 1$.

3. Correction of Cosmic Radiation. Cosmic radiation comes from the outer space and the composition of the cosmic ray is very complex. The dose of the cosmic ray absorbed by the sample is closely related to the location where the sample is collected. The special investigation indicated that the relative intensities of the cosmic rays on the earth surface are different at different altitudes and in different latitudes. Because the absorptions of cosmic rays in the atmosphere are different in different conditions, the loess itself can absorb or abate the rays: Preparation of sample and thermoluminescent measurement.

The method of acetone flotation was used to remove the fine grains of quarz (the diameter about 2–8 μm) from the loess sample, and deposit them uniformly on an aluminum plate. They were put in an evaporating box at 40 °C for 24 hours, and the samples of quartz were put on the heated plate in the furnace. As the heating rate increased, the counts of the photons emitting from the sample were recorded. The X–Y function recorder was used to plot the thermoluminescent curve (growth curve). The growth curve showed the pattern and variation of the photon numbers emitted by a sample as a function of increasing temperature. Generally, the thermoluminescence intensity of 375 °C is taken from the growth curve, and the following experimental data are obtained for a different purpose.

1. The natural storing thermoluminescent amounts of the sample, 4–8 samples were used to estimate its thermoluminescent amount and then the average amount was taken as the natural storing thermoluminescent amount TL_N.

The thermoluminescent response of samples to the a-radiation. Four samples were used and provided progressively increasing a-radiation doses $N+a_1$, $N+a_2$, $N+a_3$ and $N+a_4$ to each sample, respectively. $TL_{N+a_4} = 2TL_N$ is required in the measurement of the thermoluminescent amounts, TL_{N+a_1}, TL_{N+a_2}, TL_{N+a_3} and TL_{N+a_4}. Four TL values with average TL_N were used to plot the TL intensity, so as to obtain the thermoluminescent response of the quartz to the a-radiation,

$$S_q = CTL_{N+a}/\Delta Da \ .$$

2. The thermoluminescent response of quartz to the β-radiation. Four samples of the same weight were used to give the progressively increasing β-radiation dose, respectively. $TL_{N+\beta_4} = 2TL_N$ is also required for measuring the thermoluminescent values, $TL_{N+\beta_1}$, $TL_{N+\beta_2}$, $TL_{N+\beta_3}$ and $TL_{N+\beta_4}$. Four values with TL_N obtained were used to calculate the thermoluminescent responses of β-radiation,

$$S_p = \Delta TL_{N+a}/\Delta D\beta \ .$$

3. Thermoluminescent response of quartz to the second radiation. a_1, a_2, q_3 and a_4 were provided with four different a-radiation doses by using the measured TL samples to measure the TL intensity of each sample, TL_{2a_1}, TL_{2a_2}, TL_{2a_3} and TL_{2a_4}. The following formula can be obtained by the above method,

$$S'_q = \frac{TL_{2a}}{\Delta 2a} \ .$$

β_1, β_2, β_3 and β_4 were provided with β-radiation doses to the other four measured samples, to measure the TL amounts of the samples $TL_{2\beta_1}$, $TL_{2\beta_2}$, $TL_{2\beta_3}$ and $TL_{2\beta_4}$. The TL response to the second β-radiation can be obtained,

$$S'_p = \frac{TL_{2\beta}}{\Delta 2\beta} \ .$$

Then the sensitivity coefficients of the TL response to two kinds of radiation of the quartz in the samples are calculated.

$$\eta = \frac{S_q}{S_p} \ ,$$

$$\eta = \frac{S'_q}{S'_p} \ .$$

4. Equivalent β-radiation dose. In the experiment of TL response of quartz to β-radiation, the diagram can be used to obtain the equivalent β-radiation dose (E.D.), called the apparent value of E.D., and this is not the real value of the loess sample. Therefore, shine fading test should be used to obtain the real value. Eight identical weight samples were chosen to divide two groups, four samples in each group. The progressively increasing β-radiation doses β_1, β_2, β_3 and β_4 were provided simultaneously to the two sample groups. The four samples of one group were shone by a simulator sunlight source (an iodine-tungsten lamp with color temperature 3200 K, 1300 W manufactured by Huguan Lamp Factory, Shanghai) at 50 cm distance for 20 hours. Samples of other group were not illuminated by this light source. Then the TL intensities of the two groups of samples were measured as $TL_{N+\beta_1}$, $TL_{N+\beta_2}$, $TL_{N+\beta_3}$ and $TL_{N+\beta_4}$, and $(TL_{N+\beta_1})'$, $(TL_{N+\beta_2})'$, $(TL_{N+\beta_3})'$ and $(TL_{N+\beta_4})'$, respectively. So, two straight lines are drawn in one

figure, and the distance from the cross point of the two lines on the X-axis, X_3 to the original point is the real value of equivalent betaradiation dose E.D.

7.2.2 The Formative Age of Luochuan Loess

The sampling location of the TL from the section of Heimugou of Qingjiazhai in Luochuan region is shown in Fig. 7.2. The TL dating of loess samples in that section is listed in Table 7.3.

About 100 loess samples belonging to Malan Loess, Lishi Loess and Wucheng Loess have been measured by the thermoluminescent dating method. The data obtained showed that the samples of Malan Loess are the best, those of Lishi Loess are the second, and those of Wucheng Loess rank the third. The main reason for this is possibly due to the limitation of the TL dating method. Because the samples used in the experiments are pure quartz, the error should not be induced by the non-quartz constituents. The real reason for the error could be influenced by the TL dosimeter used for dating. Its upper limit for the TL response does not exceed 10^4Gy .Although the upper limit dose of the quartz dosimeter is higher than the $Li_2B_4O_7$ (Mn), it still seems insufficient for dating geological samples. Using the TL character of quartz in the fine fraction for dating the age of loess, the upper limit age of the samples measured could reach a million years; it is iden-

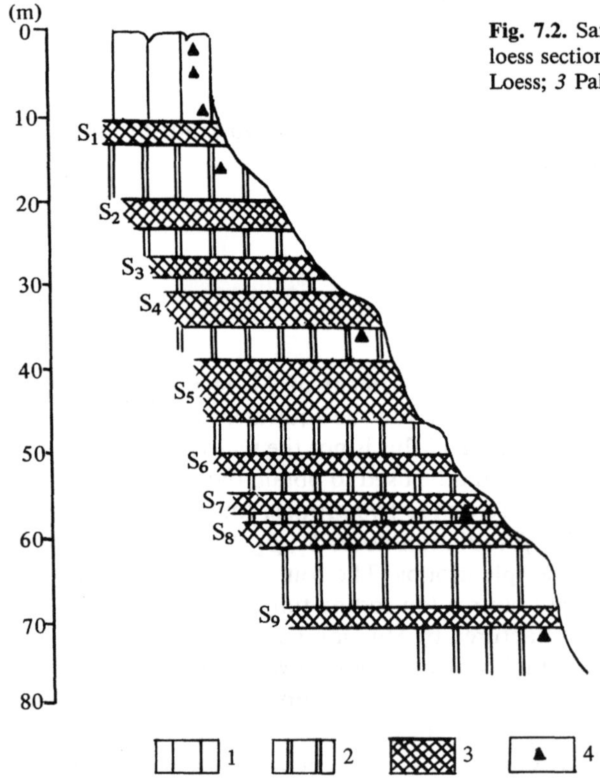

Fig. 7.2. Sampling site for TL along Qinjiazhai loess section, Luochuan. *1* Malan Loess; *2* Lishi Loess; *3* Paleosol; *4* Sampling site

Table 7.3. Age of loess samples of Heimugou section in Luochuan

Strata	Number of layers	Depth (m)	Age ($\times 10^3$ yr)
		2.5	19±2
Malan Loess	2	5	45±4
		9	103±10
	4	15	148±16
Lishi Loess	10	32	357±31
	16	52	696±60
Wucheng Loess	20	65	876±91
	85	126	1040±120

Table 7.4. Ages of Malan Loess in some sections

No.	Locality	Depth from surface (m)	Age of thermo-luminescence ($\times 10^3$ yr)	Collector and time of collection
1		3.5	39±5	Yuan Baoyin et al.
2		6.5	46±4	(1980)
3		11	103±10	
	Potou village at Heimugou, Luochuan, Shaanxi			
4		2	13±2	Bater et al. (1982)
5		5	39±9	
6		8	89±9	
7	Qinjiazhai at Heimugou,	2.5	19±2	Bater et al. (1982)
8	Luochuan, Shaanxi	5	45±4	
9		8.5	103±1	
10		1.8	7±0.6	
11	Cardening field of Caijiagou,	10.5	35±9	Dong Guangrong and
12	Yulin, Shaanxi	19.5	102±8	Li Huhou (1981)
13		20.8	124±11	
14	Milestone 215 of the highway, from Xian to Luochuan	12	135±11	Zhao Xitao and Li Huhou (1981)
15	At the foot of Gaolan Mountain, Lanzhou, Gansu	4.5	26±1	
16	Baicaoyuan, Huining, Gansu	8.8	100±6	Lu Liangcai and Zheng Honghan (1982)
17	Liugoucun, Lanzhou, Gansu	18	35±2	

tical to the age estimated by the author using the storing energy electrons existing in the quartz crystal lattice. At this dose, radiation damage to the quartz lattice has reached the maximum limit, and the age values obtained in this condition would be more erroneous.

Table 7.4 shows the age values of the samples collected from eight various locations in the Malan Loess. Numbers 1–9 samples were collected from the

Heimugou profile, Luochuan, Shaanxi Province the sampling places were different and they were only sampled at nearly similar level. The first four groups of the samples are representative samples of different level, and the latter four samples were collected at the bottom of each profile. From these age values, the following recognition can be obtained.

1. Malan Loess in northwest China was deposited about 100000 years, ago.
2. In certain sections at different locations, the ages of deposition of Malan Loess were different.
3. The age of Malan Loess can be determined by TL dating of quartz. Fixed point investigations are necesssary in specific sections, and it is indispesnable to follow rigorously all procedures prescribed for the collecting of the samples.

References

An Zhisheng, Wei Lanying (1979) The illuviation ferri-argillans and their genetic inference. Kexue Tongbao 24(8):356–359 (in Chinese, with English abstract)

An Zhisheng, Wei Lanying (1980) The fifth layer Paleosol in the Lishi Loess and their paleoclimatic significance. Acta Pedologica Sinicam 17(1):1–10

An Zhisheng, Lu Yanchou (1984) A climatostratigraphic subdivision of Late Pleistocene strata named by Malan formation in North China. Kexue Tongbao 29(9):1239–1242

An Zhisheng, Wang Juanda, Li Huamei (1977) Paleomagnctic research of the Luochuan loess section. Geochimica (4):239–249 (in Chinese, with English abstract)

An Zhisheng, Wei Lanying, Lu Yanchou, Wang Naiwen, He Xixian, Ding Shen (1979) Magnetostratigraphy of the core S-5 and the transgression in the Beijing area during the carly Matuyama Epoch. Geochimica (4):343–346 (in Chinese, with English abstract)

Andersson JG (1939) The Malan terraces of northern China. Bull Mus Far East Antiq 11:7–44

Andraea A (1913) Land- and Süßwasserschnecken aus Zentral- und Ostasien. Futterer's Durch Asien 3:43–90

Aoyagi K, Kazama T (1980) Transformational changes of clay minerals, zeolites and silica minerals during diagenesis. Sedimentology 27:179–188

Bagnold RA (1941) The physics of blown sand and desert dunes. Methuen, London

Bailey SW (1980) Summary of recommendations of AIPEA Nomenclature Committee. Clay Miner 15:85–93

Barbour GB (1930) The Paote and Chingshui physiographic stage and their significance. Bull Geol Soc China 9:45–48

Beavers AH (1957) Source and deposition of clay minerals in Peorian loess. Science 126:1285

Berg LS (1928) Zoal'nost' pokrovneikh porod Vostochnoi Evropei i proiskhordenie diossa. Zap Odessk Ova 44 (in Russian)

Berg LS (1932) Loss kak produkt veivetrivaniya i povhvoobrazovaniya. Trudei 2-i Merduiar. koifer Aichpe Vern 1 (in Russian)

Berg LS (1946) Fauna lessa. Trudy Inst Geogr Akad Nauk SSSR 37

Berg LS (1947) Klimat Zhizn Izd 2 Geogr (in Russian)

Blunt DJ (1981) Geochemistry of amino acids in sediments from Clear Lake. Calif Geol 34:380–381

Blunt DJ, Warnke D (1980) Amino acid stereochemistry in Antarctic marine sediments: Islas Orcadas Piston Cores 12-77-41 and 12-77-24. In: Hare PE, Hoering TC, King K Jr (eds) Biogeochemistry of amino acids. Wiley, New York, pp 121–128

Bognar A (1980) Distribution, properties and types of loess and loess-like sediments, in Croatic. In: Studies on loess. Akademiai Kiado, Budapest, pp 267–286

Bowler JM, Polach HA (1971) Radiocarbon analyses of soil Carbonates: An evaluation from paleosols in southeastern Australia. In: Yaalon DH (ed) Paleopedology: Origin, nature and dating of paleosols. Israel Universities Press, pp 97–108

Bradley WF (1954) X-ray diffraction criteria for the characterization of chloritic material in sediments. Proc 2nd Natl Conf Clays Clay Mines Columbia (1953):324–334

Braun A (1847) Löss. Neues Jahrb Minerol Geogr. Geol Peterfaktenkd 1:49–53

Brewer R (1964) Fabric and mineral analysis of soils. Wiley, New York, pp 205–298

Burst JF (1958) Mineral heterogeneity of "glauconite" pellets. Am Minerol 43:481–497

Cao Zhaoyuan, Yu Qinghe, Liu Lansuo, He Peiyuan (1983) Problems concerning the lower limit of Quaternary in China. Acta Geol Sinica 57:96–103 (in Chinese, with English abstract)

Central Meteorological Bureau of China (1978) Climate Atlas of the People's Republic of China. Cartogr Publ House, Beijing (in Chinese)
Chen Deniu, Lu Yanchou, An Zhisheng (1982) Snail assemblages in loess strata and their environmental implication. Proc 3rd Natl Quat Conf China, 7–15. Science Press, Beijing (in Chinese)
Chen Fangji (1980) Holocene of Beijing region and the changes of its natural environments. Sci Sinica 23(5):622–633
Chen Qingxuan, Sun Ye, Shao Yunhui (1980) Present-day activities of the tectonic systems in China. In: Scientific papers on geology for international exchange, vol I. Publ House Geol, Beijing, pp 78–88 (in Chinese)
Cheng Guoliang, Lin Jinglu, Li Suling (1978) A discussion on the ages of fossilsbearing Lantian Man Strata. Select Pap Paleontol. Science Press, Beijing, pp 151–157
Chia Langpo, Wei Chi (1976) A Palaeolithic site at Hsu-Chia Yao in Yangkao county, Shansi Province, Kaogu Shuebao (2):97–114 (in Chinese, with English abstract)
Cui Zhijiu (1982) On the significance and perspective of study on periglaciation. Proc 3rd Natl Conf Quat China. Science Press, Beijing, pp 168–175 (in Chinese)
Cressey GB (1934) China's geographic foundation. McGraw-Hill, New York
Dai Yingsheng (1980) A discussion on the water and soil erosion on the Loess Plateau in relation to the paleoclimatic environment along the middle reach of Huanghe (Yellow) River. Renmin Huanghe (4):1–8 (in Chinese)
Denisov N Ya (1946) O prirode prosadochnykdh yavienii v lessovidnykh suglinkakh. Izd-vo Sov Nauka (in Russian)
Denisov N Ya (1953a) Stroitel'nie svoistva lessa i lessovidnykh suglinkakh. Izd-vo Sov Nauka (in Russian)
Denisov N Ya (1953b) Stroitel'nie svoistva lessa i lessovidnykh suglinkov. Izd-vo II Gosstroiizdat (in Russian)
Diao Guiyi, Wen Qizhong, Y Suhua, Sun Fuqing (1982) Trace elements in Malan Loess. Proc 3rd Natl Quat Conf China. Science Press, Beijing, pp 267–268 (in Chinese)
Ding Guoyu, Lu Yanchou (1984) Map showing the relative movement of continental blocks in China (in press)
Dylik J (1954) The problem of the origine of loess in Poland. Biul Peryglacjalny 4:195–238
Dylik J (1961) The Lódź region: guide-book of excursion C. 6th INQUA Congr, Poland
Emiliani C (1955) Pleistocene temperatures. J Geol 63:538–578
Emiliani C (1966) Paleotemperature analysis of Caribbean cores. P. 6304–8 and 6304–9, and a generalized temperature curve for the past 425000 years. J Geol 74:109–126
Emiliani C (1972) Quaternary paleotemperatures and the duration of the high-temperature intervals. Science 178:398–401
Fahlbusch V (1969) Pliozäne and pleistozäne Cricetinae (Rodentia, Mammalia) aus Polen. Acta Zool Cracov 14:99–138
Flint RF (1947) Glacial geology and the Pleistocene Epoch. Wiley, New York
Flint RF (1974) Three theories in time. Quat Res 4:1–8
Gao Guorui (1981) Classification of microstructures of loess in China and their collapsibility. Sci Sinica 24(7):962–974
Geographical Department of Zhongshan Univ, Lanzhou Univ and Northwest Univ (1979) Meteorology and the climatology. People's Education Press, Beijing, pp 106–114, 128–137 (in Chinese)
Gong Shiyang, Xiong Guisu (1979) Source and distribution of silts in the Huanghe (Yellow) River. Remin Huanghe (1):1–17 (in Chinese)
Grim RE, Bray RH, Bradley WF (1937) The mica in argillaceous sediments. Am Mineral 22:813–829
Gu Xiongfei, Zheng Honghan, Han Jiamao (1980) Some problems on clay mineralogy. Geol Geochem (3):23–27 (in Chinese)
Guo Shilun, Zhou Shuhua, Meng Wu, Zhang Pengfa, Sun Shengfen, Hao Xiuhong (1980) Age determination of Peking Man by fission track dating. Kexue Tongbao 25:535
Guo Xudong, Yan Fuhua, Ye Yongyin, Sun Jianzhong (1981) Possibility of quaternary glaciation in Xi Shan, Beijing as seen from the "Glacial Remains" of Moshikou. J Glaciol Gryopedol 3(2):63–66 (in Chinese, with English abstract)
Han Jiamao (1982) A preliminary study on the clay mineralogy of the loess at Luochuan section. In: Quaternary geology and environment of China. China Ocean Press, Beijing, pp 67–72

Hays JD, Imbrie J, Shackleton NJ (1976) Variations in the earth's orbit: pacemaker of the ice ages. Science 194:1121–1132

He Bingti (1969) Loess and the agriculture of China. Chinese Univ Hong Kong (in Chinese)

Hecht AD, Barry R, Fritts H, Imbrie J, Kutzbach J, Mitchell JM, Savin SM (1979) Paleoclimatic research; status and opportunities. Quart Res 12:6–17

Heller F, Liu Tungsheng (1982) Magnetostratigraphical dating of loess deposits in China. Nature (London) 300:431–433

Heller F, Lui Tungsheng (1984) Magnetism of Chinese loess deposits (astr.) Geophys J R 77:125–141

Hilber V (1882) Recent und in loess gefundene Landschnecken aus China. Sitzungsber Math-Naturwiss Abt I K Akad Wiss Wien I 86:1883, II, 88:Ibid, 1884

Histl Res Group of Seismo Comm China seism data year book (1956) Science Publishing House (in Chinese)

Hower J, Eslinger EV, Hower ME, Perry EA (1976) Mechanism of burial metamorphism of argillaceous sediment: Mineralogical and chemical evidence. Geol Soc Am Bull 87:725–737

Hsu SC (1936) Gastropods from the Siashu Formation. Palaeontol Sinica Ser B 6(3)

Hsu KJ, Giovanoli F (1979) Messinian event in the Balck Sea. Paleogeogr Palaeoclimatool Palaeoecol 29:75–93

Huang Qiang (1963) A discussion on the assessment of foundation of subsiding loess. Tumu Gungcheng Xuebao 9(2) (in Chinese, with English abstract)

Institute of Botany, Academia Sinica and Research Group of Cenozoic Palynology of Geological Institute, Ministry of Geology (1966) Paleobotanical investigation of the Cenozoic group in the Lantian area, Shaanxi Province. Proceedings of Field Meeting on the Cenozoic Group in Lantian County, Shaanxi Province. Science Press, Beijing, pp 157–196 (in Chinese)

Institute of Geochemistry Academia Sinica (1974) Concise Handbook of Geochemistry. Science Press, Beijing (in Chinese)

Institute of Geology, State Seismological Bureau (1979) A summary instruction on "The seismotectonic map of People's Republic of China". Cartogr Publ House, Beijing (in Chinese)

Institute of Hydrogeology and Engineering Geology of the Ministry of Geology, PRC (1959) Loesses and loessy rocks in China. Publ House Geol, Beijing (in Chinese)

Institute of Pedology, Academia Sinica (1978) Soils of China. Science Press, Beijing (in Chinese)

Institute of Water Source, Shaanxi Province (1959) Characteristics of loess in Northwestern China. People's Publ House Shaanxi Province, Xian (in Chinese)

Itihara M, Kamei T (1982) The Pliocene-Pleistocene Boundary in the Osaka Group, Japan. In: Itihara M, Kuwano Y (eds) 3rd Rep the Pliocene-Pleistocene Boundary in Japan. Jpn Natl Work Group of the IGCP Project, pp 42–50

Jenkins GM, Watts DG (1968) Spectral analysis and its applications. Holden-Day, San Francisco

Jing Ke, Chen Yongzong (1983) Preliminary study of the erosion environment and rates on the Loess Plateau. Geogr Res 2(2):1–10 (in Chinese, with English abstract)

Kmet J, Makboubi E (1972) Esophageal cancer in the Caspian Littoral of Iran: Initial Study Science 175:846–853

Kes' AS (1959) K voprosu o proiskhordenii tolshi Severnogo Kitaya. Trudei Komissii po izuch. Chetvert Nepioda AN SSSR 14 (in Russian)

King SG (1926) Notes on the fossil shells of the Sanmen series. Bull Geol Soc China 5:157–160

Kodama H, Shimoda S, Sudo T (1969) Hydrous mica complexes: their structure and chemical composition. Proc Int Clay Conf Tokyo 1:185–196

Kopchenova EV (1954) Mineralogicheskii analiz shlikhov. Tosgeolizdat (in Russian)

Kriger NI (1965) Löss, ego svoistva i svyaz's geograficheskoi sredoi k Koigressu. INQUA (Ssha 1965) Izd Nauka (in Russian)

Krumbien WC (1938) Manual of sedimentary petrology. N.Y.D. Appleton-Century Company

Kubler B (1961) Sur quelques interstratifies irreguliers mica-montmorillonite. Bull Serv Carte Geol Als Lorr 14:173–178

Kubler B (1964) Les argiles, indicateurs de metamorphisme. Rev Inst Fr Petrole 19:1093–1112

Kukla GJ (1961) Quaternary sedimentation cycle – Survey of Czechoslovak Quaternary – Czwartorzed Europy srodkovej i wschodniej. INQUA VIth Int Congr Inst Geol Prace 34:145–154

Kukla GJ (1970) Correlations between loesses and deep-sea sediments. Geol Foren Stockholm Forh 92:148–180

Kukla GJ (1977) Pleistocene land-see correlations. Earth Sci Rev 13:307–374

Kyte FT, Zhou Z, Wasson JT (1981) High noble metal concentrations in a late Pliocene sediment. Nature (London) 292:417–420

Laboratory of Quaternary Palynology, Laboratory of Radiocarbone, Kweiyang Institute of Geochemistry, Academia Sinica (1978) Development of natural environment of the southern part of Liaoning Province during the last 10000 years. Sci Sinica 19(4):516–532

Larionow AK , Priklonskii VA, Anan'ev VP (1959) Lessoveie porody SSSR i ikh stroitel'nye svoistva. Gostekhzdat Lit Po Geologii Okhranie Nedr, Moskva (in Russian)

Leading Group Office on Prevention and Cure of Local Diseases, Central Committee of CPC (1984) Proceedings of scientific investigations on the Kaschen Beck disease in Yongshou County, Shaanxi Province (1979–1982). People's Hygiene Press, Beijing (in Chinese)

Li Huamei, Wang Junda (1982) Magnetostratigraphic study of several typical geologic sections in North China. In: Quaternary geology and environment of China. China Ocean Press, Beijing, pp 33–38

Li Huamei, Wang Junda (1983) Palaeomagnetic study on drill core from northern Bohai Coastal plain. Geochimica (2):196–204 (in Chinese, with English abstract)

Li Huamei, An Zhisheng, Wang Chunta (1974) Preliminary paleomagnetic study of loess from the Wucheng section, northern China. Geochimica (2):93–104 (in Chinese, with English abstract)

Li Huamei, Wang Junda, An Zhisheng (1977) Paleomagnetic investigation on drilling cores in the Hengshui area, North China Plain. Geochimica (3):210–218 (in Chinese, with English abstract)

Li Yuntong (1966) Cenozoic mollusc fauna and its stratigraphical significance in Lantian and Weinan area, Shaanxi Province. Proceedings of Field Meeting on the Cenozoic Group in Lantian County, Shaanxi Province, Science Press, Beijing, pp 225–254 (in Chinese)

Lin Zaiguan, Liang Weiming (1982) Engineering properties and zoning of loess and loess-like soils in China. Can Geotech J 19:76–91

Lui Chun (1982) Discussion of Blake event in the Datong volcanic rocks group. In: Research on geology, vol I. Inst Geol Academia Sinica, pp 179–182, Cultural Relics Publ House, Beijing (in Chinese, with English abstract)

Liu Guangxun, Ma Tingzhu, Huang Peiyu, Wang Songxian (1982) Presen-day tectonic Movement along active faults in the east of China's mainland. Seismol Geol 4(4):1–14 (in Chinese, with English abstract)

Liu Jingwen, Zhang Jiaming (1982) A preliminary discussion on the ground-fissures and earthquakes in Xian area. In: Papers of prehistoric earthquakes and Quaternary geology. Shaanxi Science and Technology Press, Xiań, pp 134–142 (in Chinese, with English abstract)

Liu Tungsheng (1958) Preliminary investigation on the loess of Shanxi and Shaanxi Provinces in the middle reaches of the Huanghe River. Quat Sinica 1(1):255–257 (in Chinese)

Liu Tungsheng (1959) Younger loess and older loess. J Geol (5):22–25 (in Chinese)

Liu Tungsheng, An Zhisheng (1984) A preliminary magnetostratigraphic study of the Beihanzhai loess section. Geochimica (2):134–137 (in Chinese, with English abstract)

Liu Tungsheng, Chang Tsunghu (1962) The Huangtu (loess) of China. Acta Geol Sinica 42:1–14 (in Chinese, with English abstract)

Liu Tungsheng, Ding Menglin (1984) A tentative chronological correlation of early human fossils horizons in China with the loess-deep sea records. Acta Anthropol Sinica 3(2):93–101 (in Chinese, with English abstract)

Liu Tungsheng, Yuan Baoyin (1982) Quaternary climatic fluctuation – A correlation of records in loess with that of the deep sea core V28–238. In: Research on geology, vol I. Inst Geol Academia Sinica, pp 113–121, Cultural Relics Publ House, Beijing (in Chinese, with English abstract)

Liu Tungsheng et al. (1964) The Loess along the Middle Reaches of the Huanghe (Yellow) River. Science Press, Beijing (in Chinese)

Liu Tungsheng et al. (1965) The Loess Deposits of China. Science Press, Beijing (in Chinese)

Liu Tungsheng et al. (1966) The composition and texture of Loess. Science Press, Beijing (in Chinese)

Liu Tungsheng, An Zhishing, Wen Chichung, Lu Yenchou, Han Chiamao, Wang Chunta, Tiao Kueiyi (1978) Geological environments of the Loess in China. Kexue Tongbao 23(1):1–9 (in Chinese, with English abstract)

Liu Tungsheng, Wen Qizhong, Zhen Honghan (1980a) Paleoclimate and evolution recorded in loess of China. In: Scientific papers on geology for international exchange, vol V., Publ House Geol, Beijing, pp 77–82 (in Chinese)

Liu Tungsheng, Chen Qingmu, Yu Zhicheng, Yuan Zhiyun (1980b) Geochemical environment problems concerning the endemic fluorine disease in China. Geochimica (1):13–22 (in Chinese, with English abstract)

Liu Tungsheng, Gu Xiongfei, An Zhisheng, Fan Yongxiang (1981) The dust fall in Beijing, China on April 18, 1980. Geol Soc Am Spec Pap 186:149–157

Liu Tungsheng, Cheng Mingyang, Li Xiufang (1982) A statellite images study on the dust storm at Beijing on April 17–21, 1980. In: Quaternary geology and environment of China. China Ocean Press, Beijing, pp 48–52

Liu Tungsheng, Dong Guangrone, An Zhisheng (1984) Desert and loess environment in Chine since the Quaternary. In: Quaternary geology and geomorphology. Proc 27th Int Geol Congr Moscow 3:185–197

Liu Tungsheng, Gu Xiongfei, Zheng Honghan (1985) Loess in Tianshan Mts. In: Geology and paleontology of the Mt. Tuomur area in Tianshan Mts. Xinjiang People's Publ House, Urumqi, pp 123–130 (in Chinese)

Liu Yulan, Wen Qizhong, Chen Qingmu (1981) F/Cl ratio in Luochuan loess section, Shaanxi Province – A geochemical index for paleoclimatic fluctuation. Geochimica (4) 388–391 (in Chinese, with English abstract)

Liu Zechun (1982) Climatostratigraphy of the sediments in the Peking Man's Cave. In: Quaternary geology and environment of China. China Ocean Press, Beijing, pp 25–51

Lozek V (1964) Quatermollusken der Tschechoslowakei. Rozpr Ustred Ustavu Geol Praha 31:374, 32tt

Lu Longhua, Zhu Fukang, Chen Xianji, Zhu Yunlai (1984) Middle-term vibration feature of Qinghai – Xizang Plateau in the summer of 1979. Proc Meteorol Examin Qingzang Plateau (II):140–151 (in Chinese)

Lu Yanchou (1981) Pleistocene climatic cycles and variation of $CaCO_3$ contents in a loess profile. Sci Geol Sinica (2):122–131 (in Chinese, with English abstract)

Lu Yanchou, An Zhisheng (1979) The quest for series of nature environmental changes during the Brunhes Epoch. Kexue Tongbao 24(5):221–224 (in Chinese, with English abstract)

Lu Yanchou, Wen Qizhong, Huang Baijun, Min Yushun, Deng Huaxing (1976) A preliminary discussion on the source of loessic materials in China – A study of the surface textures of silt quartz grains by transmission electron microscope. Geochimica (1):47–53 (in Chinese, with English abstract)

Lucas J (1963) La transformation des minéraux argileux dans la sédimentation. Etudes sur les argiles du Trias. Mem Carte Geol Als Lorr 23:202

Ma Xinghua, Qian Fang, Li Pu, Ju Shiqiang (1978) Palaeomagnetic dating of Lantian Man. Vertebrata PalAsiatica 16(4):238–243 (in Chinese, with English abstract)

Ma Yungchi (1945) Topography and soils in Tushanto Basin, Xinjiang. Geol Rev 10(3–4):175–177 (in Chinese)

Ma Yungchi (1958) Some suggestion on the origin and their paleogeographic implication of Quaternary strata. Quat Sinica 1(1):70–73 (in Chinese)

MacEwan DMC (1949) Some notes on the recording and interpretration of X-ray diagrams of soil clays. J Soil Sci 1:90–103

Mankinen EA, Dalrymple GB (1979) Revised geomagnetic polarity time scale for the interval 0–5 m.y. B.P. J Geophys Res 84:615–626

Martens EV (1879) Landschnecken von Chinesischen loess. Sitzungsbr Ges Naturforsch Fr-Berlin 73–74

Masukenkichi, Todatatsu (1944) Loess distribution in North China in Mongolic Instituteo of East Asia, 8:(in Japanese)

Mellieres F, Perez-Nieto H (1973) Les mineraux argileux des sediments recents du Golfe de Cariaco (Venezuela). Bull Gr Fr Argiles 25:65–78

Moldvay L (1962) On the governing sedimentation from colian suspension. Acta Univ Szegediensis 14:75–109

Morozov SS (1951) Novoe v reshenii problemy lessa. Uch Zap Moskovskogo Univ 149:Gruntovedenie Kn 2. (in Russian)

Obruchev VA (1933) Problema Lossa. Trudei 2-i Merdunar. Konf Aichpe veip 2, to re, v kn.: "V.A. Obruchev Izbranneie rabotei po geografii Azii", 3. Geografgiz (in Russian)

Obruchev VA (1951) Zakaspiiskaya nizmennost'. Zpa. Russk, geogr. ob-va po obshei geografii, 20, 1890; Tore, s depoln., v kn. "V.A. Obruchev, izbranneie rabotei po geografii Azii" 1. Geografgiz (in Russian)
Obruchev VA (1958) Loss kak osobneii vid pochvei, ego genezis i zadachi ego. Izucheniya v kn.: "V.A. Obruchev. Izbraiieie rabotei pogeigrafii Azii", 2. Geografgiz (in Russian)
Pei Wenzhong, Zhou Mingshen (1958) Report on finding and excavation of Paleolithic Dingcun Relics in Xiangfen County, Shanxi Province. Spec Issue, Ser A. Inst Vertebr Paleontol. Academia Sinica. No 2, Beijing (in Chinese)
Peng Lin (1983) Soil resources for dionght-resistent crope in loess areas and their reasonable ultilization. Collect Pap Soc Pedol, Shaanxi Province
Perry E, Hower J (1970) Burial diagenesis in gulf coast pelitic sediments. Clays Clay Miner 18:165–177
Pewe TL (1981) Desert dust: An overview. Geol Soc Am Spec Pap 186:1–10
Pewe TL, Péwe EA, Péwé RH, Jounaux A, Slatt RM (1981) Desert dust: Characteristics and rate of deposition in central Arizona. USA. Geol Soc Am Spec Pap 186:169–190
Piechocki A (1977) The Late Pleistocene and Holocene Mollusca of the Kunow Region (N-E, Margin of the Swiefokrzyskie Mts.). Folia Quat 49:Krakow
Ping Chi (1929) Fossil terrestrial gastropods from North China. Palaeontol Sinica Ser B 6(5)
Ping Chi (1930) Two new species of the Cretaceous fresh-water gastropoda in Mongolia. Am Mus Novitates 473
Ping Chi (1931) Tertiary and Quaternary non-marine gastropods of North China. Palaeontol Sinica Ser B 6(6)
Popov VV (1957) Klassifikatsiya lossoveikh porod. Trudei In-ta Geol Nauk An SSSR, Ser Geolmorfol Chvert Geol Veip (in Russian)
Qian Fang, Ma Xinghua, Wu Xihao (1983) Preliminary study on the Quaternary magnetic stratigraphy of China. Mar Geol Quat Geol 3:17–29 (in Chinese, with English abstract)
Qinghai-Gansu Comprehensive Expedition, Academia Sinica (1962) Engineering geological characteristics of loesses in eastern Gansu along arterial channel of diverted Taohe River: Physical, hydrographical, mechanical and chemical properties. In: Engineering geological problem of diverted Taohe River (II). Science Press, Beijing (in Chinese)
Qiu Shihua, Chai Lianzheng, Chen Tiemai, Yuan Sixun (1983) Stabilishing of "standard of sugar carbons in China" for radiocarbon dating. Kexue Tongbao 28(3):170–174
Rahn KA, Borys RD, Snaw GE (1981) Asian desert dust over Alaska: Anatomy of an Arctic haze episode. Geol Soc Am Spec Pap 186:592–596
Reynolds RC Jr, Hower J (1970) The nature of interlayering in mixedlayer illitemont-morillonite. Clays Clay Miner 18:25–36
Richthofen F v (1877) China 1. Berlin
Richthofen F v (1882) On the mode of origin of the loess. Geol Mag 9:Ser 2, 293–305
Ruhe RV, Olson CG (1980) Clay-mineral indicators of glacial and nonglacial sources of Wisconsinan loesses in southern Indiana, USA. Geoderma 24:283–297
Ruhe RV, Miller GA, Vreeken WJ (1971) Paleosols, loess sedimentation and soil stratigraphy. In: Yallon DH (ed) Paleopedology: Origin, nature and dating of paleosols, pp 41–46
Russel RJ (1944) Lower Mississippi valley loess. Bull Geol Soc Am 55:1–40
Schmithener H (1932) Chinesische Lösslandschaft und Städte. Stuttgart
Sedletsky ID, Ananev VP (1953) Zolovoe preoiskhordenke i sostav mineralov lossa Severnogo Kitaya. DAN SSSR 90:651–654 (in Russian)
Shackleton NJ, Opdyke ND (1973) Oxygen isotope and paleomagnetic stratigraphy of Equatorial Pacific core V28-238: Oxygen isotope temperatures and ice volume on a 10^5 year and 10^6 year scale. Quat Res 3:39–55
Shackleton NJ, Opdyke ND (1976) Oxygen isotope and paleomagnetic stratigraphy of Pacific core V28–239: late Pliocene to latest Pleistocene. Geol Soc Am Mem 145:449–464
Shackleton NJ, Opdyke ND (1977) Oxygen isotope and paleomagnetic evidence for early Northern Hemisphere glaciation. Nature (London) 270:216–219
Shi Yuanchun (1958) Loess and its origin in western Shanxi Province. Quat Sinica 1(1):252–253 (in Chinese)
Singh G, Opdyke ND, Bowler JM (1981) Late Cainozoic stratigraphy palaeomagnetic chronology and vegetational history from lake George, NSW. J Geol Soc Aust 28:435–452

Smalley IJ (1971) "In-situ" theories of loess formation and the significance of the calcium-carbonate content of loess. Earth Sci Rev 7:67–86
Smalley IJ, Cabrera JG (1970) The shape and surface texture of loess particles. Geol Soc Am Bull 81:1591–1595
Smaley IJ, Vita-Finzi C (1968) The formation of fine particles in sandy deserts and the nature of "desert" loess. J Sediment Petrol 28:766–774
Song Daquan (1950) Study on paleopedology of Xiasu Formation in Nanjing – Feature and development rule of several representative profiles. Bull Pedol Soc China 1(3–4):141–153 (in Chinese)
Su Lianyi, Bao Yunying, Tian Guoguang, Qiu Yijian, Xie Hongxiang (1959) Analysis of loess particles in the middle reaches of the Huanghe (Yellow) River. Sci Geol Sinica (8):251–255 (in Chinese)
Sun Dianqing, Wu Xihao, Pu Qingyu (1982) Some suggestions on the work concerning Quaternary glacial geology. Proc 3rd Natl Quat Conf China. Science Press, Beijing, pp 155–161 (in Chinese)
Swineford A, Frye JC (1951) Petrography of the Peoria loess in Kansas. J Geol 59:306–322
Swineford A, Frye JC (1955) Petrographic comparison of some loess samples from Western Europe with Kansas loess. J Sediment Petrol 25:3–23
Tang Keli (1981) Micromorphology and genesis of the paleosols in loess, Wugong, Shaanxi. Kexue Tongbao 26:177–179 (in Chinese)
Tang Songliu, Sun Jingxin, Tu Shude, Qian Qinfang, Wang Yuqi, Chen Bingru, Huang Yanchu, Jiang Zhachun (1980) Rare earth elements in some soils from Guangzhou. Acta pedol Sinica 17:299–307 (in Chinese, with English abstract)
Tang Yingjun, Ji Hongxiang (1983) A Pliocene-Pleistocene transitional fauna from Yuxian, Northern Hebei. Vertebrata PalAsiatica 21:245–254 (in Chinese, with English abstract)
Teilhard de Chardin P, Licent E (1930) Geological observations in N. Manchuria and Barga (Hailar). Bull Geol Soc China 9:23–35
Teilhard de Chardin P, Peveteau J (1930) Les manniferes fossiles de Nihowan (China). Ann Paleontol 19
Teilhard de Chardin P, Young CC (1930) Preliminary observations on the preloessic and post-Pontian formations in Western Shansi and Northern Shensi. Mem Geol Surv China Ser A 8
Teilhard de Chardin P, Young CC (1931) Fossil mammals from the late Cenozoic of northern China. Palaeontol Sinica Ser C 9
Teruggi ME (1957) The nature and origin of Argentine loess. J Sediment Petrol 27:322–332
Thorp J (1935) Soil profile studies as an aid to understanding recent geology. Bull Geol Soc China 14:359–392
Thorp J (1936) Geography of the soil of China. Geol Surv China 1–552
Torii M, Yue Leping, Hayashida A, Maenaka K, Yokoyama T, Wang Yongyan, Sasajima S (1983) Preliminary results of magnetostratigraphy of Chinese loess in Luochuan area, Shaanxi, China. In: Sasajima S, Wang Yongyan (eds) Some problems on the quaternary chronology of Chinese Loess with special emphasis on Luochuan Loess. (in Chinese, withe English abstract)
Wang Dezbi (1974) Optical mineralogy. People's Publ House of Shanghai, Shanghei, pp 34–356 (in Chinese)
Wang Jiayin (1965) On the loess problems of historic period. Quat Sinica 4(1):1–8 (in Chinese)
Wang Tingtai, Wang Pinhsien (1980) Relationship between sea-level changes and climatic fluctuations in East China since late Pleistocene. Acta Geog Sinica, vol 35, No 4, p 299 (in Chinese)
Wang Junda, Li Huamei (1982) Paleomagnetic studies on a loess section in western Gansu. Geochimica (1):85–89 (in Chinese, with English abstract)
Wang Kelu, Pei Jingxian (1964) Mineral composition of loess at Wucheng, Xixian County, Shanxi Province. The problems of quaternary geology. Science Press, Beijing, pp 89–110 (in Chinese)
Wang Naiwen, He Xixian (1982) Discovery of calcareous nannofossils in Beijing Plain and its significance for stratigraphy, palaeogeography and paleoclimatology. Kexue Tongbao 27:805–808
Wang Xianjue, Chen Yuwei, Lei Jianquan, Wu Mingqing, Zhao Yiyang (1982) REE geochemistry in sea-floor sediments in the continental shelf of East China Sea. Geochimica (1):56–65 (in Chinese, with English abstract)
Wang Yongyan, Hsueh Hsianghsi, Ho Juchang, Chang Kuowei (1966) Division of the Quaternary strata in the loess district of north Shanxi and east Kansu. Acta Geol Sinica 46(1):102–117 (in Chinese, with English abstract)
Wang Yongyan, Yue Leping, Wu Zaibao, Chen Xiling, Duan Wen (1980) On the stratigraphic problems of the loess on the plateau north of Weihe River, Shaanxi Province in accordance with the palaeomagnetic data. Geol Rev 26(2):141–147 (in Chinese, with English abstract)

Wang Yongyan et al. (1982) Loess and quaternary geology, 1976–1980. People's Publ House Shaanxi Province, Xian (in Chinese)
Weaver CE (1960) Possible uses of clay minerals in search for oil. Clays Clay Miner 8th Natl Clay Conf (1959):214–227
Weber K (1972) Notes on determination of illite crystallinity. Neues Jahrb Mineral Abh 46:267–276
Wells KL, Riecken FF (1969) Regional distribution of potassium in the B horizon clay of some prairie loess soil of the Midwest. Soil Sci Soc Am Proc 33:583–587
Wen Qizhong, Yu Suhua (1979) REE in sedimentary process. Geol Geochem (5):55–61 (in Chinese)
Wen Qizhong, Zhu Haizhi, Guo Jinluan (1964) Some geochemical problems of the loess in the middle reaches of the Huanghe (Yellow) River. The problems of quaternary geology. Science Press, Beijing, pp 111–125 (in Chinese)
Wen Qizhong, Sun Fuqing, Diao Guiyi, Yu Suhua (1981a) Ratios of oxides and relative values of weathering leaching or accumulation in Luochuan loess section, Shaanxi Province and their geological significance. Geochimica (4):381–387 (in Chinese, with English abstract)
Wen Qizhong, Yu Suhua, Gu Xiongfei, Lei Jianquan (1981b) A preliminary investigation of REE in loess. Geochimica (2):151–157 (in Chinese, with English abstract)
Wen Qizhong, Zheng Honghan, Han Jiamao, Wang Junda, Lin Shaomeng, Qiao Yulou, Wei Lanying, Diao Guiyi (1982) Loess of Longxi Basin in Gansu Province. Sci Geogr Sinica 2(3):202–209 (in Chinese, with English abstract)
Wen Qizhong, Yu Suhua, Sun Fuqing, Wang Yuqi, Chen Bingru, Tu Shude, Sun Jingxin (1984) Rare-earth elements in Luochuan loess section, Shaanxi Province. Geochimica (2):126–133 (in Chinese, with English abstract)
Wu Jukang (1964) Mandible of the Sinanthropus-type discovered at Lantian, Shensi. Vertebrata PalAsiatica 8(1):1–17 (in Chinese, with English abstract)
Wu Jukang (1966) The hominid skull of Lantian, Shensi. Vertebrata PalAsiatica 10(1):1–12 (in Chinese, with English abstract)
Wu Zirong, Yuan Baoyin, Sun Jianzhong (1980) Subdivision and recognition of the Nehewan strata. Sci Geol Sinica (1):87–95 (in Chinese, with English abstract)
Xia Yumei, Wang Peifang, Wang Manhua (1983) Preliminary research on sporo-pollen association of the section of Huang Shan in Harbin. Sci Geogr Sinica 3(1):183–187 (in Chinese, with English abstract)
Xinjing Comprehensive Expedition, Academia Sinica, Institute of Geography, Academia Sinica and Geomorphological Research Group of the Xinjiang Comprehensive Expedition (1978) Geomorphology of Xinjiang Uygur Autonomous Region. Science Press, Beijing (in Chinese)
Xiong Yi, Wen Qixiao (1953) How to improve soils in Northwest China. Kexue Tongbao (10):48–54 (in Chinese)
Yan Zhen (1960) Stratigraphic division of the Quaternary sediments in the Baoji ara. Quat Sinica 3(1–2):88–103 (in Chinese)
Yang Huairen, Xu Xin (1980) Quaternary environmental changs in eastern China. Journal of Nanjing University (natural science edition) 1:127–144
Yen Tengchien (1936) Some Quaternary gastropoda from eastern Szechuan. Bull Geol Soc China 15(3):331–342 (in Chinese)
Yen Tengchien (1938) Eine neue Gatting Pseudophysa in gen. aus dem Pliozan von Yünnan (China). Arch Mollusk 70:279–280
Yen Tengchien (1943) Review and summary of Tertiary and Quaternary non-marine mollusks of China. Proc Acad Natl Sci Phil 95:267–309
Yih LF (1920) The geology of Hsi-Shan or the Western Hills of Peking. Mem Geol Sur China Ser A (1)
Yu Wen (1963) Gastropoda Fauna in China. Science Press, Beijing (in Chinese)
Yuan Sixun, Chen Tiemei, Gao Shijun (1983) Uranium series dating of "Ordos Man" and "Sjara Osso-Gol Culture". Acta Anthropol Sinica 2(1):90–94
Zhai Lisheng (1983) An outline of construction engineering geology of collapsing loess region in China. Science Press, Beijing (in Chinese)
Zhang De'er (1982) Analysis of dust rain in the historic times of China. Kexue Tongbao 27(5):294–297
Zhang De'er (1984) Synoptic-climatic studies of dust fall in China since the historic times. Sci Sinica 27(8):825–836
Zhang Shaozhen (1958) Preliminary study on clay minerals of loess in Sanmen Gorge and Lanzhou. Inf Loess (2):179–194 (in Chinese)

Zhang Shuyuan (1965) Detritus minerals of loess in the area near Gulang, Gansu Province. Quat Sinica 4(1):32–46 (in Chinese)
Zhang Zonghu (1962) Collapse and permeability of loess-like soils in China. Sci Geol Sinica (12):1–8 (in Chinese)
Zhang Zonghu (1983) Lithological and stratigraphical analysis on the loess profiles of the Loess Plateau in China. Mar Geol Quat Geol 3(3):14–28 (in Chinese, with English abstract)
Zhang Zonghu, Hu Huiming, Zhai Rongting, Wang Yunsheng, Yao Zuijing (1959) Comprehensive investigation of geological and engineering properties of loesses in eastern Gansu. In: Selected works on hydrogeology and engineering geology, vol III. Publ House Geol, Beijing, pp 3–97 (in Chinese)
Zhao Shusen, Liu Mingling, Ma Zhibang (1982) U-series dating of Pleistocene speleothems. In: Research on geology, vol I. Inst Geol, Academia Sinica. Cultural Relics Publ House, Beijing, pp 294–299 (in Chinese, with English abstract)
Zhao Songling, Chin Yunshan (1982) Transgressions and sea-level changes in the eastern coastal region of China in the last 300000 years. Quaternary geology and environment of China. China Ocean Press, Beijing, pp 147–154
Zhao Songling, Yang Guangfu, Cang Shuxi, Zhang Hongcai, Huang Qingfu, Xia Dongxing, Wang Yongjie, Liu Fushou, Liu Chengfu (1978) On the marine stratigraphy and coastlines of the western coast of the gulf of Hebei. Oceanol Limnol 9(1):15–25 (in Chinese, with English abstract)
Zhen Heqing (1965) Study on detritus minerals and texture of loess in the middle reaches of the Huanghe (Yellow) River. Quat Sinica 4(1):47–61 (in Chinese)
Zheng Honghan (1982) Paleoclimate events recorded in clay minerals in Loess of China. Quaternary geology and environment of China. China Ocean Press, Beijing, pp 59–66
Zheng Honghan (1984) Holocene loess in the Huanghe (Yellow) River. Geochimica (3):238–245 (in Chinese, with English abstract)
Zheng Yanwu (1982) The collapsing property of loess in China. Publ House Geol, Beijing (in Chinese)
Zhou Kunshu, Liang Xiulong, Ye Yongying, Wang Wenlin (1960) Sporo-pollen and plant relics of buried soils in older loess at Chenjiaya of Wangjiagou, Lishi County, Shanxi Province. Quat Sinica 1(1):104–111
Zhou Kunshu, Liang Xiulong, Liu Ruiling (1982) Preliminary exploation on the paleoclimate recorded in loess of China. Proc 3rd Natl Quat Conf China. Science Press, Beijing, pp 134–138 (in Chinese)
Zhou Kunshu, Liang Xiulong, Ye Yongying (1984) Sporo-pollen analysis of loess at Liushugou of Wucheng, Shanxi Province. Quaternary sporo-pollen analysis and the paleoenvironment. Science Press, Beijing, pp 109–114 (in Chinese)
Zhu Haizhi (1959) Statistic analysis on grain size of loess under microscope. Sci Geol Sinica (3) (in Chinese)
Zhu Haizhi (1963) Some characteristics of the grain size and texture of Malan Loess in the middle reaches of the Huanghe (Yellow) River. Sci Geol Sinica (2):88–102 (in Chinese, with English abstract)
Zhu Haizhi (1982) Research on some signs of prehistoric seismic activities. Papers of prehistoric earthquakes and quaternary geology. Shaanxi Science and Technology Press, Xian, pp 17–29 (in Chinese, with English abstract)
Zhu Haizhi, Wen Qizhong, Guo Jinluan (1964) Preliminary study on the characteristics of loess texture. The problems of quaternary geology. Science Press, Beijing, pp 140–150 (in Chinese)
Zhu Kezhen (1973) A preliminary study on the climate fluctuations during the last 5000 years in China. Sci Sinica 16(2):291–296 (in Chinese, with English abstract)
Zhu Xianmo (1958) Discussion on the problems of red layers intercalated in the loess section. Quat Sinica 1(1):74–82 (in Chinese)
Zhu Xianmo (1965) Paleosols in loessic sediments of China. Quat Sinica 4(1):9–19 (in Chinese)
Zhu Xianmo (1981) Land resources of Shaanxi Province and their resonable use. Shaanxi Science and Technology Press, Xian (in Chinese)
Zhu Xianmo, Wang Jingqin, Chen Junqing (1982) Preliminary studies on the "Litholac" soaking on the Gobi rock surface in Gobi desert (Abstract). Quaternary geology and environment of China. China Ocean Press, Beijing, p 94
Zhu Zenda, Wu Zheng, Liu Shu, Di Xingming (1980) Introduction to deserts of China. Science Press, Beijing (in Chinese)

Subject Index

MIX
Papier aus verantwortungsvollen Quellen
Paper from responsible sources
FSC® C105338

If you have any concerns about our products,
you can contact us on
ProductSafety@springernature.com

In case Publisher is established outside the EU,
the EU authorized representative is:
Springer Nature Customer Service Center GmbH
Europaplatz 3, 69115 Heidelberg, Germany

Printed by Libri Plureos GmbH
in Hamburg, Germany